Endothelial Cells

Volume II

Editor

Una S. Ryan, Ph.D.
Professor of Medicine
Department of Medicine
University of Miami
Miami, Florida

CRC Press, Inc.
Boca Raton, Florida

Library of Congress Cataloging-in-Publication Data

Endothelial cells/editor, Una S. Ryan.
    p. cm.
  Includes bibliographies and indexes.
  ISBN 0-8493-4988-5 (set)
  1. Endothelium — Cytology. I. Ryan, Una, S., 1941-
  [DNLM: 1. Endothelium — anatomy & histology. 2. Endothelium —
physiology. QS 532.5.E7 E566]
QP88.45.E53 1988
599'.087 — dc 19
DNLM/DLC

                                                              87-32542

This book represents information obtained from authentic and highly regarded sources. Reprinted material is quoted with permission, and sources are indicated. A wide variety of references are listed. Every reasonable effort has been made to give reliable data and information, but the author and the publisher cannot assume responsibility for the validity of all materials or for the consequences of their use.

Direct all inquiries to CRC Press, Inc., 2000 Corporate Blvd., N.W., Boca Raton, Florida, 33431.

© 1988 by CRC Press, Inc.

International Standard Book Number 0-8493-4988-5 (set)
International Standard Book Number 0-8493-4990-7 (v.1)
International Standard Book Number 0-8493-4991-5 (v.2)
International Standard Book Number 0-8493-4992-3 (v.3)

Library of Congress Card Number 87-32542
Printed in the United States

PREFACE

In the Foreword to a book on Yankee epitaphs called *Over Their Dead Bodies** the authors remark " . . . on the rich chronicle they offered of contemporary ideas and events". This is precisely the mission of this book, although I doubt that any of the authors would wish their chapters to represent their "last words".

The endothelium itself is far from a morbid subject. It has come alive in recent years and its story is told in these three volumes. It has risen from being a ghostly substance, scarcely showing in early pathology or histology texts, of no known properties save that of lining the blood vessels, to being a collection of cells endowed with pores that could be mathematically modeled but never seen, to possessing a rich array of enzymatic and processing properties. Its stature has grown to that of a metabolically active and responsive tissue endowed with a diversity of enzymes, receptors, and transport molecules. It can be grown in culture and manipulated into postures it may never have to succumb to in vivo, its gene products have been cloned, and it has been made to reveal its relationships with other cells and molecules both near neighbors or distance targets. It is claimed as a regulator of blood pressure, a team player in hemostasis, a sparring partner with various blood cell types, and the dancing partner of the vascular smooth muscle cell. At one time seen as the innocent victim of inflammatory attack we now know that it frequently calls the tune. It is both a target and a source of hormones, growth factors, vasoactive substances, hemostatic factors, and oxygen radicals.

It binds complement components, can express receptors for immune reactions, presents antigens, and can engulf and kill microorganisms. It can be activated, excited, and primed. Activated endothelium represents a remarkable amplification surface for local immune and inflammatory reactions and is able to initiate events that lead to closing off a vessel. Activation of endothelium plays a key role in the host response yet when inappropriately expressed can underlie much of vascular pathology. In fact, it is likely that all diseases have a vascular etiology.

Proud though we are of the extraordinary accomplishments of the last 2 decades of research we know that the endothelium has not yet yielded up all its mysteries. I hope this book will stand as a monument to the success of its brilliant band of contributors and a stepping stone for the next runners in the relay race to capture the prize of understanding the vascular endothelial cell.

Una S. Ryan
Miami, October 1987

---

* Mann, T. C. and Greene, J., *Over Their Dead Bodies: Yankee Epitaphs and History,* Stephen Greene Press, Brattleboro, Vermont, 1962.

THE EDITOR

Una S. Ryan, Ph.D., is Professor of Medicine at the University of Miami School of Medicine, Chief of the Division of Vascular Cell Biology, and Director of the Hybridoma Facility.

Dr. Ryan obtained her training in England and received a B.Sc. degree from the University of Bristol in 1963 and a Ph.D. from the University of Cambridge in 1968. She was a Howard Hughes Investigator and Director of the Laboratory for Ultrastructure Studies at the Howard Hughes Medical Institute, Miami, from 1967 to 1971. She served as Instructor in Medicine from 1967 to 1972, Assistant Professor of Medicine from 1972 to 1977, Associate Professor from 1977 to 1980, Research Professor from 1980 to 1986, and Professor of Medicine from 1986.

Dr. Ryan is a member of the American Society for Cell Biology, Society for Neuroscience, Tissue Culture Association, American Heart Association, Council on Basic Research, Council on Circulation, Cardiopulmonary Council, American Physiological Society, Microcirculatory Society, European Society for Microcirculation, American Thoracic Society, New York Academy of Sciences, International Society for Heart Research, and the Royal Society of Medicine.

She was an Established Investigator of the American Heart Association from 1972 to 1977. She received the Louis Artur Lucian Award for Research in Circulatory Diseases in 1984, and the Lamport Lectureship in 1985. She received a MERIT Award from the National Institutes of Health in 1986, she is President-Elect of the Sigma Xi Miami chapter, and was elected to the Royal Society of Medicine in 1986.

She has been the recipient of numerous research grants from the National Institutes of Health and has served on and chaired many review and advisory panels, including the Pulmonary Diseases Advisory Committee, 1972 to 1976 Committee A. She is currently a member of Pathology A Study Section.

Dr. Ryan is editor of *Tissue & Cell* (an international cell biology journal) and an advisory or reviewing editor for a large number of other journals. She has edited 3 books, authored over 180 papers and over 160 abstracts. Her major research interest centers on the cell biology of the pulmonary endothelium.

CONTRIBUTORS

Susan J. Braunhut, Ph.D.
Research Associate
Departments of Surgery and
  Ophthalmology
Harvard Medical School
Children's Hospital
Boston, Massachusetts

Wilson Hales Burgess, Ph.D.
Senior Scientist
Laboratory of Molecular Biology
Jerome H. Holland Laboratory for the
  Biomedical Sciences
American Red Cross
Rockville, Maryland

Patricia A. D'Amore, Ph.D.
Assistant Professor of Pathology and
Research Associate in Surgery
Departments of Surgery and Pathology
Harvard Medical School
Children's Hospital
Boston, Massachusetts

Peter F. Davies, Ph.D.
Associate Professor of Pathology
Department of Pathology
Brigham and Women's Hospital
Harvard Medical School
Boston, Massachusetts

Paul E. DiCorleto, Ph.D.
Staff Member
Department of Brain and Vascular
  Research
Cleveland Clinic Research Institute
Cleveland, Ohio

Paul L. Fox, Ph.D.
Project Scientist
Department of Brain and Vascular
  Research
Cleveland Clinic Research Institute
Cleveland, Ohio

David E. Gannon, M.D.
Research Fellow
Department of Pathology
University of Michigan
Ann Arbor, Michigan

Marco Gardinali, M.D.
Physician and Researcher
Department of Medicine
University of Milan
Milan, Italy

Harry L. Goldsmith, Ph.D.
Professor
Department of Medicine
Division of Experimental Medicine
McGill University
Montreal, Quebec, Canada

Avrum I. Gotlieb, M.D., C.M.
Associate Professor of Pathology
Department of Pathology
University of Toronto
Toronto General Hospital
Toronto, Ontario, Canada

John M. Harlan, M.D.
Associate Professor of Medicine
Department of Medicine
University of Washington
Seattle, Washington

Ronald L. Heimark, Ph.D.
Research Assistant Professor
Department of Pathology
University of Washington
Seattle, Washington

Peter M. Henson, Ph.D.
Executive Vice President for Biomedical
  Affairs
Office for Biomedical Affairs
National Jewish Center for Immunology
  and Respiratory Medicine
Denver, Colorado

Tony E. Hugli, Ph.D.
Member
Department of Immunology
Research Institute of Scripps Clinic
La Jolla, California

Takeshi Karino, Ph.D.
Associate Professor
Department of Medicine
Division of Experimental Medicine
McGill University
Montreal, Quebec, Canada

Michael Klagsbrun, Ph.D.
Associate Professor
Departments of Biological Chemistry and
  Surgery
Harvard Medical School
Boston, Massachusetts

Dale C. Lien, M.D.
Research Associate
Department of Pediatrics
National Jewish Center for Immunology
  and Respiratory Medicine
Denver, Colorado

Claes Lundberg, Ph.D.
Researcher
Department of Pharmacology
PHARMACIA
Uppsala, Sweden

Thomas Maciag, Ph.D.
Head of the Molecular Biology
  Laboratory
Jerome H. Holland Laboratory for the
  Biomedical Sciences
American Red Cross
Rockville, Maryland

Francois Marceau, M.D., Ph.D.
Researcher
Department of Immunology
Le Center Hospital de'University LAVAL
  (CHUL)
Quebec, Canada

Jordan S. Pober, M.D., Ph.D.
Associate Professor
Department of Pathology
Harvard Medical School and Brigham and
  Women's Hospital
Boston, Massachusetts

Una S. Ryan, Ph.D.
Professor of Medicine
Department of Medicine
University of Miami
Miami, Florida

Duane R. Schultz, Ph.D.
Professor of Medicine
Department of Medicine
University of Miami School of Medicine
Miami, Florida

Barbara R. Schwartz, M.S.
Researcher
Department of Medicine
Division of Hematology
University of Washington
Seattle, Washington

Stephen M. Schwartz, M.D., Ph.D.
Professor of Pathology and
Adjunct Professor of Bioengineering
Department of Pathology and
  Bioengineering
University of Washington
Seattle, Washington

Marcia G. Tonnesen, M.D.
Chief
Dermatology Service
Veterans Administration Medical Center
Denver, Colorado

James Varani, Ph.D.
Associate Professor
Department of Pathology
University of Michigan
Ann Arbor, Michigan

Peter A. Ward, M.D.
Professor and Chairman
Department of Pathology
University of Michigan
Ann Arbor, Michigan

Michael K. K. Wong, Ph.D.
Medical Student
Department of Pathology
Faculty of Medicine
University of Toronto
Toronto, Ontario, Canada

G. Scott Worthen, M.D.
Staff Physician
Department of Medicine
National Jewish Center for Immunology
  and Respiratory Medicine
Denver, Colorado

Bruce R. Zetter, Ph.D.
Associate Professor
Department of Physiology and Surgery
Harvard Medical School
Boston, Massachusetts

TABLE OF CONTENTS

Volume I

TABLE OF CONTENTS

Volume II

TABLE OF CONTENTS

Volume III

Growth Factors and Growth Control

Chapter 13

THE STRUCTURAL AND FUNCTIONAL PROPERTIES OF ENDOTHELIAL CELL GROWTH FACTOR AND ITS RECEPTOR

Thomas Maciag and Wilson Hales Burgess

TABLE OF CONTENTS

I. INTRODUCTION

Endothelial cells in vivo form provide a nonthrombogenic monolayer surface which lines the lumen of blood vessels and provides a cellular interface between blood and tissue.[1] The ubiquitous presence of the endothelial cell within the vascular tree highlights the importance of this cell type in the maintenance of tissue homeostasis.[1] The strategic localization of the endothelial cell in vivo endows this cell type with the ability to participate as a modulator of cellular homeostatsis during wound repair and tissue remodeling.[2,3,9] Likewise, the endothelium, as a quiescent population of cells in vivo,[4] can be activated to participate in a variety of pathophysiological situations[1-5] which include tumor development, atherogenesis, arthritis, bone remodeling, psoriasis, diabetic retinopathy, and chronic inflammatory conditions.

The endothelial cell also plays a fundamental role in the formation of new blood vessels.[1-3,5,8,9] The process of neovascularization involves organized events which include the controlled migration and proliferation of the endothelial cell.[3,5-8] These events are orchestrated by presumed biochemical modulators[2,5] which control both the proliferative[2,8,10] and nonproliferative[5] aspects of angiogenesis and utilize the endothelial cell as the target for these hormonal agents.[2,3,5-7,10] Thus, the identification of polypeptide signals which modulate endothelial cell migration and division are critical to the elucidation of the biochemical events which underlie the mechanism of angiogenesis.

In the past decade, two polypeptide modulators of endothelial cell division have been identified.[10-12] The mitogens, the basic polypeptide, fibroblast growth factor (FGF), and the acidic polypeptide, endothelial cell growth factor (ECGF), have both been characterized as potent polypeptide regulators of endothelial cell division in vitro[13] and stimulators of angiogenesis in vivo,[14] The elucidation of the biochemical structures for these polypeptides is a hallmark in the field of experimental neovascularization because these structures provide a rigorous basis for characterization and future experimentation. Although both polypeptide families are assumed to be equally important as modulators of angiogenesis, this review will be mainly concerned with the biological function of the acidic family of polypeptide growth factors.

II. THE BIOLOGICAL PROPERTIES OF ECGF

Although reliable methods had been established for the isolation of human endothelial cells from large blood vessels,[1,15] these cells have been difficult to serially propagate in vitro. Human endothelial cells, seeded in vitro at low cell density, do not respond to serum,[5,10,11,16] a rich source of mitogens for other mammalian cell types.[17] The identification of basic FGF from bovine brain[10] and the preparation of crude extracts of ECGF from bovine neural tissue[11] yielded sufficient biological activity for the propagation of human endothelial cells in vitro.[16,18]

The biological activity of ECGF is dependent upon the presence of serum[16,18] and acts in a synergistic manner with the hormones and growth factors contributed by serum.[16] Although a reliable serum-free culture system for human endothelial cells has not been established, it was known that human endothelial cells do not respond to platelet-derived growth factor (PDGF),[19,20] a polypeptide serum-derived mitogen for other mesenchymal cells.[17] Thus, it was clear from these early studies that plasma did not contain a mitogen for human endothelial cell proliferation but instead required a polypeptide growth factor derived from a tissue source.[10,11]

A number of biological attributes are associated with ECGF and each is consistent with the ability of ECGF to stimulate human endothelial cell proliferation. ECGF significantly delays the premature senescence of human endothelial cells. Populations of human cells are able to achieve between 50 and 60 cumulative population doublings when the cell culture conditions include ECGF and serum.[5,18] This is consistent with the number of population doublings achieved by human fibroblasts serially propagated in the presence of serum.[21] ECGF also reduces the amount of serum required to generate human endothelial cell division and when grown on a suitable extracellular matrix component (fibronectin or collagen), ECGF supports the clonal growth of human endothelial cells.[5,16,18] ECGF is also a potent chemoattractant for human endothelial cells in vitro,[22] stimulating chemotaxis at concentrations similar to those required for the stimulation of DNA synthesis.

The mitogenic[18,23] and chemotactic[22] activity of ECGF is potentiated by the glycosaminoglycan heparin. This property is unique to ECGF and can be utilized as a criterion for distinction between ECGF and basic-FGF.[13,24,25] Heparin also potentiates the mitogenic activity of crude preparations of ECGF[18] and alters immunological and receptor binding epitopes on the polypeptide.[23] Although the mechanism by which heparin potentiates the biological activity of ECGF is not clear, the mechanism may involve a structural interaction between ECGF and heparin.[23-26] Consistent with this suggestion is the ability of ECGF to bind to immobilized heparin.[26] Although heparin can be considered as a cofactor for ECGF, it is not clear whether heparin itself possesses intrinsic activity as a result of down-regulation of the ECGF receptor by the heparin:ECGF complex.

III. THE PURIFICATION OF ECGF

ECGF is an anionic, single-chain polypeptide mitogen derived from extracts of bovine neural tissue prepared at a neutral pH.[11] The biological activity of the polypeptide is destroyed by heat, strong alkali, or acid.[5,11] Like most other growth factors, ECGF possesses two forms:[27] a high molecular weight (M_r) and a low-M_r form. These forms are related since it is possible to generate low M_r-ECGF from high M_r-ECGF by dissociation of the high M_r-ECGF complex with weak acids.[27] Alteration of the dielectric constant and ionic strength of the high M_r-ECGF complex also results in the formation of low M_r-ECGF.

ECGF also binds avidly to immobilize heparin.[26] This property has found utility in the construction of purification schemes for representative polypeptides which belong to the acidic and basic families of the endothelial cell mitogens (see Reference 24 for an exhaustive review). Using neutral pH extraction, the high M_r to low M_r transition, and the avid biological affinity for heparin, a rapid purification procedure has been established.[28] The biological activity of ECGF which elutes from heparin-Sepharose between 0.9 and 1.1 M NaCl is biologically active as a human endothelial cell mitogen in the low nanogram-per-milliliter range and contains a mixture of two forms of low M_r-ECGF. The elution from heparin-Sepharose is characteristic of ECGF and distinguishes it from other heparin-binding growth factors such as PDGF which elutes at lower concentrations of NaCl and basic FGF which elutes at 2.0 M NaCl.[24,29] This purification procedure is very rapid and results in a preparation of ECGF which is heterogeneous by sodium dodecyl sulfate polyacrylamide gel electrophoresis (SDS-PAGE).

The post-heparin-Sepharose ECGF fraction can be further purified by reversed-phase high pressure liquid chromatography (HPLC) which discriminates two biologically active forms of low M_r-ECGF.[28] These polypeptides have been named alpha- and beta-ECGF in reference to their retention properties.[28] Both forms are biologically active below the nanogram-per-milliliter range in endothelial cell growth and DNA synthesis assays.[28] The two forms of ECGF are further related by heparin affinity, radioreceptor binding, and antibody recognition as well as amino acid composition.[28] However, differences between alpha- and beta-ECGF can be resolved by SDS-PAGE; alpha-ECGF possesses a M_r 17,000 while beta-ECGF possesses a M_r 20,000. Digestion of alpha- and bent-ECGF with CNBr generates two peptide fragments from both forms of the single-chain polypeptide.[28] The larger of the two fragments obtained by CNBr digestion of both alpha- and beta-ECGF possesses similar M_rs and amino acid compositions, while the smaller of the two fragments remain dissimilar by SDS-PAGE.[28] Analysis of alpha- and beta-ECGF CNBr digests by SDS-PAGE in the presence and absence of reducing agents suggest that both forms of ECGF are single-chain polypeptides containing a methionine residue positioned between at least one disulfide bond.[28]

IV. THE STRUCTURE OF BOVINE ECGF

The availability of a rapid and efficient purification procedure for both bovine alpha- and beta-ECGF generated sufficient quantities of these polypeptides for structural analysis.[28] Automated Edman degradation of intact bovine alpha-ECGF generates a sequence starting the amino terminus with NH_2-NYKKPK-COOH.[30] In contrast, bovine beta-ECGF does not yield any detectable phenylthiohydantion amino acid derivatives after ten cycles of automated Edman degration. Digestion of bovine beta-ECGF with CNBr followed by automated Edman degration yields the sequence NH_2-DTDGLL-COOH. In contrast, Edman degration of CNBr-modified bovine alpha-ECGF yields two sequences, the amino terminal sequence NH_2-NYKKPK-COOH and a second sequence, NH_2-DTDGLL-COOH, which is identical to the single sequence derived from CNBr-modified bovine beta-ECGF. These results suggest that bovine alpha- and beta-ECGF are structurally related,[30] and bovine beta-ECGF contains a blocked amino terminus.

Digestion of alpha- and beta-ECGF with trypsin followed reversed-phase HPLC yielded elution profiles consistent with the suggestion that the two forms of bovine ECGF are structurally related. The profile observed for bovine beta-ECGF contains all of the trypsin-derived fragments observed in the bovine alpha-ECGF profile in addition to two unique peptides.[28,30]

The trypsin-digested fragments obtained from intact bovine alpha-ECGF were subjected to automated Edman degration. The carboxyl terminus of bovine beta-ECGF was assigned to the trypsin fragment which did not contain either a lysine or arginine residue by amino acid composition. Since this fragment was observed in both the beta- and alpha-ECGF trypsin digestion elution profile, it was concluded that bovine alpha- and beta-ECGF contain the same carboxyl terminal sequences. Thus, it became apparent that these data limited the differences between bovine beta- and alpha-ECGF to the amino terminus of the polypeptides and further suggested that bovine alpha-ECGF contains an amino terminal extension.[30]

The complete sequence of bovine alpha-ECGF was derived from overlapping sequence derived from intact alpha-ECGF and fragments derived from CNBr, trypsin, and thermolysin digestion of bovine alpha-ECGF. The sequence NH_2-NGRSK-COOH was assigned by homology with the sequence reported for bovine acidic FGF,[14,31] a potent endothelial cell polypeptide mitogen and angiogenesis factor.[32]

V. BOVINE BETA-ECGF IS A PRECURSOR OF BOVINE ACIDIC-FGF AND ALPHA-ECGF

The sequence of the amino terminal region unique to bovine beta-ECGF was obtained by isolation of the two unique trypsin-derived peptide fragments. Although one fragment was blocked at the amino terminus, the second fragment yielded the sequence NH_2-FNLPLG-COOH. This sequence, which is the amino terminal sequence of bovine acidic-FGF,[14,31] suggests that acidic-FGF is a precursor form of alpha-ECGF.[30] Limited sequence of the blocked amino terminal trypsin-derived peptide fragment was obtained by sequencing peptides independently isolated after thermolysin digestion and partial acid hydrolysis of the blocked amino terminal trypsin-derived peptide.[30] Confirmation of the sequence of the blocked amino terminal trypsin-derived beta-ECGF fragment and the identification of the blocking group as Acl-beta-ECGF was obtained by fast atom bombardment mass spectrometry.[30] These data further suggest that bovine beta-ECGF is a precursor form of bovine acidic-FGF and alpha-ECGF.

The origin of bovine acidic-FGF and alpha-ECGF from bovine beta-ECGF most likely involves specific proteolytic processing. Cleavage at the K_{14}-F_{15} position in beta-ECGF yields acidic-FGF, a cleavage site which may involve the enzymatic activity of a brain-derived serine protease. Cleavage at the G_{20}-N_{21} position in beta-ECGF is unique and may involve processing by an unknown protease with unusual specificity. The identification of these related but amino-terminal-truncated polypeptides also calls attention to the extraction conditions used early in the purification proecdures for acidic-FGF and beta-ECGF. Acidic-FGF is purified from bovine brain by extraction of the tissue at pH 4.5,[12,14,31] conditions which would favor the enzymatic activity of a serine protease. In contrast, beta-ECGF is extracted near pH 7.0 in the presence of EDTA,[28] conditions which should not favor the protcolytic activity of a serine and/or calcium-activated protease(s) and serve to protect beta-ECGF from proteolytic digestion. However, these conditions do favor the activity of a protease with specificity for the G_{20}-N_{21} bond in beta-ECGF. These data also argue that the first 20 amino acid residues of the amino terminal extension of bovine beta-ECGF do not significantly contribute to the mitogenic activity of the polypeptide because acidic-FGF (des 1-14, beta-ECGF) and alpha-ECGF (des 1-20, beta-ECGF) are potent polypeptide mitogens.[14,28,31]

VI. THE CLONING OF THE GENE ENCODING HUMAN BETA-ECGF

The amino acid sequence of bovine alpha-ECGF was used for the design and synthesis of synthetic radiolabeled oligonucleotide probes. These probes were used to screen a human cDNA library constructed with poly A-selected mRNA purified from human brain stem tissue. Positive cDNA clones were purified and sequenced.[33] Analysis revealed a sequence of 465 nucleotides which represents the open reading frame encoding human beta-ECGF. This open reading frame is flanked by termination codons.[33] The amino acid sequence deduced from the cDNA sequence[33] agrees well with the amino acid sequence described for bovine beta-ECGF[30] and human acidic-FGF.[31] The sequence, 5′-ACCAUGG-3′, which includes the proposed translation initiation codon and is found upstream of the open reading frame,[33] also agrees well with the consensus sequence which is optimal for the initiation of translation in eukaryotes.[34] These data argue that human ECGF is synthesized as intact beta-ECGF.

The gene which encodes ECGF is transcribed as a 4.3-kilobase mRNA transcript, and the ECGF mRNA contains lengthy 5′ and 3′ untranslated sequences. Examination of the amino acid sequence of human beta-ECGF reveals the absence of a sequence which could serve as a signal peptide. The lack of an apparant signal peptide suggests that biosynthetic ECGF, a polypeptide mitogen which exerts its biological effect through an extracellular, high-affinity receptor,[23,25] does not exit the cell in a traditional manner. This property is also shared by the growth factors, basic-FGF[35] and interleukin-1 (IL-1),[36] two other polypeptide mitogens which act through membrane receptors. Although the significance of this observation is presently not clear, it is of interest that beta-ECGF and basic-FGF, two structurally and biologically related polypeptides which contain sequences homologies with IL-1, share this common feature. In this regard, it is important to note that ECGF has not been found in either plasma or serum.[23]

The gene which encodes human beta-ECGF has been localized to chromosome 5 in the region 5q31.3-33.2 by *in situ* hybridization.[33] In contrast, basic-FGF has been localized to another human chromosome.[37] Human beta-ECGF resides on the long arm of chromosome 5 between the receptor for PDGF and the c-fms proto-oncogene, the receptor for colony-stimulating Factor 1. The significance of this growth factor and receptor-rich region is unknown.

VII. BOVINE BETA-ECGF AND BOVINE BASIC-FGF ARE STRUCTURALLY RELATED POLYPEPTIDES

The polypeptide growth factor basic-FGF,[10] a potent inducer of angiogenesis in vivo and promoter of endothelial cell division in vitro, has been purified from bovine brain and its structure determined.[38] Basic-FGF is structurally related to beta-ECGF with the more prominent homologies occurring in the COOH- and NH_2-terminal domains of the polypeptide. The amino terminal sequence of basic-FGF was derived from the cDNA sequence of the bovine gene.[35] Basic-FGF, unlike beta-ECGF, contains short, imperfect sequence repeats and contains multiple sequences related to the tetrapeptide RGDS, a sequence found in fibronectin, fibrinogen, and von Willebrand factor.[33] The RGDS sequence is involved in the extracellular attachment of these proteins. Since the structural homology between bovine beta-ECGF and basic-FGF is approximately 55%, it is reasonable to suggest that these polypeptides have evolved from a common ancestral gene.[14]

Beta-ECGF is also structurally homologous to the polypeptides IL-1-alpha and IL-1-beta. Although the structural homology is approximately 20% between ECGF and IL-1-beta and 16% between ECGF and IL-1-alpha,[14] beta-ECGF and IL-1 do not share functional similarities. However, the absence of an apparent signal peptide sequence within the structure of these three polypeptide growth factors argues that ECGF and basic-FGF may have evolved from a common ancestral gene which also included IL-1.

The lack of a signal peptide sequence in ECGF and basic-FGF suggests that the mechanism of biosynthesis and endocytotic transport of these polypeptide mitogens may be unique. In this regard, it is important to stress the ability of both polypeptides to avidly bind heparin and the observation that the basic-FGF contains sequences which suggest an association with the extracellular matrix. Furthermore, basic-FGF has been found to be tightly associated with the extracellular matrix in vivo.[38] Whether these polypeptide mitogens are indeed extracellular matrix-associated growth factors and are released in association with cell lysis are areas which must be explored before these polypeptides can be assigned an autocrine or paracrine function.

VIII. HEPARIN MODULATES THE BIOLOGICAL ACTIVITY OF ECGF

The early observations of Thornton and colleagues[18] demonstrated that the glycosaminoglycan heparin potentiates the mitogenic activity of crude preparations of bovine ECGF. In addition, heparin also induces an ECGF-dependent increase in human endothelial cell density at confluence.[18] Although heparin possesses no intrinsic biological activity as a mitogen, the potentiation of ECGF mitogenic activity by heparin is unique because other glycosaminoglycans do not substitute for heparin in this regard.

Heparin also potentiates the biological activity of the purified polypeptide alpha-ECGF both as a mitogen[23,25] and chemoattractant.[22] It has been suggested that the potentiation by heparin involves conformational changes in the structure of ECGF which can be detected by immunological and radioreceptor methods.[23] Monoclonal antibodies prepared against ECGF bind the antigen better in the presence of heparin.[23] Similarly, ECGF binds tighter to its high-affinity receptor present on the endothelial cell surface in the presence of heparin.[23] It is, therefore, likely that the potentiation of ECGF biological activity by heparin involves the induction by the glycosaminoglycan of a favored structural polypeptide conformation.

The ability of heparin to restore biological activity to inactive preparations of purified ECGF without an alteration in structure[23] is also consistent with this suggestion.

The augmentation of the mitogenic activity of ECGF by heparin is also dependent upon the concentration of the glycosaminoglycan. Half-maximum saturation is achieved at approximately 10 μg of heparin per milliliter in the presence of 1 nM ECGF.[23] These data are significant because if one assumes a stoichiometry of 1:1 between the association of the polypeptide and the glycosaminoglycan, the dose dependence data with heparin argues that a relatively minor component of the rather crude heparin preparation is responsible for the biological attribute of heparin. It will be of interest to determine whether the heparan sulfate proteoglycan can bind ECGF and substitute for heparin as a potentiator of the biological activity of ECGF.

IX. THE ECGF RECEPTOR

The polypeptide alpha-ECGF can be efficiently radioiodinated using the lactoperoxidase method,[23,25,39] with a majority of the radiolabel incorporated into the second residue in the amino-terminal region (NH$_2$-NYK-COOH) of the polypeptide.[41] Analysis of binding data using ^{125}I-ECGF demonstrates the presence of a high-affinity receptor which is present on the cell surface of murine, bovine, rabbit, and human endothelial cells.[23] The K_d for ligand affinity is approximately 1 nM with approximately 30,000 ECGF receptors per cell present on the endothelial cell surface.[23] Binding studies performed at 37 and 4°C as a function of time have revealed that the ECGF receptor is rapidly down-regulated by ligand occupancy; a process observed with other growth factors.[17]

Ligand occupancy of the ECGF receptor is also required for signal transduction to the nucleus. Evidence for this statement was provided by studies performed with monoclonal antibodies against ECGF.[25] Monoclonal ECGF antibodies, which bind and inhibit the biological activity of ECGF, also prevent ^{125}I-ECGF from binding to its receptor. In contrast, monoclonal antibodies, which bind ECGF but do not inhibit the biological activity of the polypeptide, do not inhibit receptor binding.

Since the ECGF receptor presents itself at the endothelial cell surface as a stable binding domain at 4°C,[23] the cross-linking agent disuccinimidyl suberate (DDS) was used to determine the biochemical nature of the ECGF receptor. It was possible to covalently couple ^{125}I-ECGF to the surface of the endothelial cell with DSS at 4°C.[39] The cross-linked ^{125}I-ECGF:receptor complex possessed a M_r 170,000 by SDS-PAGE.[39] The cross-linking reaction was specific for ^{125}I-ECGF because the radioactivity associated with the M_r 170,000 species was lost when the incubation of the endothelial cells with ^{125}I-ECGF was performed with a 100-fold molar excess of unlabeled ECGF.[39] Furthermore, other polypeptide hormones and growth factors do not compete for ^{125}I-ECGF in the DSS-mediated cross-linking reaction,[39] an observation which is consistent with binding data.[23]

Chemical and enzymatic analysis of the ^{125}I-ECGF cross-linked species suggests that the ECGF receptor is a single-chain polypeptide.[39] The M_r of the ^{125}I-I-ECGF:receptor complex is sensitive to cleavage with CNBr and trypsin. Furthermore, the electrophoretic mobility of the cross-linked ^{125}I-ECGF:receptor complex does not change in the presence or absence of reducing agents. The M_r of the ECGF receptor after subtraction of the M_r of the ligand is approximately 150,000.[39]

The M_r 150,000 ECGF receptor polypeptide is also internalized as a result of ligand occupancy,[39] an observation which is consistent with binding data.[23] Endothelial cells incubated with ^{125}I-ECGF for 1 hr at 37°C revealed a significant decrease in ECGF receptor number[23] which correlates well with a significant decrease in the intensity of the radiolabeled ^{125}I-ECGF:receptor cross-linked species.[39] In contrast, endothelial cells incubated with ^{125}I-ECGF for 1 hr at 4°C do not demonstrate a significant decrease in the number of ECGF receptors[23] or a decrease in the intensity of the cross-linked ligand:receptor complex.[39] In both situations, the specificity of the ECGF receptor cross-linked species was demonstrated by competition with a 100-fold molar excess of cold ECGF.[23,39]

The presence of the ECGF receptor is not a unique property of endothelial cells.[23] Fibroblasts and Balb/C 3T3 cells also contain the M_r 150,000 polypeptide receptor and can utilize ECGF a mitogen.[14,23] Both cell types also possess approximately 30,000 receptor per cell with a K_d of approximately 1 nM. A comparison of the K_d for receptor affinity with the concentration of ECGF required to stimulate half-maximum proliferation for both endothelial cells and fibroblasts suggests that only partial receptor occupancy is required for ECGF-induced mitogenic signal transduction.[23] This correlation has also been observed with other growth factors.[17]

X. FUTURE DIRECTIONS

A new family of polypeptide growth factors has been defined on the basis of structural information deduced from amino acid[14,24] and nucleotide sequence[33,35] information. This family of growth factors encompasses the acidic class of heparin-binding polypeptides[14] which induce DNA synthesis in endothelial cells and fibroblasts.[14,23] Although the cloning of the gene which encodes this polypeptide family has provided a rigorous structural basis for the classification of this polypeptide family, these data have also presented a paradox. How can a growth factor exert an autocrine or paracrine response without containing a signal peptide for extracellular transport after biosynthesis? The resolution of this question promises to provide fundamental information applicable to human development, physiology, and disease. However, at this point, the paradox calls attention to the importance of cell death and lysis as a modulator of cellular function and growth and the potential role for glycosaminoglycans as regulators and/or protectors of these ligand-induced functions. Likewise, studies concerning the mechanism of receptor-mediated signal transduction for this growth factor family has also been initiated. Although a recent report provides exciting evidence for ligand-induced receptor phosphorylation,[40] future directions must provide a structural basis for this activity.

REFERENCES

1. **Gimbrone, M. A., Jr.,** *Prog. Hemostasis Thromb.,* 3, 1, 1976.
2. **Folkman, J.,** *Adv. Cancer Res.,* 19, 331, 1984.
3. **Auerbach, R., Kubai, L., and Sidky, Y.,** *J. Cancer Res.,* 36, 3435, 1976.
4. **Gimbrone, M. A., Jr., Cotran, R. S., Leapman, S. B., and Folkman, J.,** *J. Natl. Cancer Inst.,* 52, 413, 1974.
5. **Maciag, T.,** *Prog. Hemostasis Thromb.,* 7, 167, 1982.
6. **Auerbach, R. and Joseph, J.,** in *Biology of Endothelial Cells,* Jaffe, E., Ed., Martinus Nijhoff, The Hague, 1984, 393.
7. **Risau, W.,** *Proc. Natl. Acad. Sci. U.S.A.,* 83, 3855, 1986.
8. **Folkman, J.,** *Ann. Intern. Med.,* 82, 96, 1975.
9. **Folkman, J. and Cotran, R.,** *Int. Rev. Exp. Pathol.,* 16, 207, 1976.
10. **Gospodarowicz, D.,** *Nature (London),* 249, 123, 1974.
11. **Maciag, T., Cerundolo, J., Ilsley, S., Kelley, P. R., and Forand, R.,** *Proc. Natl. Acad. Sci. U.S.A.,* 76, 5674, 1979.
12. **Lemmon, S. K., Riley, M. C., Thomas, K. A., Hoover, G. A., Maciag, T., and Bradshaw, R. A.,** *J. Cell Biol.,* 95, 162, 1982.
13. **Gospodarowicz, D., Massoglia, S., Cheng, J., and Fujii, D. K.,** *J. Cell Phys.,* 127, 121, 1986.
14. **Thomas, K. A.,** *Comments Mol. Cell Biophys.,* 3, 1, 1985.

15. **Jaffe, E. A., Nichman, R. I., Becker, L. G., and Minick, C. R.,** *J. Clin. Invest.,* 52, 2745, 1973.
16. **Maciag, T., Hoover, G. A., Stemerman, M. B., and Weinstein, R.,** *J. Cell Biol.,* 91, 420, 1981.
17. **James, R. and Bradshaw, R. A.,** *Annu. Rev. Biochem.,* 83, 259, 1984.
18. **Thornton, S. C., Mueller, S. N., and Levine, E. M.,** *Science,* 222, 623, 1983.
19. **Wall, R. T., Harker, L. A., Quadracci, L. J., and Stricker, G. E.,** *J. Cell. Physiol.,* 96, 203, 1978.
20. **Thorgienson, G. and Robertson, A. L., Jr.,** *Atherosclerosis,* 21, 259, 1975.
21. **Hayflick, L. and Moorehead, P. S.,** *Exp. Cell Res.,* 25, 585, 1961.
22. **Terranova, V. P., DiFlorio, R., Lyall, R. M., Hic, S., Friesel, R., and Maciag, T.,** *J. Cell Biol.,* 101, 2330, 1985.
23. **Schreiber, A. B., Kenney, J., Kowalski, W. J., Friesel, R., Mehlman, T., and Maciag, T.,** *Proc. Natl. Acad. Sci. U.S.A.,* 82, 6138, 1985.
24. **Lobb, R. Y., Harper, J. W., and Fett, J. W.,** *Anal. Biochem.,* 154, 1, 1986.
25. **Schreiber, A. B., Kenney, J., Kowalski, J., Thomas, K. A., Gimenez-Gallego, G., Rios-Candelore, M., DiSalvo, J., Barritault, D., Courty, J., Courtois, Y., Moenner, M., Loret, C., Burgess, W. H., Mehlman, T., Friesel, R., Johnson, W. V., and Maciag, T.,** *J. Cell Biol.,* 101, 1623, 1985.
26. **Maciag, T., Mehlman, T., Friesel, R., and Schreiber, A. B.,** *Science,* 225, 932, 1984.
27. **Maciag, T., Hoover, G. A., and Weinstein, R.,** *J. Biol. Chem.,* 257, 5333, 1982.
28. **Burgess, W. H., Mehlman, T., Friesel, R., Johnson, W. V., and Maciag, T.,** *J. Biol. Chem.,* 260, 11389, 1985.
29. **Klagsbrun, M. and Shing, Y.,** *Proc. Natl. Acad. Sci. U.S.A.,* 82, 805, 1985.
30. **Burgess, W. H., Mehlman, T., Marshak, D. R., Fraser, B. A., and Maciag, T.,** *Proc. Natl. Acad. Sci. U.S.A.,* 83, 7216, 1986.
31. **Gimenez-Gallego, G., Rodkey, J., Bennett, C., Riso-Candelore, M., DiSalvo, J., and Thomas, K. A.,** *Science,* 130, 1385, 1985.
32. **Thomas, K. A., Rios-Chandelore, M., Gimenez-Gallego, G. D., Salvo, J., Bennett, C., Rodkey, J., and Fitzpatrick, S.,** *Proc. Natl. Acad. Sci. U.S.A.,* 82, 6409, 1985.
33. **Jaye, M., Howk, R., Burgess, W. H., Ricca, G. A., Chiu, I.-M., Ravera, M., O'Brien, S. J., Maciag, T., and Drohan, W. N.,** *Science,* 233, 541, 1986.
34. **Kozak, M.,** *Cell,* 44, 283, 1986.
35. **Abraham, J. A., Mergia, A., Whang, J. L., Tumulo, A., Friedman, J., Hjerrild, K. A., Gospodarowicz, D., and Fiddes, J. C.,** *Science,* 233, 545, 1986.
36. **Clark, S. C., Ayra, S. K., Wong-Staal, F., Matsumoto-Kobayaski, M., Kay, R. M., Kaufman, R. J., Brown, E. L., Soemaker, C., Copeland, S., Orogzland, S., Smith, K., Sarngadharan, M. G., Lindneer, S., and Gallo, R. C.,** *Proc. Natl. Acad. Sci. U.S.A.,* 81, 2543, 1984.
37. **Fiddes, J.,** personal communication, 1986.
38. **Esch, F., Baird, A., Ling, N., Ueno, N., Hill, F., Denoray, L., Klepper, R., Gospodarowicz, D., Bohlen, P., and Guillemin, R.,** *Proc. Natl. Acad. Sci. U.S.A.,* 82, 6507, 1985.
39. **Friesel, F., Burgess, W. H., Mehlman, T., and Maciag, T.,** *J. Biol. Chem.,* 261, 7581, 1986.
40. **Huang, S. S. and Huang, J. F.,** *J. Biol. Chem.,* 261, 9568, 1986.
41. **Burgess, W. H., Friesel, R., Mehlman, T., and Maciag, T.,** unpublished observations.

Chapter 14

STIMULATORY AND INHIBITORY FACTORS IN VASCULAR GROWTH CONTROL

Patricia A. D'Amore and Susan J. Braunhut

TABLE OF CONTENTS

I. INTRODUCTION

A. Endothelial Cell (EC) Turnover

The proliferation of vascular EC appears to be very stringently controlled. Labeling studies have documented the turnover rate of vascular EC in normal tissues of adults to be consistently low, with 0.01 to 0.1% of the large- and small-vessel EC labeled following a 24-hr ^{3}H-thymidine pulse.[1-3] This closely controlled growth has been shown to be maintained by EC in culture.[4] Using thymidine incorporation and autoradiography, Haudenschild and co-workers[4] demonstrated that confluent cultures of large-vessel EC did not respond to freshly added sera. In contrast, in cultures of 3T3 cells, which are also "contact-inhibited", virtually every cell responded to the addition of fresh nutrients by the synthesis of DNA.

B. Potential Regulators

Although the mechanisms responsible for vascular growth control have not been elucidated, studies in this area indicate several potential regulators including growth factors, the extracellular matrix, cell-cell interactions, and external mechanical forces. This chapter will focus on the first of these — soluble factors that either stimulate or inhibit vascular EC growth. The major emphasis will be on factors demonstrated to modulate EC proliferation in vitro and angiogenesis in vivo.

C. Background

In studying the process of angiogenesis, it might be assumed that factors with a demonstrated effect on neovascularization would have a direct action on EC growth. It is also possible, however, that neovascularization, which is comprised of a series of successive events, may have more than one effector, and the effectors, in turn, are coordinated in some fashion. The delineation of some of the steps of angiogenesis came from a series of morphological studies of developing blood vessels. Ausprunk and Folkman documented that during the process of neovascularization, at least three levels of activity are apparent.[5] The point of origin of the new vascular bed, usually a venule, is characterized by distinct alterations in the basement membrane. This event has been postulated to occur as a result of protease release from EC.[6] Immediately distal to this area of basement membrane dissolution is a population of rapidly proliferating EC. Forward of this region lies a wave of

Table 1
SMALL-MOLECULAR-WEIGHT GROWTH FACTORS

Factor	Neovascularization		Mitogenicity		Migration	
	Eye	CAM	BAE	BCE	AE	CE
Polyamines	ND[a]	+	+	ND	ND	ND
Nucleosides	ND	ND	+	ND	ND	ND
Adipocyte-derived factor	ND	−	+	+	−	−
	ND	+	−	−	+	+
Histamine	ND	ND	+	+	ND	+
Selenium	+	+	+	−	−	+
Adenosine diphosphate	+	+	±	−	±	−
Heparin	−	− (low dose)	−	−	−	+
	+	+ (high dose)				
Copper	+	−	−	−	+	−
Ceruloplasmin	+	−	−	−	+	−
Heparin-Cu^{2+}	ND	+	ND	ND	ND	ND
Gly-his-lys-Cu^{2+}	ND	+	ND	ND	ND	ND
Prostaglandins	ND	+	ND	ND	ND	ND
TAF	+	+	+	+	ND	ND
ECGF-2a-2b	ND	ND	HUVE	+	ND	ND
Lactic acid	+	ND	ND	ND	ND	ND
Hyaluronic acid	ND	+	ND	ND	ND	ND
Fibronectin-Heparin	ND	ND	ND	ND	ND	+

[a] ND = not determined.

migratory EC, the foremost of which form lumens and anastamose, and secrete new basement membrane components. These processes — remodeling of the extracellular matrix, mobilization, and proliferation of the endothelium — are all part of the process of angiogenesis, but it is not known whether these represent discrete events or are simply parts of an interwoven cascade. This chapter will examine some effectors of angiogenesis and will attempt to clarify what specific function they might serve in the process of neovascularization.

Two types of assays are commonly used to assess angiogenic activity. The anterior eye chamber or corneal pocket assay[7] and the chick chorioallantoic membrane (CAM) assay[8] measure an angiogenic response to a test factor in vivo. These two assays can be evaluated histologically for inflammatory cell involvement but cannot distinguish among the actions of effectors in terms of specific aspects of the angiogenic process (e.g., proliferation vs. migration). Factors can be examined in vitro for their effect on individual steps of angiogenesis: migration, proliferation, and protease release. Furthermore, using in vitro assays, factors can be tested for a direct action on a particular vascular cell. Aortic, umbilical vein, and capillary EC, as well as pericytes and smooth muscle cells, can now be cultured and compared for their responsiveness to specific factors. Conflicting observations pertaining to the action of a test substance are often simply the result of different culture conditions or cell strains. Therefore, the method of analysis and the source and species of the target cell must be taken into account when making direct comparisons of factors and their activities.

II. GROWTH FACTORS

A. Small Molecular Weight Growth Factors

A number of small molecular weight factors have been demonstrated to participate in the angiogenic process and/or are implicated in promoting growth of EC (Table 1). These factors can be arbitrarily divided into two categories: (1) factors that are angiogenic and have been examined directly for their effect on EC proliferation, and (2) factors that are angiogenic, but function to influence basement membrane remodeling or EC mobilization (chemotaxis).

1. Mitogenic/Angiogenic Factors

In this first category, there are a number of factors that are known to induce EC proliferation in vitro and/or stimulate angiogenesis in vivo.

a. Polyamines

The polyamines spermine, spermidine, and putrescine have been shown to stimulate angiogenesis in the CAM assay and the growth of human EC in vitro. When human EC are stimulated to proliferate by polyamines, there is a concomitant increase in the production of specific EC proteins and mRNAs.[148] It is unclear what contribution these polyamines make to spontaneous angiogenesis. It may be significant that these same polyamines become cytotoxic for EC if they are oxidized. This event may play some role in regulating the action of the polyamines during wound healing, where an hypoxic microenvironment has been shown to be important in the healing process.[9]

b. Nucleosides

In a systematic study examining aortic EC growth in a variety of medias supplemented with different combinations of nutrients, it was demonstrated that pyrimidine nucleosides in combination with serum increased cell growth by 85 to 100% over controls.[10] These studies proved useful in defining optimal growth conditions for EC and in revealing a synergistic effect of deoxythymidine and deoxycytidine on EC growth. It was recognized by the authors of this study that the pyrimidine nucleosides probably satisfy a nutritional requirement of EC and do not act as true mitogens. Application of nucleoside antimetabolic drugs as a potential antiangiogenic therapy may still prove valid.[11]

c. Adipocyte-Derived Angiogenic Factor

Adipocytes in culture secrete a potent mitogen for both aortic and capillary EC.[12] The relative amount of mitogenic activity associated with cells that differentiated into adipocytes has been shown to be at least tenfold higher than undifferentiated cells.[12] Differentiated adipocytes do not require serum to secrete the mitogenic factor, and aortic EC do not require serum to respond to this factor. Preliminary characterization of this factor revealed that it is not dialyzable and is insensitive to heating, trypsin, or chymotrypsin enzymatic treatment. The mitogenic activity found in association with preadipocyte-conditioned media is destroyed by enzyme treatment, indicating that preadipocytes release a distinct factor or that protease inhibitors are present only in the differentiated cell-conditioned media that protects the factor from degradation. The adipocyte-derived mitogenic activity was separated from other an-giogenic factors secreted by differentiated adipocytes by reverse phase chromatography (C18 column). The mitogenic activity did not bind this column, whereas, three distinct angiogenic/chemotactic factors were retained.[13] The mitogen is nonproteinous and, although it does not chromatograph like a prostanoid lipid, its synthesis is blocked by prostaglandin synthesis inhibitors. It is neither angiogenic in the CAM nor chemotactic for EC.[13]

During embryogenesis, adipose tissue becomes highly vascularized by the ingrowth of numerous small vessels, an event which precedes full development of a fat pad. The observation that a potent EC mitogen is produced as a function of adipocyte differentiation provides a system that will allow investigation of the role of growth factors in development. The finding of different factors, each having a distinct activity (mitogenic vs. chemotactic factors), supports the concept that different factors may modulate the distinct steps of neovascularization, at least in this particular system.

d. Histamine

Histamine has been shown to be a mitogenic for both bovine aortic EC[14] and human capillary EC.[15] Mast cells, a major source of histamine in vivo, are known to participate in inflammatory processes but also have been noted in association with growing and regenerating tissues in the absence of other hallmarks of inflammation. The mitogenic effect of histamine is blocked by H_1 but not H_2 antagonists, implicating a specific interaction of histamine with a H_1 receptor.[15] Heparin, another mast cell releasate, has been demonstrated to be chemotactic for capillary EC,[16] but this action appears to be separate from the effects of other mast cell products, such as histamine.

Little is known about the signals that control degranulation of mast cells. It would be of interest to know if granules can be differentially released since it appears that different mast cell products relate to different steps in the angiogenic process. Finally, it appears that mast cell products facilitate, but are not required, for angiogenesis, since animals deficient in mast cells are able to mount an angiogenic response.[17]

e. Selenium

Selenium is a nonmetal compound that has been reported to be angiogenic in the CAM and corneal pocket assays.[18] Histological analysis reveals that the angiogenesis occurred without inflammation.[18] Three seleno compounds — selenomethionine, selenocystine, and selenious acid — were examined for their ability to induce migration and proliferation of bovine capillary and aortic EC in vitro.[18] All three compounds induced the migration of capillary (but not aortic) EC (at doses of 10^{-6} to 10^{-8} M). In contrast, they stimulated a marked increase in aortic EC proliferation, but only marginally increased capillary EC proliferation. This is somewhat surprising since a majority of the factors that induce capillary EC proliferation also cause their migration. The growth factor activity of these compounds was not due to the amino acid portion of the molecules, as the same amino acids at an equivalent molar ratio were not mitogenic or chemotactic. These same studies conducted using fibronectin-coated plates and four different growth medias, yielded identical results, indicating that differences in nutritional and plating requirements between the two cell types was not responsible for the different responsiveness of the cells.[18]

These studies are relevant in several regards. First, the observed differences between large and microvascular EC in their functional response to selenium is intriguing since potential differences between EC from different sources is of continued interest to vascular cell biologists. Second, selenium is a small compound that could easily escape detection in an "angiogenic factor" preparation. With potency at 10^{-8} M, it potentially could be a contaminant of some of the preparations currently being characterized. Last, these studies may help to explain the cardiovascular lesions observed in small mammals relating to selenium-deficient diets.[19]

2. Chemotactic/Angiogenic Factors

The second category of angiogenic factors consists of those that evoke angiogenesis but are not mitogenic for EC. Rather, these factors have been shown to participate in other events of the angiogenic process, such as EC mobilization and/or basement membrane remodeling.

a. Adenosine Diphosphate (ADP)

ADP has been shown to be angiogenic in the CAM[20,21] as well as in the corneal pocket[22] assays. It does not, however, stimulate proliferation of either aortic or capillary EC or migration of capillary EC in vitro, but does cause aortic EC migration.[23] It has recently been shown that the angiogenic response to ADP in the corneal pocket assay is accompanied by a substantial infiltrate of eosinophils and it is now suspected that ADP is chemotactic for a subpopulation of leukocytes that, in turn, mediates the angiogenic response.[22]

b. Heparin

Heparin, at high doses (100 μg/mℓ), has been reported to be angiogenic both in the CAM[20] and corneal pocket assays.[24] Corticosteroids abolish the latter response, however, indicating that inflammation may be involved.[24] Heparin is not mitogenic for bovine adrenal capillary or aortic EC in vitro but potentiates the mitogenic activity of acidic fibroblast growth factor (FGF) for human umbilical vein EC.[25] Others have shown that heparin stimulates capillary but not aortic EC migration.[16] It has, thus, been suggested that the angiogenic effect of heparin in vivo is via its chemoattractant activity. It is interesting to note that low doses (10 μg/mℓ) of heparin become angiogenic in the CAM and corneal pocket assays when administered in combination with copper.[24]

c. Copper

Copper, over a range of doses, induces angiogenesis in the corneal pocket assay[26] but not in the CAM.[22] It causes migration of bovine aortic but not of bovine capillary EC and is nonmitogenic for EC.[27] Ceruloplasmin exhibits effects identical to those observed for copper.[27]

A review of this literature reveals some additional, intriguing parallels. Heparin, ceruloplasmin, and gly-his-lys all take on new activity when coupled with copper; in combination with copper, they all prove angiogenic in the CAM assay.[22] It was previously shown, using the corneal pocket assay, that copper accumulates locally during an angiogenic response, and that rabbits placed on a copper-deficient diet are unable to mount a complete angiogenic response.[26] Although ceruloplasmin is the native carrier of serum copper, all three factors have been shown to bind molecules of copper, albeit, in varying amounts. Gly-his-lys has been shown to facilitate copper uptake by hepatoma cells,[28] but it is not known if this occurs in EC. Gly-his-lys bound to copper has been shown to have superoxide dismutase activity which may explain the accelerated wound healing observed in animals treated with this compound.[29] The gly-his-lys-Cu^{2+} complex has demonstrated effects on neurite outgrowth — both dendritic and axonal — inhibitory activity on platelet aggregation, and potent chemoattractant activity for EC and mast cells.[29] All of these functions are thought to play a role in wound healing and full restoration of the wound site. Gly-his-lys does not, however, evoke the migration of fibroblasts or influence their collagen production and, thus, exhibits some specificity for cells involved in the angiogenic process. It would be interesting to know if heparin or ceruloplasmin-Cu^{2+} complexes have these wound healing functions in common with the gly-his-lys-Cu^{2+} complex. It remains to be determined if copper bound to heparin, ceruloplasmin, and gly-his-lys mediates neovascularization via the same mechanisms.

d. Prostaglandins

Prostaglandins, type E prostaglandins (PGE) in particular, have long been suspected to play a role in angiogenesis. It has been noted that growing tumors, both spontaneous and induced, have elevated levels of prostaglandins in fluid associated with the tumor.[30] Prostaglandins of the E series have been demonstrated to evoke angiogenesis in the CAM assay.[31] The angiogenic activity of the various prostaglandins now has been shown to be directly correlated with their ability to increase local copper concentrations and can be abolished by placing animals on a copper-free diet.[26] The exact role of copper in the angiogenic process is not known, nor is it clear if copper requires the carriers outlined above to be effective.

e. Tumor-Derived Factors

The concept of a tumor angiogenesis factor (TAF) was first described by Folkman as a diffusible substance from tumor cells that causes new blood vessel growth.[32] Factors that fit this description have been subsequently reported by a number of investigators[33-37] from a variety of tumors, but also from normal tissues including cat retina[36] and synovial fluid.[38] It is presumed that these factors are related, if not identical, to each other based on their molecular weight, charge, and the specificity of the reported target cells; however, the definitive characterization of these tumor angiogenic factors has not been reported. These factors are anionic, have low molecular weights (200 to 800), and have ultraviolet absorption peaks at 260 nm. There have been conflicting reports of the action of TAF. The various investigators agree that TAFs are capable of evoking angiogenesis in the CAM and corneal pocket assays. However, one group reported that tumor extracts failed to elicit EC proliferation in vitro,[39] whereas others reported that the tumor factors stimulated capillary but not aortic EC division.[34,37] Fenselau and colleagues have demonstrated that a highly purified factor derived from the Walker 256 rat tumor stimulates fetal bovine aortic as well as capillary EC proliferation in the absence of serum components without the use of a collagen substrate.[35] Corneal fibroblasts and smooth muscle cells were unresponsive to the activity. Fenselau et al. found that platelet-derived growth factor was inhibitory for aortic EC and thus speculated that culture conditions might explain the discrepancy between their findings and the findings of others who conducted proliferation studies in the presence of platelet factors.[35]

ECGF-2A and ECGF-2B, two low-molecular-weight proteins purified from human hepatoma cell-conditioned media, stimulate the proliferation of human umbilical vein EC but not smooth muscle cells, fibroblast, or hepatoma cells in serum-free media.[40] ECGF-2A has an estimated molecular weight of 6500, an isoelectric point of 6.0, and an amino acid sequence that is homologous with the first 25 amino acids of the NH_2-terminus of human pancreatic trypsin inhibitor. In addition to its growth-promoting activity, ECGF-2A exhibits inhibitory activity for pancreatic trypsin. ECGF-2B has a molecular weight of 27,000, is very acidic (pI below 4.0), is homologous with the first 49 NH_2-terminal residues of a urinary proteinase inhibitor, and inhibits pancreatic trypsin. It has not been determined if the growth-promoting activity of these two proteins is related to their protease inhibitory activity.

It is not clear at this time what relationship, if any, these small molecular weight tumor-derived factors bear to the heparin-binding growth factors (see Section B below). In particular, the factor derived from retina has been claimed to resemble, in some of its biochemical characteristics,[36] the bovine retinal angiogenic factor described by D'Amore et al.[41] Heparin-binding EC mitogen from retina exists in multiple forms with molecular weights of 16,000 to 20,000.[42] The latter is not reduced to lower molecular weight forms even after treatment with ethanol,[41] which was used in the purification of the small molecular weight molecule from retina.[36] This does not rule out the possibility that the smaller molecule is an active fragment of the larger heparin-binding mitogen, generated by the homogenization of the tissue during the initial extraction procedure.[36]

f. Lactic Acid

In one study, lactic acid (in particular, the l-lactate form) was reported to induce neovascularization in the corneal pocket assay.[43] It is known that the development of hypoxic conditions precedes neovascularization in a variety of tissues, and that lactic acid increases locally as a by-product of anaerobic glycolysis. Comparable swelling of the anterior chamber, by injection of silicon oil or intracorneal injection of other, more caustic agents, was not sufficient to evoke neovascularization to the same extent as lactic acid. Lactic acid has been reported to increase the proliferation of fibroblasts in vitro and to promote the synthesis of hyaluronic acid and other extracellular matrix constituents by these cells.[44] It is not clear by what mechanism — direct or indirect — lactic acid stimulates neovascularization.

g. Hyaluronic Acid

Sodium hyaluronate has been demonstrated to induce angiogenesis in the CAM assay.[45] The angiogenic activity of partially degraded hyaluronate has recently been shown to be retained by highly purified fragments of sodium hyaluronate ranging from 4 to 25 disaccarides in length.[149] Based on histological analysis, inflammation does not appear to be involved in this neovascular growth.

h. Fibronectin/Heparin Fragments

Fragments of heparin and fibronectin, when tested in combination, have been reported to be chemotactic for bovine capillary EC in vitro.[46] Bovine plasma fibronectin fragments (10,000 to 30,000 daltons), prepared by trypsin digest, were tested for their EC chemotactic activity alone and in combination with a 4500-dalton fragment of heparin. When assayed separately (10 to 200 μg/mℓ), neither stimulated the mobilization of EC. However, in combination (fibronectin at 5 μg/mℓ heparin 50 μg/mℓ), the fragments stimulated capillary EC migration. It is not clear from these studies the purity of the reagents used and there is no indication whether proliferation of EC was observed.

The authors of these studies have speculated that in nonpathologic angiogenesis, such as wound healing and tissue remodeling, where basement membrane and extracellular matrix material is broken down, generation of such fragments may be all that is required for a full angiogenic response. In these cases, they postulate that specific "endothelial cell mitogens" may not be involved. It is not clear that the data support this hypothesis. In some cases, these factors have not been tested for their mitogenic activity, and yet the authors presume their primary action is as chemoattractants. If the factors are shown to be nonmitogenic in future studies, it would appear that an additional effector would be required to induce the necessary expansion of the EC population. If these factors prove to be mitogenic in initial studies, particularly in the case of extracellular matrix (ECM) components, it must be demonstrated that these are highly purified reagents, free of possible contaminating EC mitogens, which have been shown to reside bound to the ECM.[47]

B. Peptide Growth Factors

1. Heparin-Binding Growth Factors (HBGF)

Following the successful culture of vascular EC in the early 1970s, there were numerous reports of cell, tissue, and organ extracts that stimulated the proliferation of the cells (for review see Reference 48). These reports accumulated for nearly a decade with little progress in the purification and characterization of these activities and with no information on the relationship of these "factors" to one another. A breakthrough in the purification of EC growth factors came with the discovery by Shing and co-workers that a tumor-derived EC mitogen bound with high affinity to the glycosaminoglycan heparin.[49] Since that time, there have been many reports on the purification and characterization of heparin-binding EC from a variety of tissues (Table 2).

There are two other chapters (Chapters 13 and 14) in this publication that deal in detail with HBGFs. Therefore, the treatment of the subject in this chapter will be neither extensive nor inclusive.

a. Purification

The first EC mitogen to be purified by taking advantage of its heparin affinity was a cationic polypeptide factor from a rat chondrosarcoma.[49] This factor had an isoelectric point of between 9 and 10, eluted from heparin with 1.8 M NaCl and had a molecular weight of 18,400. The initial observation of heparin binding of the cationic tumor-derived factor was surprising. In light of the basic nature of the activity, one would expect the molecule to bind heparin. It was the strength of binding, however, that was unexpected. Whereas platelet-derived growth factor, a peptide growth factor of similar charge, bound heparin and was eluted with 0.6 M NaCl, 1.8 M NaCl was required to elute the tumor-derived EC mitogen.

EC stimulators that were acidic molecules (isoelectric points of 4 to 5) had also been identified in a number of tissues (see Section II.B.c below). Because of the anionic nature

Table 2
PEPTIDE GROWTH FACTORS

Factor	Neovascularization		Mitogenicity		Migration	
	Eye	CAM	BAE	BCE	BAE	BCE
Acidic FGF	+	+	+	+	+	+
Basic FGF	+	+	+	+	ND[a]	ND
Angiogenin	ND	+	ND	ND	ND	ND
ECGF	+	ND	+	+	ND	ND
Insulin	ND	ND	±	+	ND	ND
Lipoproteins	ND	ND	+	+	ND	ND
Macrophage-derived factors	+	ND	ND	−	ND	+
Platelet-derived growth factor	ND	ND	−	−	ND	ND
Thrombin	ND	ND	−	ND	ND	ND

[a]   ND = not determined.

of these factors, it was not expected that they would bind to heparin, a highly sulfated molecule. However, these factors did bind to immobilized heparin[42,50] and were eluted with 1.0 M NaCl. Subsequent purification, combining heparin and reverse phase or ion-exchange chromatography, revealed at least two forms of the acidic factor with molecular weights of 16,000 to 18,000. Amino acid sequence analysis indicated a 35% homology with interleukin-1.[51]

b. Characterization

Rapid progress in the characterization of these two classes of EC mitogens has been made in the past 2 years. Amino acid sequence determination has revealed a 53% homology between the two molecules.[52,53] These data indicated that the two classes of heparin-binding EC mitogens (acidic and basic FGFs; HBGF and FGF are used interchangeably) are the products of separate genes that arose from a single ancestral one. This idea has been confirmed by recent reports of the cloning and sequencing of the genes for both acidic[54] and basic[55] FGF.

Heterogeneity has been reported in the molecular weights of both the acidic and basic HBGF. High (>50,000) and low-(15,000 to 20,000) molecular weight forms of the EC mitogens have been demonstrated in the retina[41,42] and brain.[56] Further, microheterogeneities have been demonstrated among the lower molecular weight forms in both the acidic[57] and basic FGFs.[58,59] Although the exact significance of the various-sized molecules is not precisely known, it is generally thought that the variability is an artifact of purification procedures.[58,59]

c. Distribution

Cationic FGF has been demonstrated in virtually every tissue examined (for examples see References 60 to 63). Human tumor cells[58] and cultured EC[47] have been shown to synthesize cationic FGF, and the latter have been demonstrated to deposit the activity into the sub-endothelial basement membrane.[47]

Anionic FGF has been primarily isolated from neural tissues including retina,[42,64,65] brain,[50,63] and hypothalamus.[66] More recently, acidic FGF has been purified from human tumors[67] and bone.[68]

d. Activity

Anionic and cationic HBGF have been shown to be mitogenic[42,50,66] and chemotactic[69-71] for both large- and small-vessel EC. Additionally, these factors have been shown to stimulate the growth and/or differentiation of a variety of other cell types including neurites.[72,73] Cationic HBGF has been shown to be angiogenic (300 to 600 ng per assay) in the CAM assay and the rat corneal pocket.[74] Histological analysis revealed that the angiogenic response was noninflammatory, that is, that the purified factor induces new vessel formation directly not via a second (inflammatory) cell. Acidic HBGF have also been demonstrated to be angiogenic in the same assay systems.[75] Although histological analysis was not reported in the latter study, it is suspected that acidic HBGF would not prove inflammatory but would also act as a direct angiogenic stimulus.

e. Receptors

Receptors for HBGF have been demonstrated on a variety of cell types including BHK-21,[76] lens epithelial cells,[77] lung capillary EC,[78] fibroblasts,[79] skeletal muscle cells,[80] and 3T3 cells.[80] The estimate of the molecular mass of the receptors in these studies ranges from 125,000 to 165,000 daltons. Heparin has been shown to inhibit the binding of cationic FGF in BKH cells[76] and enhance the binding of acidic FGF to pulmonary capillary EC.[79]

f. Role of HBGF

At the outset of this chapter, we emphasized the fact that under normal conditions, vascular EC have very low turnover rates. Thus, it is paradoxical that virtually every tissue examined to date contains at least one of the two identified classes of heparin-binding EC mitogens. Recently reported DNA sequence information reveals that neither of the classes of HBGF possesses classic signal peptides, suggesting that these molecules are not constitutively secreted proteins. However, the observations of Vlodavsky and co-workers of significant levels of HBGF associated with the extracellular matrix indicates that some activity is released from the cell, although the mechanism by which this occurs is not known.[47] Thus it is possible that the factors remain cell associated until they are "needed". Nearly every condition in which neovascularization occurs is associated with ischemia. One explanation, therefore, is that the alterations in the microenvironment that occur with ischemia (e.g., hypoxia and acidosis) stimulate the release of these factors (from the cells and/or the matrix). The mechanism underlying the release could be relatively nonspecific, i.e., changes in membrane permeability due to cell damage. Recent reports of the differentiation-dependent production of acidic FGF in brain[81] and kidney[82] further suggests a role for these factors in vascular development.

2. Nonheparin-Binding Peptide Growth Factors
a. Angiogenin

Angiogenin, a 14,400-dalton, cationic protein isolated from a human adenocarcinoma cell line, has been reported to be angiogenic in the CAM assay.[83] The amino acid and gene sequences for angiogenin were determined,[84,85] and it was found that the protein is 35% homologous with human pancreatic ribonuclease. Though the initial studies indicated that angiogenin did not have ribonuclease activity,[85] a more recent paper reports endonucleolytic activity for both 28S and 18S RNA.[86]

b. Epidermal Growth Factor (EGF)

Human vascular EC from umbilical vein,[87] iliac artery, and thoracic aorta[88] have been demonstrated to be stimulated to proliferate in response to EGF. The mitogenic activity on human EC was reported to be potentiated by the addition of thrombin.[87] On the other hand, neither thrombin, EGF, nor a combination of the two had any effect on the proliferation of bovine vascular EC.[87] This latter finding is in contrast with the EGF stimulation of bovine capillary EC proliferation reported by McAuslan and co-workers.[89]

Additionally, EGF has been reported to be angiogenic in the rabbit corneal pocket assay.[90]

However, the relatively high concentration of EGF required to obtain reproducible neovascularization (10 μg per implant) and the observation of leukocyte infiltration at the implant site raises the question as to whether the vessel growth was the direct result of EGF or secondary to the presence of leukocytes.

c. Insulin and Insulin-Like Growth Factors

There are numerous reports characterizing and quantitating receptors for insulin and insulin-like factors on EC from all levels of the vasculature. Insulin receptors have been demonstrated on cultured EC from bovine adipose tissue, human umbilical vein, and bovine pulmonary arteries and aortas,[91-93] as well as freshly isolated liver EC[94] and myocardial EC in vivo.[95] In addition, both large-vessel and capillary EC have been found to have receptors for insulin-like growth factor I (IGF-I) and II and multiplication-stimulating activity (MSA).[92,96,97] Use of anti-insulin antibodies did not change the binding of labeled MSA, but did reduce the binding of IGF-I by 30 to 40%, suggesting "overlapping determinants between insulin and IGF-I receptors that were not present on MSA receptors".[92] Interestingly, when compared to human umbilical vein EC, bovine cells bound four to ten times more MSA.[92]

In spite of the extensive characterization of insulin receptors on EC, relatively little has been reported about the biological action of insulin on EC. King and co-workers have shown that insulin is mitogenic for cultured retinal capillary EC but not for large-vessel (aortic) EC.[98] These results are in contrast to reports that insulin stimulates ^{3}H-thymidine incorporation into bovine aortic EC DNA[99] and is mitogenic for human arterial EC.[87] One explanation for the apparent lack of biological effect of insulin on EC comes from the work of King and Johnson who, using a dual-chambered vessel, demonstrated that aortic EC transport insulin across the monolayer by a receptor-mediated and energy-dependent system, leaving a majority of the insulin intact.[100] A recent report by the same group demonstrates that both large- and small-vessel EC bind and process insulin and IGF-I by degradative and nondegradative pathways with the latter predominating.[101] Thus, it is possible that the major function of insulin receptors on EC is to facilitate the shuttling of insulin to the parenchymal cells of the tissue and the other cells of the blood vessel.

d. Macrophage-Derived Growth Factors

There are a number of reports that macrophages contain materials that modulate EC function in vitro and induce angiogenesis in vivo. Banda and co-workers have demonstrated that wound fluid contains an activity that stimulates new vessel growth in the corneal pocket assay and is chemotactic but not mitogenic for capillary EC.[102,103] These investigators demonstrated that macrophages could be stimulated to secrete angiogenic activity by culturing them under conditions that simulated a wound environment (e.g., hypoxia or high lactate).[104,105] The chemotactic factor identified in these studies was reported to have a molecular weight between 2,000 and 14,000.[103]

A molecule of similar molecular size (7,000 to 10,000) has been isolated from sarcoid macrophage-epitheloid cells. However, unlike the molecule described by Banda et al., this factor stimulates the proliferation of aortic EC.[106] Basic FGF has been demonstrated in macrophages and was suggested by those investigators to account for the macrophage-associated EC mitogenic activity.[62] However, the finding that activity isolated from the sarcoid macrophage cells has an acidic isoelectric point, does not adhere to heparin-Sepharose, and is destroyed by dithiothreitol treatment differentiates it from basic FGF.

Tumor-associated macrophages have similarly been shown to produce factors mitogenic for EC and angiogenic in vivo.[107] Interestingly, tumor cell suspensions that had been depleted of macrophages were less effective inducers of angiogenesis, leading the authors to speculate that tumor neovascularization is mediated, at least in part, by macrophages. The convincing demonstration of macrophage-derived mitogenic/angiogenic activity, together with the known association of macrophages with tumors, inflammation, wound repair and defense mechanisms, points to a role for these cells in angiogenesis.

e. Platelet-Derived Growth Factors

Although none of the platelet-derived EC growth factor activities have been purified to homogeneity, there are a few reports that provide sufficient characterization to permit comparison. Clemmons and colleagues have isolated from platelets a dialyzable factor (MW 700), stable to acid and boiling, that stimulates the proliferation of aortic EC.[108] The authors speculated that since FGF was mitogenic only in the presence of nondialyzed serum or plasma plus this moleucle, this factor may be the component of serum that supports EC growth.

A larger molecular weight molecule has been partially purified from platelets and shown to stimulate ^{3}H-thymidine incorporation by umbilical vein EC.[109] The activity was demonstrated to be heat labile and to elute from Sephacryl-200 with apparent molecular weights of 65,000 and 135,000. Since the heparin affinity of this factor has not been reported, its relationship to the heparin-binding EC mitogens described above is unclear. The fact that the activity appears to have a molecular weight larger than that of the FGFs does not eliminate this possibility since the FGFs have a tendency to aggregate, may exist as larger precursors, and/or have carrier molecules associated with them.[41,56]

Mitogenic activity for EC has been partially purified from platelets by Miyazono et al.[110] Although the heparin affinity of this factor was not reported, its characteristics — including an isoelectric point of 4.0, a molecular weight of 20,000, and insensitivity to dithiothreitol treatment —strongly suggest identity with acidic FGF.

f. Thrombin

There are conflicting data on the action of thrombin on EC. Thrombin alone was shown to have no effect on the proliferation of human EC;[87,111] whereas, a recent study reports a nearly twofold stimulation of human EC proliferation using similar thrombin concentrations (1 to 2 μg/mℓ).[83] When added in conjunction with FGF, thrombin potentiated the mitogenic effect of FGF on human umbilical vein EC but not bovine vascular EC.[87] It is important to note that since this work predates the development of methods for the purification of FGFs to homogeneity, the nature of the FGF used in these experiments (acidic, basic, or a mixture) is unknown. Though pure platelet-derived growth factor (PDGF) was not mitogenic for human vein EC,[87,111] the addition of thrombin and PDGF was reported to stimulate for EC growth.[111]

In support of these data is the report of thrombin-binding sites on human EC. Using ^{125}I-thrombin, Awbrey and co-workers demonstrated two sets of receptors, a small population (3300 per cell) of high-affinity receptors and a larger population of low-affinity receptors.[112] As of this writing, bovine cells have not been examined for thrombin receptors. Lack of binding sites on these cells might be expected in light of the absence of potentiating effects.

g. Lipoproteins

It has been reported that subconfluent cultures of human umbilical vein EC and smooth muscle cells are growth-inhibited by low-density lipoproteins (LDL) and that at low doses of LDL, this effect could be blocked by the addition of high-density lipoproteins (HDL).[113] It has now been demonstrated that HDL are mitogenic for EC, and are one of the essential components required for maintaining EC in a chemically defined, serum-free media.[114,115] Corneal, adrenal, corpus luteal, and cortical brain EC, when grown in the presence of HDL and transferrin, exhibit growth equal to that in full serum.[115-117] Cells grown on plastic or gelatin-coated dishes did require HBGF; whereas, cells grown on EC-derived extracellular matrix did not. EC cultured under these defined conditions exhibited morphology typical of

EC, and proliferation rates and life spans similar to EC cultured in serum. Thus, it appears that FGF, HDL, and transferrin are the minimal supplements for maintenance of optimal EC growth.

HDL are composed of complex aggregates of apolipoproteins and lipids. Initial attempts to fractionate the active component of HDL has revealed that apolipoprotein C_1, the lowest molecular weight apolipoprotein of HDL (6331 MW), has a mitogenic effect on aortic EC equal to that of intact HDL. However, this effect could be demonstrated only in lipoprotein-deficient serum and not in serum-free media, and the HDL fraction that remained after total removal of apolipoprotein C_1 was also mitogenic.[118]

A 27,000-dalton protein has been identified as a potential candidate for mediating the intracellular events of HDL-induced mitogenesis.[117] Using ^{32}P-labeling, Darbon et al.[117] demonstrated dose-dependent serine and threonine phosphorylation in response to HDL treatment. This 27 kdalton protein was also phosphorylated in response to thrombin but not in response to the addition of insulin, EGF, or PDGF.

HDL has also been shown to activate a regulatory enzyme of cholesterol metabolism, 3-hydroxy-3-methylglutaryl coenzyme A (HMG-CoA).[119] This enzyme is required for EC growth and survival; EC treated with a reductase inhibitor, compactin, do not survive. Actively dividing cells express 50-fold higher levels of HMG-CoA than quiescent cells and the activity of the enzyme has been shown to increase in a dose-dependent fashion in response to HDL treatment.[119]

III. INHIBITORY FACTORS

A. Nonpeptide Inhibitory Factors

1. Phorbol Esters

Doctrow and Folkman have reported that phorbol esters inhibit the growth of bovine capillary aortic EC in vitro.[120] The inhibition of growth is concomitant with a change in the shape of the EC, characterized by a rounded cell body and long, cytoplasmic cell processes (Figure 1a). The proliferation of bovine capillary EC was inhibited in a dose-dependent fashion with the addition of phorbol 12, 13-dibutyrate (PDBu) to the media. PDBu also inhibited proliferation of bovine capillary EC in response to acidic and basic HBGF (Figure 1b). One possible mechanism of this growth inhibition was revealed by a comparison of the action of five different phorbol compounds in these studies. The ability of the phorbol esters to inhibit EC growth was directly correlated with their ability to activate protein kinase C. Protein kinase was detected in the cytosol of capillary cells, making it feasible that the inhibition is mediated by the activation of protein kinase C.

Montesano and Orci[121] found that a phorbol ester, phorbol myristate acetate (PMA), induced a shape change, independent of DNA or protein synthesis, in both sparse and confluent cultures of bovine capillary EC. When capillary EC were treated with PMA on hydrated collagen gels, they migrated into the gel and formed branching cords that anastomosed below the surface of the gel, a morphology that was reminiscent of a capillary network.[5] These in vitro observations may reflect events that occur in vivo during angiogenesis. Factors that effect protein kinase C activation may signal the formation of new capillaries by rendering capillary EC unresponsive to mitogens, causing the cells to align and organize into a capillary network.

2. Retinoids

Retinoids have proven growth inhibitory in a number of transformed and normal cell types. In many instances of growth inhibition, the cells were reported to be altered phenotypically, undergoing changes representative of a more differentiated cell type. There are, as yet, very few reports that examine the effects of retinoids on vascular cells.

Melnykovych and Clowes reported that bovine aortic EC are growth inhibited by the

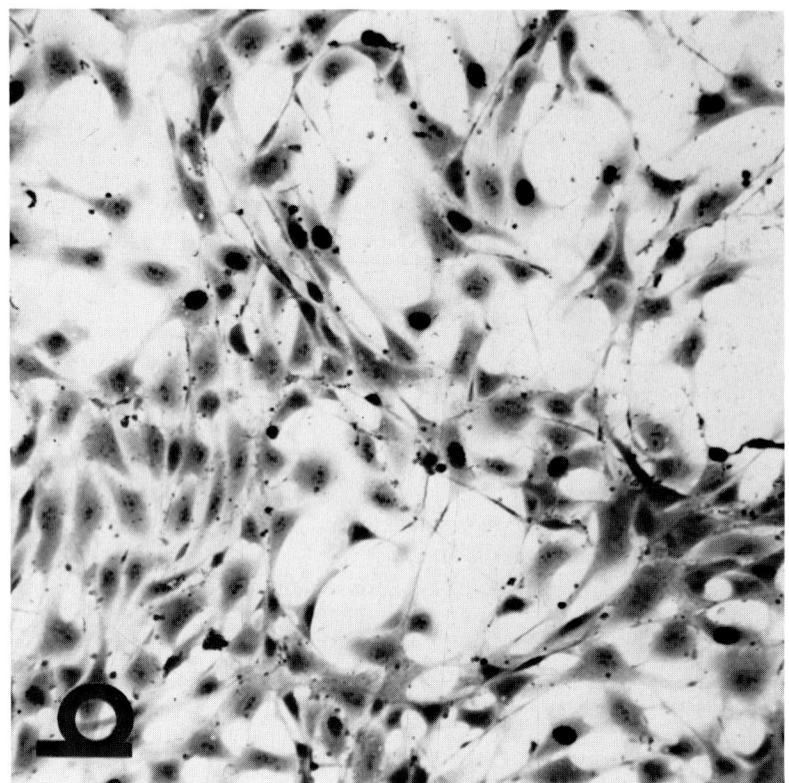

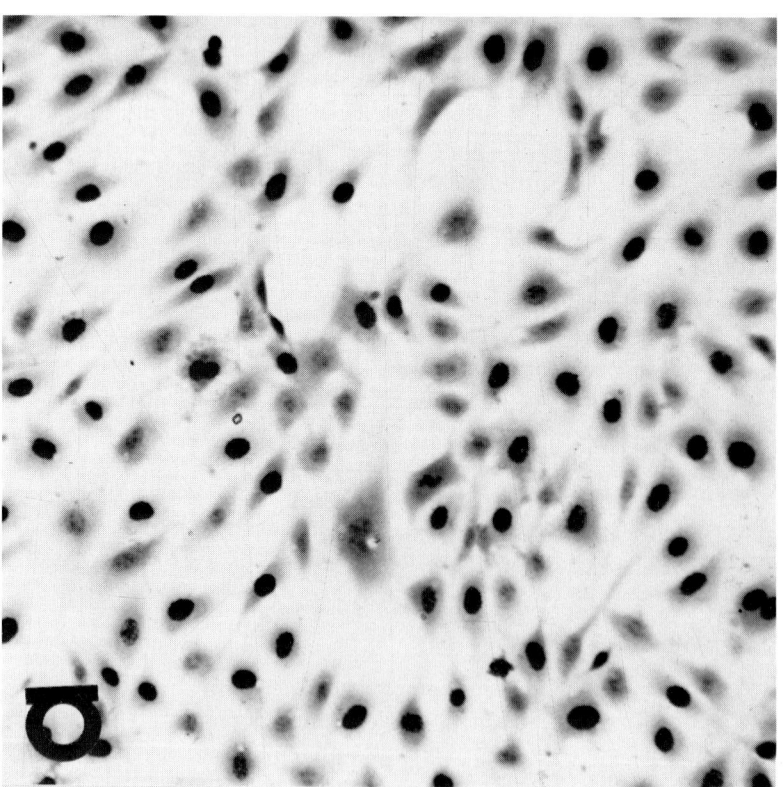

FIGURE 1.   Autoradiography of bovine capillary EC incubated for 24 hr in media containing 1% calf serum supplemented with 4ng/mℓ   HBGF in the (a) absence or (b) presence of 150 n*M* of the phorbol ester, PDBu.   Cells were pulsed for 3 hr with ³H-thymidine, fixed, counterstained with methylene blue, and processed for tritium autoradiography.   Cultures of EC treated with phor-bol esters exhibit a markedly reduced number of labeled nuclei and undergo a morphologic change from control cells incubated in 1% calf serum alone (Figure 2a) or in media containing HBGF (Figure 1a).   (Magnification × 200.)   (Photograph courtesy of Dr. Susan Doctrow.)

addition of retinol, with 20% inhibition in 10% fetal calf serum (FCS); 38% in 2% FCS.[122] The effect of retinol on aortic EC was examined in parallel cultures of cells grown in 10% delipidized serum. (Delipidization removes all of the serum lipids but also denatures up to 30% of the serum proteins.) Under these conditions, the addition of retinol was shown to stimulate bovine aortic EC growth. However, the control cells (grown in delipidized serum in the absence of retinol) did not divide over the time course of the experiment. These data suggest that delipidization removes components that are rate limiting for EC growth and causes one to question the growth stimulation observed with the retinoids.

Our own observations on the effect of retinoids on bovine microvascular cells have been more conclusive.[123] Two forms of Vitamin A — retinol and retinal acetate (1 μM)—were growth inhibitory for bovine capillary EC grown in the presence of either 2 or 5% calf serum (CS). The inhibition (compared to untreated controls) was 33% over an initial 72-hr period, and 100% in a subsequent 72-hr period. The observed growth inhibition was accompanied by a dramatic change in the morphology of the cell noted by 24 hrs (Figure 2). The cells had long processes and the cell bodies retracted and appear more rounded after treatment with either retinol or retinyl acetate (Figure 2b). The observed EC shape change was reminiscent of that reported after treatment with phorbol esters[120,121] (Figure 1) and tumor necrosis factor,[124] treatments that also resulted in EC growth inhibition. The inhibition was reversible and thus not cytotoxic for EC. Retinoic acid, another form of the vitamin, had no effect on capillary EC growth. Retinol and retinal acetate, which inhibited microvascular EC growth, did not effect retinal pericyte growth. Aortic EC, assayed under similar conditions, were marginally growth inhibited (19%) by the addition of 1 μM retinol or 1 μM retinyl acetate. Growth inhibition was not observed in studies using higher serum concentrations (10%). This may be due to the presence of mitogens in the serum.

3. *Heparin — Heparin/Steroids*

Rosenbaum et al. reported that human umbilical vein EC growth is inhibited by up to 30% when heparin is added to growth media containing low amounts of serum (1%), but this effect is not observed at higher concentrations of serum.[125] Gospodarowicz and co-workers reported that heparin (10 $\mu g/m\ell$) from several commercial sources strongly inhibits the growth of bovine adrenal and brain cortex capillary EC but not human umbilical vein EC, in response to serum (10%) or acidic and basic FGF.[115] We have repeatedly observed the dose-dependent inhibition of adrenal capillary EC by heparin alone and strong potentiation of the mitogenic effect of acidic FGF by heparin.[150] The only differences that are immediately apparent between the studies of Gospodarowicz and ourselves is that our cells are cultured in gelatin-coated dishes and supplemented with a crude extract of retina (containing both acidic and basic FGF); whereas, Gospodarowicz and colleagues maintain cells in basic FGF and conduct their studies using cells grown on plastic or EC extracellular matrix.[115] It is not known which of these differences, if any, account for the contradictory observations.

Folkman and co-workers have reported that the administration of heparin (or a heparin fragment) along with cortisone resulted in the regression of tumor masses.[126] Strong evidence indicates that the tumor regression was due to the inhibition of angiogenesis. A recent report by Ingber et al. suggests that the mechanism of this antiangiogenic effect is the induction of basement membrane dissolution.[127]

B. Peptide Inhibitory Factors
1. *Transforming Growth Factor-Beta (TGF-B)*

TGF-β is a platelet releasate that has been reported to stimulate wound healing by inducing vessel ingrowth.[128,129] TGF-β, a 25,000-dalton peptide, has been demonstrated to inhibit the growth of bovine aortic EC. This growth inhibition was correlated with an increase in the ability of aortic EC to form colonies in soft agar, a measurement of anchorage-independence.

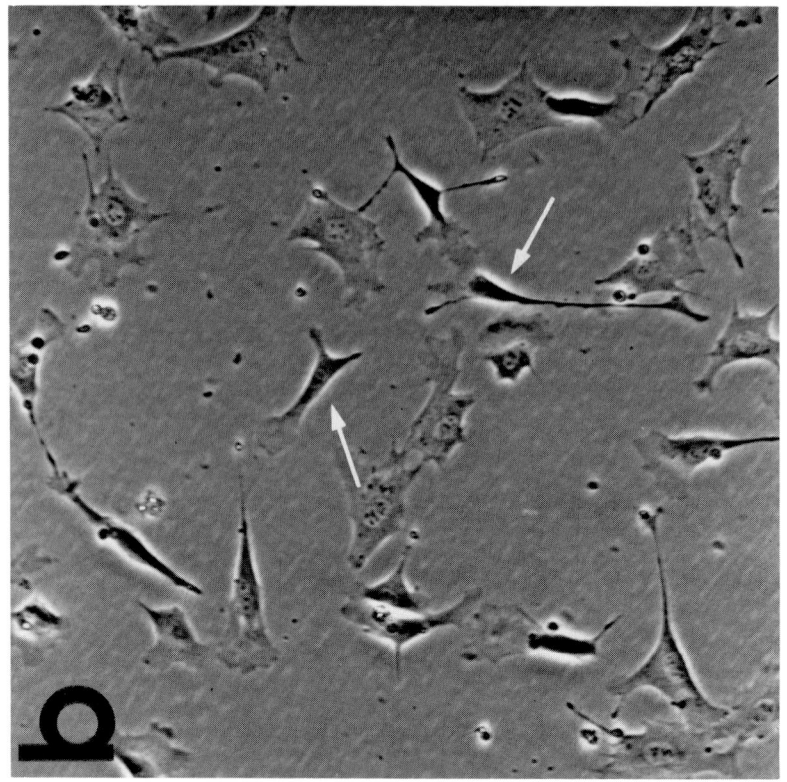

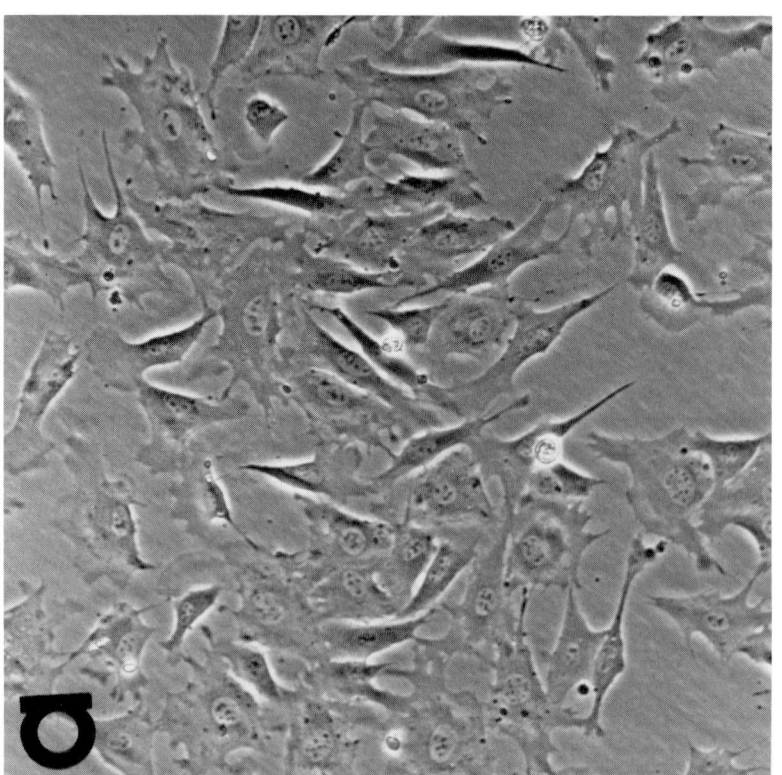

FIGURE 2.   Phase contrast microscopy of bovine capillary EC grown in media containing (a) 5% calf serum alone, or (b) with 1 μM retinol. Cells incubated with retinol (b) display an altered morphology as compared to the control cells incubated in 5% calf serum (a). (Magnification × 93.) (Photograph provided courtesy of Dr. Susan Doctrow.)

More recently, TGF-β was shown to inhibit the proliferation and migration of EC in an in vitro wound model.[130] The inhibition was transient and the cells recovered and healed the wound by 48 hr. The basis for the apparent paradox — that TGF-β is angiogenic in vivo yet inhibits proliferation and migration in vitro — is unclear.

2. Tumor Necrosis Factor (TNF)

TNF was first described by Carswell and colleagues in 1975 as a substance derived from the serum of zymosan-treated rodents which caused the necrosis of certain tumors.[131] TNF causes hemorrhage, necrosis, and regression of a variety of transplantable mouse and human tumors in vivo. It is not known if the rejection of these various tumors is via the same or different mechanisms. TNF has been shown to be directly cytotoxic for some human tumor cells,[132] but its main effect is thought to be directed toward the vascular bed of the tumor. Angiogenesis is associated with the growth of solid tumors. Tumor vessels differ from their normal counterpart by lacking innervation, exhibiting a more robust proliferation of EC, and having weak and leaky walls. The effect of TNF on solid tumors is thought to be due to its direct action on the endothelium of these tumor vessels. Evidence for this comes from a series of in vitro studies, demonstrating that TNF is cytotoxic for bovine capillary EC and growth inhibitory for both bovine aortic and human umbilical vein EC.[119] A monoclonal antibody raised to a 16,000 to 18,000-MW isolate neutralized the growth-inhibitory effects.

3. Gamma-Interferon

Gamma-interferon has been reported by several investigators to inhibit the proliferation of EC.[133,134] Human leukocyte interferon was demonstrated to have two effects on capillary EC: it inhibited capillary EC growth in response to mouse sarcoma-conditioned media, and prevented the migration of capillary EC in response to tumor-chemotactic factors.[133] Gamma-interferon also has been shown to inhibit the migration of bovine aortic EC.[133]

4. Inhibitors from Tissue Extracts
a. Vitreous

Vitreous is one of several poorly vascularized tissues in the adult that has been reported to have antiangiogenic activity.[135] Vitreous extracts directly inhibit bovine aortic EC proliferation in vitro[136] and mitogen-induced angiogenesis in the CAM assays in vivo.[137] It now appears that vitreous, derived from a number of species, has at least two inhibitors, one low (< 10,000) and one high (>50,000) molecular weight.[136] The high molecular weight species has antitrypsin activity, and to a lesser extent, antichymotrypsin activity. The low molecular weight forms do not exhibit protease inhibitor activity but are distinct from other low molecular weight inhibitors, such as protamine,[138] by their inability to adhere to heparin. In the vitreous, the high molecular weight inhibitor fraction is enriched with hyaluronic acid, a glycosaminoglycan with demonstrated antiangiogenic activity in the CAM.[137] Prior treatment of this fraction[136] or whole vitreous[137] with hyaluronidase does not effect the inhibition of aortic EC growth or inhibition of acidic HBGF-induced neovascularization in the CAM, indicating that the inhibition is not due to contamination with hyaluronic acid.

b. Cartilage

Cartilage extracts have been demonstrated to inhibit solid tumor growth, presumably via their ability to prevent angiogenesis.[139] Sorgente and Dorey fractionated cartilage extracts. One fraction had antitrypsin activity and did not inhibit bovine aortic EC proliferation in vitro, and another low molecular weight constituent exhibited no antitrypsin activity but inhibited thymidine incorporation in bovine aortic EC.[140] Langer and Murray further purified this activity using HPLC and identified two peaks (16,000 and 30,000 MW) with inhibitory activity in the rabbit corneal assay.[141] Takigawa et al. have recently described a high molecular weight (100,000 daltons) factor, isolated from cartilage that inhibited the growth of pulmonary artery EC.[142] This factor has not been evaluated for its in vivo effects or its antitrypsin activity. The relationship that these factors bear to one another remains to be clarified.

c. Aorta

Aortic tissue has been reported to have antiangiogenic activity and to inhibit the proliferation of bovine aortic EC in vitro.[143,144] Two low-molecular-weight forms have been shown to have inhibitory effects on the growth of large-vessel EC, one with and one without antitrypsin activity.

d. Lens

Bovine and human lens extracts have demonstrated inhibitory activity for bovine aortic EC. Two distinct species have been described, one greater than and one less than 50,000 daltons.[145] The inhibitory effect in aortic EC was reversible and specific in that no inhibition of smooth muscle cell proliferation was observed. The authors speculate that the inhibitor present in the lens may act to prevent iris vascularization. The neovascularization of iris that occurs when the lens is removed may result from the removal of this natural inhibitor.

IV. CONCLUSIONS

This chapter has sought to review what is currently known about the soluble factors that act to either stimulate or inhibit EC growth. The modulating effects of these factors on EC cannot, however, be taken out of the context of the cell state at the time of the signal. Cell shape, extracellular matrix, and cell-cell contact are but a few of the other variables that can modulate the actual responsiveness of EC to the soluble factors. Several of these modulating forces will be reviewed in other chapters.

A central question concerning the mechanism of growth modulation by stimulatory and inhibitory molecules is how these effects might be mediated in vivo. Any model that seeks to explain the in vitro observations should have relevance not only to the regulation of single cells but to the coordination and regulation of populations of cells within a tissue. One of the theoretical models, which proposes to explain how physical forces and extracellular matrix can act to regulate cell shape, differentiation, and growth, defines the cell as a tensegrity structure.[146] In architectural terms, a tensegrity structure is composed of compression-resistant or nonelastic elements that do not touch but are continously interconnected by elastic or tensile components. In this model, stress fibers and microtubules are the rigid, compression-resistant units of the cell, held together by contractile elements, with the microfilaments within an elastic cytoplasm. The internal structure is linked to a smaller, separate tensegrity system within the nucleus and externally to the plasmalemma and the basement membrane. Thus, the spatial arrangement of scaffold-like elements could mediate such actions as closing or opening nuclear pores or clustering of transmembrane molecules such as growth factor receptors.[147]

Cell shape and the nature of the substratum have already been shown to dramatically influence gene expression, cell growth, and differentiation in a number of other cell systems,[147] providing reason to believe that the growth of the endothelium may be similarly modulated by the combined actions of the extracellular matrix, growth factors/inhibitors, other cells, and mechanical forces.

V. SUMMARY

A wide variety of growth stimulators and inhibitors have been shown to modulate EC function in vitro. In general, these factors have not been systematically investigated so the specificity of the effect with regard to cell type (large- vs. small-vessel EC, vascular vs. nonvascular cell types) and effect (proliferation, migration, etc.) are not known. Furthermore, the role of the factors in vivo is virtually uninvestigated and represents an important area for future efforts.

REFERENCES

1. **Schwartz, S. M. and Benditt, E.,** Aortic endothelial cell replication. I. Effects of age and hypertension in the rat, *Cir. Res.,* 41, 248, 1977.
2. **Hobson, B. and Denekamp, J.,** Endothelial proliferation in tumours and normal tissues: continuous labelling studies, *Br. J. Cancer,* 49, 405, 1984.
3. **Engerman, R. L., Pfaffenbach, D., and Davis, M. D.,** Cell turnover of capillaries, *Lab. Invest.,* 17, 738, 1967.
4. **Haudenschild, C., Zahniser, D., Folkman, J., and Klagsbrun, M.,** Human vascular endothelial cells in culture: lack of response to serum growth factors, *Exp. Cell. Res.,* 98, 175, 1976.
5. **Ausprunk, D. H. and Folkman, J.,** Migration and proliferation of endothelial cells in preformed and newly formed blood vessels during tumor angiogenesis, *Microvasc. Res.,* 14, 53, 1977.
6. **Gross, J. L., Moscatelli, D., and Rifkin, D. B.,** Increased capillary endothelial cell protease activity in response to angiogenic stimuli in vitro, *Proc. Natl. Acad. Sci. U.S.A.,* 80, 2623, 1983.
7. **Gimbrone, M. A., Jr., Cotran, R. S., Leapman, S. B., and Folkman, J.,** Tumor growth and neovascularization: an experimental model using the rabbit cornea, *J. Natl. Cancer Inst.,* 52, 413, 1974.
8. **Auerbach, R., Kubai, L., Knighton, D., and Folkman, J.,** A simple procedure for the long-term cultivation of chick embryos, *Dev. Biol.,* 41, 391, 1974.
9. **Silver, I. A.,** The measurement of oxygen tension in healing tissue, *Prog. Resp. Res.,* 3, 124, 1969.
10. **Fenselau, A., Kaiser, D., and Wallis, K.,** Nucleoside requirements for the in vitro growth of bovine aortic endothelial cells, *J. Cell. Physiol.,* 108, 375, 1981.
11. **Fenselau, A., Wallis, K., and Kaiser, D.,** Studies on a new class of potential antiangiogenic agents: nucleoside antimetabolite drugs, *Microvasc. Res.,* 22, 58, 1981.
12. **Castellot, J. J., Jr., Karnovsky, M. J., and Spiegelman, B. M.,** Potent stimulation of vascular endothelial cell growth by differentiated 3T3 adipocytes, *Proc. Natl. Acad. Sci. U.S.A.,* 77, 6007, 1980.
13. **Dobson, D. E., Castellot, J. J., and Spiegelman, B. M.,** Angiogenesis stimulated by 3T3-adipocytes is mediated by prostanoid lipids, *J. Cell Biol.,* 101, 109a, 1987.
14. **D'Amore, P. A. and Shepro, D.,** Stimulation of growth and calcium influx in cultured bovine, aortic endothelial cells by platelets and vasoactive substance, *J. Cell. Physiol.,* 92, 177, 1977.
15. **Marks, R. M., Roche, W. R., Czerniecki, M., Penny, R., and Nelson, D. S.,** Mast cell granules cause proliferation of human microvascular endothelial cells, *Lab. Invest.,* 55, 289, 1986.
16. **Azizkhan, R. G., Azizkhan, J. C., Zetter, B. R., and Folkman, J.,** Mast cell heparin stimulates migration of capillary endothelial cells in vitro., *J. Exp. Med.,* 152, 931, 1980.
17. **Fraser, R. H. and Simpson, J. G.,** Role of mast cells in experimental tumour angiogenesis, in *Development of the Vascular System,* CIBA Foundation Symposium, O'Connor, N. J., Ed., Pitman Books, London, 1983.
18. **McAuslan, B. R. and Reilly, W.,** Selenium-induced cell migration and proliferation: relevance to angiogenesis and microangiopathy, *Microvas. Res.,* 32, 112, 1986.
19. **Grant, E. A.,** Morphological and etiological studies of diabetic microangiopathy in pigs, *Acta Vet. Scand. Suppl.,* 3, 2, 1961.
20. **Fraser, R. A., Ellis, M., and Stalker, A. L.,** Experimental angiogenesis in the chorioallantoic membrane, in *Current Advances in Basic and Clinical Microcirculatory Research,* Lewis, D. H., Ed., S. Karger, Basel, 1979, 25.
21. **Dusseau, J. W., Hutchins, P. M., and Malbasa, D. S.,** Stimulation of angiogenesis by adenosine on the chick chorioallantoic membrane, *Circ. Res.,* 59, 163, 1986.

22. **McAuslan, B. R., Reilly, W. G., Hannan, G. N., and Gole, G. A.,** Angiogenic factors and their assay: activity of formyl methionyl leucyl phenylalanine, adenosine diphosphate, heparin, copper, and bovine endothelium stimulating factor, *Microvas. Res.,* 26, 323, 1983.

23. **Tenscher, E. and Weidler, V.,** Adenosine nucleotides, adenosine and adenine as angiogenesis factors, *Biomed. Biochim. Acta,* 44, 493, 1985.

24. **McAuslan, B. R., Hannan, G. N., and Reilley, W.,** Signals causing change in morphology, phenotype, growth mode and gene expression of vascular endothelial cells, *J. Cell. Physiol.,* 112, 96, 1982.

25. **Thornton, S. C., Mueller, S. N., and Levine, E. M.,** Human endothelial cells: use of heparin in cloning and long-term serial cultivation, *Science,* 222, 623, 1983.

26. **Ziche, M., Jones, J., and Gullino, P.,** Role of prostaglandin E_1 and copper in angiogenesis, *J. Natl. Cancer Inst.,* 69, 475, 1982.

27. **McAuslan, B. R. and Reilly, W.,** Endothelial cell phagokinesis in response to specific metal ions, *Exp. Cell Res.,* 130, 147, 1980.

28. **Pickart, L., Freedman, J. H., Loker, W. J., Peisach, J., Perkins, C. M., Stenkamp, R. E., and Weinstein, B.,** Growth-modulating plasma tripeptide may function by facilitating copper uptake into cells, *Nature (London),* 288, 715, 1980.

29. **Pickart, L.,** The biological effects and mechanism of action of the plasma tripeptide glycly-L-histidly-L-lysine, *Lymphokines,* 8, 425, 1983.

30. **Owen, K., Gomolka, D., and Droller, M. J.,** Production of prostaglandin E_2 by tumor cells in vitro, *Cancer Res.,* 40, 3167, 1980.

31. **Form, D. M. and Auerbach, R.,** PGE_2 and angiogenesis, *Proc. Soc. Exp. Biol. Med.,* 172, 1983.

32. **Folkman, J.,** Tumor angiogenesis, *Adv. Cancer Res.,* 19, 331, 1974.

33. **Phillips, P., Steward, J. K., and Kumar, S.,** Tumour angiogenesis factor (TAF) in human and animal tumours, *Int. J. Cancer,* 17, 549, 1976.

34. **Schor, A. M., Schor, S. L., Weiss, J. B., Brown, R. A., Kumar, S., and Phillips, P.,** Stimulation by a low-molecular-weight angiogenic factor of capillary endothelial cells in culture, *Br. J. Cancer,* 41, 790, 1980.

35. **Fenselau, A., Watt, S., and Mello, R. J.,** Tumor angiogenic factor: purification from the Walker 256 rat tumor, *J. Biol. Chem.,* 256, 9605, 1981.

36. **Kissun, R. D., Hill, C. R., Garner, A., Phillips, P., Kumar, S., and Weiss, J. B.,** A low-molecular-weight angiogenic factor in cat retina, *Br. J. Ophthalmol.,* 66, 165, 1982.

37. **Keegan, A., Hill, C., Kumar, S., Phillips, P., Schor, A., and Weiss, J.,** Purified tumour angiogenesis factor enhances proliferation of capillary, but not aortic, endothelial cells in vitro, *J. Cell Sci.,* 55, 261, 1982.

38. **Brown, R. A., Weiss, J. B., Tomlinson, I. W., Phillips, P., and Kumar, S.,** Angiogenic factor from synovial fluid resembling that from tumours, *Lancet,* 29, 682, 1980.

39. **McAuslan, B. R. and Hoffman, H.,** Endothelium stimulating factor from Walker carcinoma cells: relation to tumour angiogenesis factor, *Exp. Cell Res.,* 119, 181, 1979.

40. **McKeehan, W. L., Sakagami, Y., Hoshi, H., and McKeehan, K. A.,** Two apparent human endothelial cell growth factors from human hepatoma cells are tumor-associated proteinase inhibitors, *J. Biol. Chem.,* 261, 5378, 1986.

41. **D'Amore, P. A., Glaser, B. M., Brunson, S. K., and Fenselau, A. F.,** Angiogenic activity from bovine retina: partial purification and characterization, *Proc. Natl. Acad. Sci. U.S.A.,* 78, 3068, 1981.

42. **D'Amore, P. A. and Klagsbrun, M.,** Endothelial cell mitogens derived from retina and hypothalamus: biochemical and biological similarities, *J. Cell Biol.,* 99, 1545, 1984.

43. **Imre, G.,** The role of increased lactic acid concentration in neovascularizations, *Acta Morphol. Acad. Sci. Hung.,* 32, 97, 1984.

44. **Comstock, J. P. and Udenfriend, S.,** Effect of lactate on collagen proline hydroxylase activity in cultured L929 fibroblasts, *Proc. Natl. Acad. Sci. U.S.A.,* 66, 552, 1970.

45. **West, D. C., Hampson, I. N., Arnold, F., and Kumar, S.,** Angiogenesis induced by degradation products of hyaluronic acid, *Science,* 228, 1324, 1985.

46. **Ungari, S., Katari, R. S., Alessandri, G., and Gullino, P. M.,** Cooperation between fibronectin and heparin in the mobilization of capillary endothelium, *Invasion Metastasis,* 5, 193, 1985.

47. **Vlodavsky, I., Folkman, J., Sullivan, R., Fridman, R., Ishai-Michaeli, R., Sasse, J., and Klagsbrun, M.,** Endothelial cell-derived basic fibroblast growth factor: synthesis and deposition into subendothelial extracellular matrix, *Proc. Natl. Acad. Sci. U.S.A.,* 84, 2292, 1987.

48. **Shepro, D. and D'Amore, P. A.,** Physiology and biochemistry of the vascular wall endothelium, in *Handbook of Physiology . The Cardiovascular System,* Vol. IV (Part 1), Renkin, E. M. and Michel, C. C., Eds., Waverly Press, Baltimore, 1984, 165.

49. **Shing, Y., Folkman, J., Sullivan, R., Butterfield, C., Murray, J., and Klagsbrun, M.,** Heparin affinity: purification of a tumor-derived capillary endothelial cell growth factor, *Science,* 223, 1296, 1984.

50. **Maciag, T., Mehlman, T., Friesel, R., and Schreiber, A.,** Heparin binds endothelial cell growth factor, the principal mitogen in the bovine brain, *Science*, 225, 932, 1984.

51. **Thomas, K. A., Rios-Candelore, M., Gimenez-Gallego, G., DiSalvo, J., Bennett, C., Rodkey, J., and Fitzpatrick, S.,** Pure brain-derived acidic fibroblast growth factor is a potent angiogenic vascular endothelial cell mitogen with sequence homology to interleukin 1, *Proc. Natl. Acad. Sci. U.S.A.*, 82, 6409, 1985.

52. **Gimenez-Gallego, G., Rodkey, J., Bennett, C., Rios-Candelore, M., DiSalvo, J., and Thomas, K.,** Brain-derived acidic fibroblast growth factor: complete amino acid sequence and homologies, *Science*, 230, 1385, 1985.

53. **Esch, F., Ueno, N., Baird, A., Hill, F., Denoroy, L., Ling, N., Gospodarowicz, D., and Guillemin, R.,** Primary structure of bovine brain acidic fibroblast growth factor (FGF), *Biochem. Biophys. Res. Commun.*, 133, 554, 1985.

54. **Jaye, M., McConathy, E., Drohan, W., Tong, B., Deuel, T., and Maciag, T.,** Modulation of sis gene transcript during endothelial cell differentiation in vitro, *Science*, 228, 882, 1985.

55. **Abraham, J. A., Mergia, Wheng, J. L., Tumolo, A., Freidman, J., Hjerrild, K. A., Gospodarowicz, D., and Fiddes, J. C.,** Nucleotide sequence of a bovine clone encoding the angiogenic protein, basic fibroblast growth factor, *Science*, 233, 545, 1986.

56. **Maciag, T., Hoover, G. A., and Weinstein, R.,** High and low molecular weight forms of endothelial cell growth factor, *J. Biol. Chem.*, 257, 5333, 1982.

57. **Burgess, W. H., Mehlman, T., Freisel, R., Johnson, W. V., and Maciag, T.,** Mutiple forms of endothelial cell growth factor, *J. Biol. Chem.*, 260, 11389, 1985.

58. **Klagsbrun, M., Sasse, J., Sullivan, R., and Smith, J. A.,** Human tumor cells synthesize an endothelial cell growth factor that is structurally related to basic fibroblast growth factor, *Proc. Natl. Acad. Sci. U.S.A.*, 83, 2448, 1986.

59. **Ueno, N., Baird, A., Esch, F., Ling, N., and Guillemin, R.,** Isolation of an amino terminal extended form of basic fibroblast growth factor, *Biochem. Biophys. Res. Commun.*, 138, 580, 1986.

60. **Gospodarowicz, D., Cheng, J., Lui, G-M., Baird, A., and Bohlen, P.,** Isolation by heparin-Sepharose affinity chromatography of brain fibroblast growth factor: identity with pituitary fibroblast growth factor, *Proc. Natl. Acad. Sci. U.S.A.*, 81, 6983, 1984.

61. **Sullivan, R. and Klagsbrun, M.,** Purification of cartilage-derived growth factor by heparin affinity chromatography, *J. Biol. Chem.*, 260, 2399, 1985.

62. **Baird, A., Mormede, P., and Bohlen, P.,** Immunoreactive fibroblast growth factor in cells of peritoneal exudate suggest its idenity with macrophage-derived growth factor, *Biochem. Biophys. Res. Commun.*, 126, 358, 1985.

63. **Lobb, R. R. and Fett, J. W.,** Purification of two distinct growth factors from bovine neural tissue by heparin affinity chromatography, *Biochemistry*, 23, 6295, 1984.

64. **Baird, A., Esch, F., Gospodarowicz, D., and Guillemin, R.,** Retina and eye derived endothelial cell growth factors: partial molecular characterization and identity with acidic and basic fibroblast growth factors, *Biochemistry*, 24, 7855, 1986.

65. **Courty, J., Loret, C., Moenner, M., Chevallier, B., Lagente, O., Courtois, Y., and Barritault, D.,** Bovine retina contains three growth factor activities with different affinity to heparin: eye-derived growth factor I, II, III, *Biochimie*, 67, 265, 1985.

66. **Klagsbrun, M. and Shing, Y.,** Heparin affinity of anionic and cationic capillary endothelial cell growth factors: analysis of hypothalamus-derived growth factors and fibroblast growth factors, *Proc. Natl. Acad. Sci. U.S.A.*, 82, 805, 1985.

67. **Lobb, R. R., Rybak, S. M., Sinclair, D. K., and Fett, J. W.,** Lysates of two established human tumor lines contain heparin binding growth factors related to bovine acidic fibroblast growth factor, *Biochem. Biophys. Res. Commun.*, 139, 861, 1986.

68. **Hauschka, P., Mavarakos, A. E., Iafrati, M. D., Doleman, S. E., and Klagsbrun, M.,** Growth factors in bone matrix, *J. Biol. Chem.*, 261, 12665, 1986.

69. **Azizhkan, J., Sullivan, J., Azizkhan, R., Zetter, B. R., and Klagsbrun, M.,** Stimulation of increased capillary endothelial cell motility by chondrosarcoma cell-derived factors, *Cancer Res.*, 43, 3281, 1983.

70. **Herman, I. M. and D'Amore, P. A.,** Capillary endothelial cell migration: loss of stress fibres in response to retina-derived growth factor, *J. Muscle Res. Cell Motility*, 5, 697, 1984.

71. **Terranova, V. P., DiFlorio, R., Lyall, R. M., Hic, S., Friesel, R., and Maciag, R.,** Human endothelial cells are chemotactic to endothelial cell growth factor and heparin, *J. Cell Biol.*, 101, 2330, 1985.

72. **Wagner, J. A. and D'Amore, P. A.,** Neurite outgrowth induced by an endothelial cell mitogen isolated from retina, *J. Cell Biol.*, 103, 1363, 1986.

73. **Togari, A., Dickens, G., Kuzuya, H., and Guroff, G.,** The effect of fibroblast growth factor on PC12 cells, *J. Neurosci.*, 5, 307, 1985.

74. **Shing, Y., Folkman, J., Haudenschild, C., Lund, D., Crum, R., and Klagsbrun, M.,** Angiogenesis is stimulated by a tumor-derived endothelial cell growth factor, *J. Cell. Biochem.,* 29, 275, 1985.

75. **Lobb, R. R., Alderman, E. M., and Fett, J. W.,** Induction of angiogenesis by bovine brain derived class 1 heparin-binding growth factor, *Biochemistry,* 24, 4869, 1985.

76. **Neufeld, G. and Gospodarowicz, D.,** The identification and partial characterization of the fibroblast growth factor receptor of baby hamster kidney cells, *J. Biol. Chem.,* 260, 13860, 1985.

77. **Moenner, M., Chevallier, B., Badet, J., and Barritault, D.,** Evidence and characterization of the receptor to eye-derived growth factor I, the retinal form of basic fibroblast growth factor, on bovine epithelial lens cells, *Proc. Natl. Acad. Sci. U.S.A.,* 83, 5024, 1986.

78. **Friesel, R., Burgess, W. H., Mehlman, T., and Maciag, T.,** The characterization of the receptor for endothelial cell growth factor by covalent ligand attachment, *J. Biol. Chem.,* 261, 7581, 1986.

79. **Schreiber, A. B., Kenney, J., Kowalski, W. J., Friesel, R., Mehlman, T., and Maciag, T.,** Interaction of endothelial cell growth factor with heparin: characterization by receptor and antibody recognition, *Proc. Natl. Acad. Sci. U.S.A.,* 82, 6138, 1985.

80. **Olwin B. B. and Hauschka, S. D.,** Identification of the fibroblast growth factor receptor of Swiss 3T3 cells and mouse skeletal muscle myoblasts, *Biochemistry,* 25, 3487, 1986.

81. **Risau, W.,** Developing brain produces an angiogenesis factor, *Proc. Natl. Acad. Sci. U.S.A.,* 83, 3855, 1986.

82. **Risau, W. and Ekblom, P.,** Production of heparin-binding angiogenesis factor by the embryonic kidney, *J. Cell Biol.,* 103, 1101, 1986.

83. **Fett, J. W., Strydom, D. J., Lobb, R. R., Alderman, E. M., Bethune, J. L., Riordan, J. F., and Vallee, B. M.,** Isolation and characterization of angiogenin, an angiogenic protein from human carcinoma cells, *Biochemistry,* 24, 5480, 1985.

84. **Strydom, D. J., Fett, J. W., Lobb, R. R., Alderman, E. M., Bethune, J. L., Riordan, J. F., and Vallee, B. L.,** Amino acid sequence of human tumor derived angiogenin, *Biochemistry,* 24, 5486, 1985.

85. **Kurachi, K., Davie, E. W., Strydom, D. J., Riordan, J. F., and Vallee, B. L.,** Sequence of the cDNA and gene for angiogenin, a human angiogenesis factor, *Biochemistry,* 24, 5494, 1985.

86. **Shapiro, R., Riordan, J. F., and Vallee, B. L.,** Characteristic ribonucleolytic activity of human angiogenin, *Biochemistry,* 25, 3527, 1986.

87. **Gospodarowicz, D., Brown, K. D., Birdwell, C. R., and Zetter, B. R.,** Control of proliferation of human vascular endothelial cells, *J. Cell Biol.,* 77, 774, 1978.

88. **Weinstein, R. and Weng, K.,** Growth factor responses of human arterial endothelial cells in vitro, *In Vitro Cell. Dev. Biol.,* 22, 549, 1986.

89. **McAuslan, B. R., Bender, V., Reilly, W., and Moss, B. A.,** New functions of epidermal growth factor: stimulation of capillary endothelial cell migration and matrix dependent proliferation, *Cell Biol. Int. Rep.,* 9, 175, 1985.

90. **Gospodarowicz, D., Bialecki, H., and Thakral, T. K.,** The angiogenic activity of the fibroblast and epidermal growth factor, *Exp. Eye Res.,* 28, 501, 1979.

91. **Bar, R. S., Hoak, J. C., and Peacock, M. L,** Insulin receptors in human endothelial cells: identification and characterization, *J. Clin. Endocrinol. Metab.,* 47, 699, 1978.

92. **Bar, R. S., Goldsmith, J. C., Rechler, M. M., Peacock, M. L., and Nissley, S. P.,** Interactions of multiplication-stimulating activity with bovine endothelium: comparative studies in primary, passaged, and cloned cells cultured from the pulmonary and systemic circulations, *Endocrinology,* 110, 990, 1982.

93. **Bar, R. S., Dolash, S., Dake, B. L, and Boes, M.,** Cultured capillary endothelial cells from bovine adipose tissue: a model for insulin binding and action in microvascular endothelium, *Metabolism,* 35, 317, 1986.

94. **Soda, R. and Tavassoli, M.,** Distribution of insulin receptors in liver cell suspensions using a minibead probe, *Exp. Cell Res.,* 145, 389, 1983.

95. **Bar, R. S., Boes, M., and Sandra, A.,** Receptors for insulin-like growth factor-I (IGF-I) in myocardial capillary endothelium of the intact perfused heart, *Biochem. Biophys. Res. Commun.,* 133, 724, 1985.

96. **Bar, R. S. and Boes, M.,** Distinct receptors for IGF-I, IGF-II, and insulin are present on bovine capillary endothelial cells and large vessel endothelial cells, *Biochem. Biophys. Res. Commun.,* 124, 203, 1984.

97. **Jialal, I., Crettaz, M., Hachiya, H. L., Kahn, C. R., Moses, A. C., Buzney, S. M., and King, G. L.,** Characterization of the receptors for insulin and the insulin-like growth factors on micro- and macrovascular tissues, *Endrocinology,* 117, 1222, 1985.

98. **King, G. L., Johnson, S. M., and Jialal, I.,** Processing and transport of insulin by vascular endothelial cells. Effects of sulfonylureas on insulin receptors, *Am. J. Med.,* 79, 43, 1985.

99. **Kaiser, N., Vlodavsky, I., Tur-Sinai, A., Fuks, Z., and Cerasi, E.,** Insulin binding and degradation in vascular endothelial cells: modulation by cell growth and culture organization, *Endocrinology,* 113, 228, 1983.

100. **King, G. L. and Johnson, S.,** Receptor-mediated transport of insulin across endothelial cells, *Science,* 227, 1583, 1985.

101. **Banskota, N. K., Carpentier, J.-L., and King, G. L.,** Processing and release of insulin and insulin-like growth factor-I by macro- and microvascular endothelial cells, in press.
102. **Banda, M. J., Knighton, D. R., Hunt, T. K., and Werb, Z.,** Isolation of a nonmitogenic angiogenesis factor from wound fluid, *Proc. Natl. Acad. Sci. U.S.A.,* 79, 7773, 1982.
103. **Banda, M., Dwyer, J. S., and Beckman, A.,** Wound fluid angiogenesis factor stimulates the directed migration of capillary endothelial cells, *J. Cell Biochem.,* 29, 183, 1985.
104. **Knighton, D. R., Hunt, T. K., Scheuenstuhl, H., Halliday, B. J., Werb, Z., and Banda, M. J.,** Oxygen tension regulates the expression of angiogenesis factor by macrophages, *Science,* 221, 1283, 1983.
105. **Jensen, J. A., Hunt, T. K., Scheuenstuhl, H., and Banda, M.,** Effect of lactate, pyruvate, and pH on secretion of angiogenesis and mitogenesis factors by macrophages, *Lab. Invest.,* 54, 574, 1986.
106. **Okabe, T. and Takaku, F.,** A macrophage factor that stimulates the proliferation of vascular endothelial cells, *Biochem. Biophys. Res. Commun.,* 134, 344, 1986.
107. **Polverini, P. J. and Leibovich, S. J.,** Induction of neovascularization in vivo and endothelial proliferation in vitro by tumor-associated macrophages, *Lab. Invest.,* 51, 635, 1984.
108. **Clemmons, D. R., Isley, W. L., and Brown, M. T.,** Dialyzable factor in human serum of platelet origin stimulates endothelial cell replication and growth, *Proc. Natl. Acad. Sci. U.S.A.,* 80, 1641, 1983.
109. **King, G. L. and Buchwald, S.,** Characterization and partial purification of an endothelial cell growth factor from human platelets, *J. Clin. Invest.,* 73, 392, 1984.
110. **Miyazono, K., Okabe, T., Urabe, A., Yamanaka, M., and Takaku, F.,** A platelet factor that stimulates the proliferation of vascular endothelial cells, *Biochem. Biophys. Res. Commun.,* 126, 83, 1985.
111. **Zetter, B. R. and Antoniades, H. N.,** Stimulation of human vascular endothelial cell growth by a platelet-derived growth factor and thrombin, *J. Supramol. Struct.,* 11, 361, 1979.
112. **Awbrey, B. J., Hoak, J. C., and Owen, W. G.,** Binding of human thrombin to cultured human endothelial cells, *J. Biol. Chem.,* 254, 4092, 1979.
113. **Hessler, J. R., Robertson, A. L., and Chislom, G. M.,** LDL-induced cytotoxicity and its inhibition by HDL in human vascular smooth muscle cells in culture, *Atherosclerosis,* 32, 213, 1979.
114. **Tauber, J-P., Cheng, J., and Gospodarowicz, D.,** Effect of high and low density lipoproteins on proliferation of bovine vascular endothelial cells, *J. Clin. Invest.,* 66, 696, 1980.
115. **Gospodarowicz, D., Massoglia, S., Cheng, J., and Fujii, D. K.,** Effect of fibroblast growth factor and lipoproteins on the proliferation of endothelial cells derived from bovine adrenal cortex, brain cortex, and corpus luteum capillaries, *J. Cell Physiol.,* 127, 121, 1986.
116. **Tauber, J. P., Cheng, J., Massoglia, S., and Gospodarowicz, D.,** High density lipoproteins and the growth of vascular endothelial cells in serum-free medium, *In Vitro,* 17, 519, 1981.
117. **Darbon, J. M., Tournier, J. P., Tauber, J. P., and Bayard, F.,** Possible role of protein phosphorylation in the mitogenic effect of high density lipoproteins on cultured vascular endothelial cells, *J. Biol. Chem.,* 261, 8002, 1986.
118. **Tournier, J.-F., Bayard, F., and Tauber, J. P.,** Rapid purification and activity of apolipoprotein Cl on the proliferation of bovine vascular endothelial cells in vitro, *Biochim. Biophys. Acta,* 804, 216, 1984.
119. **Cohen, D. C., Massoglia, S. L., and Gospodarowicz, D.,** Correlation between two effects of high density lipoprotein on vascular endothelial cells, *J. Biol. Chem.,* 257, 9429, 1982.
120. **Doctrow, S. R. and Folkman, J.,** Protein kinase C activators suppress stimulation of capillary endothelial cell growth by angiogenic endothelial mitogens, *J. Cell Biol.,* 104, 679, 1987.
121. **Montesano, R. and Orci, L.,** Tumor-promoting phorbol esters induce angiogenesis in vitro, *Cell,* 42, 469, 1985.
122. **Melnykovych, G. and Clowes, K. K.,** Growth stimulation of bovine endothelial cells by vitamin A, *J. Cell Physiol.,* 109, 265, 1981.
123. **Braunhut, S. J., Doctrow, S. R., and D'Amore, P. A.,** The modulation of endothelial cell growth by retinoids, *J. Cell Biol.,* 103, 196a, 1986.
124. **Sato, N., Goto, T., Haranaka, K., Satomi, N., Nariuchi, H., Mano-Hirano, Y,. and Sawasaki, Y.,** Actions of tumor necrosis factor on cultured vascular endothelial cells: morphologic modulation, growth inhibition, and cytotoxicity, *J. Natl. Cancer Inst.,* 76, 1113, 1986.
125. **Rosenbaum, J., Toblem, G., Molho, P., Barzu, T., and Caen, J. P.,** Modulation of endothelial cells growth induced by heparin, *Cell Biol. Int. Rep.,* 10, 437, 1986.
126. **Folkman, J., Langer, R., Lindhardt, R. J., Haudenschild, C., and Taylor, S.,** Angiogeneisis inhibition and tumor regression caused by heparin or a heparin fragment in the presence of cortisone, *Science,* 221, 719, 1983.
127. **Ingber, D. E., Madri, J. A., and Folkman, J.,** A possible mechanism for inhibition of angiogenesis by angiostatic steroids: induction of capillary basement membrane dissolution, *Endocrinology,* 119, 1768, 1986.
128. **Sporn, M. B., Roberts, A., Shull, J. H., Smith, J. M., Ward, J. M., and Sodbk, J.,** Polypeptide transforming growth factors isolated from bovine sources and used for wound healing in vivo, *Science,* 219, 1329, 1983.

129. **Roberts, A. B.,** New class of transforming growth factors potentiated by EGF: isolation from non-neoplastic tissue, *Proc. Natl. Acad. Sci. U.S.A.,* 78, 5339, 1981.

130. **Heimark, R. L., Twardzik, D. R., and Schwartz, S. M.,** Inhibition of endothelial regeneration by type-beta transforming growth factor from platelets, *Science,* 233, 1078, 1986.

131. **Carswell, E. A., Old, L. J., Kassel, R. L, Green, S., Fiore, N., and Williamson, B.,** An endotoxin-induced serum factor that causes the necrosis of tumors, *Proc. Natl. Acad. Sci. U.S.A.,* 72, 3666, 1975.

132. **Haranaka, K., Satomi, N., and Sakurai, A.,** Differences in tumor necrosis factor productive ability among rodents, *Br. J. Cancer,* 50, 471, 1984.

133. **Brouty-Boye, D. and Zetter, B. R.,** Inhibition of cell motility by interferon, *Science,* 208, 516, 1980.

134. **Eldor, A., Fridman, R., Vlodavsky, I., Hy-Am, E., Fuks, Z., and Panet, A.,** Interferon enhances prostacyclin production by cultured vascular endothelial cells, *J. Clin. Invest.,* 73, 251, 1984.

135. **Brem, S., Brem, H., Folkman, J., Finkelstein, D., and Patz, A.,** Prolonged tumor dormancy by prevention of neovascularization in the vitreous, *Cancer Res.,* 36, 2807, 1976.

136. **Jacobson, B., Dorfman, T., Basu, P. K., and Hasany, S. M.,** Inhibition of vascular endothelial cell growth and trypsin activity by vitreous, *Exp. Eye Res.,* 41, 581, 1985.

137. **Lutty, G. A., Thompson, D. C., Gallup, J. Y., Mello, R. J., Patz, A., and Fenselau, A.,** Vitreous: an inhibitor of retinal extract induced neovascularization, *Invest. Ophthalmol. Visual Sci.,* 24, 52, 1983.

138. **Taylor, S. and Folkman, J.,** Protamine is an inhibitor of angiogenesis, *Nature (London),* 297, 307, 1982.

139. **Langer, R., Conn, H., Vacanti, J., Haudenschild, C., and Folkman, J.,** Control of tumor growth in animals by infusion of an angiogenesis inhibitor, *Proc. Natl. Acad. Sci. U.S.A.,* 77, 4331, 1980.

140. **Sorgente, N. and Dorey, C. K.,** Inhibition of endothelial cell growth by a factor isolated from cartilage, *Exp. Cell Res.,* 128, 63, 1980.

141. **Langer, R. and Murray, J.,** Angiogenesis inhibitors and their delivery systems, *Appl. Biochem. Bioeng.,* 8, 9, 1983.

142. **Takigawa, M., Shira, E., Enomoto, M., Hiraki, Y., Fukuya, M., Suzuki, F., Shiio, T., and Tugari, Y.,** Cartilage-derived anti-tumor factor (CATF) inhibits the proliferation of endothelial cells in culture, *Cell Biol. Int. Rep.,* 9, 619, 1985.

143. **Eisenstein, R., Goren, S. B., Schumacher, B., and Chromokos, E.,** The inhibition of corneal neovascularization with aortic extracts in rabbits, *Am. J. Ophalmol.,* 88, 1005, 1979.

144. **Eisenstein, R., Harper, E., and Kuettner, K. E.,** Growth regulators in connective tissues. II. Evidence for the presence of several growth inhibitors in aortic extracts, *Arterial Wall,* 5, 163, 1979.

145. **Williams, G. A., Eisenstein, R., Schumacher, B., and Grant, D.,** Inhibitor of vascular endothelial cell growth in the lens, *Am. J. Ophthalmol.,* 97, 366, 1984.

146. **Ingber, D. E. and Jamieson, J. D.,** Cells as tensegrity structures: architectural regulation of histodifferentiation by physical forces transduced over basement membrane, in *Gene Expression during Normal and Malignant Differentiation,* Anderson, L. C., Gahmberg, G. C., and Ekblom, P., Eds., Academic Press, New York, 1985, 13.

147. **Bissel, M. J., Hall, H. C., and Parry, C.,** How does extracellular matrix direct gene expression?, *J. Theor. Biol.,* 99, 31, 1982.

148. **Morgan, D. M. L.,** personal communication.

149. **West, D. L.,** personal communication.

150. **D'Amore, P. A. and Braunhut, S. J.,** unpublished observations.

Chapter 15

ANGIOGENESIS FACTORS

Michael Klagsbrun

TABLE OF CONTENTS

I. INTRODUCTION

Proliferation of blood vessels is a process necessary for the normal growth and development of tissue. In the adult, angiogenesis occurs infrequently. Exceptions are found in the female reproductive system, where angiogenesis occurs in the follicle during its development, in the corpus luteum during ovulation, and in the placenta after pregnancy. These periods of angiogenesis are relatively brief and tightly regulated. Normal angiogenesis also occurs as part of the body's repair processes, e.g., in the healing of wounds and fractures. By contrast, uncontrolled angiogenesis can contribute to serious disease. As examples, the growth of solid tumors is dependent on vascularization, and in diabetic retinopathy vascularization of the retina often leads to blindness.

Some of the earliest insights into the mechanisms of angiogenesis came from studies of tumor vascularization (for reviews, see References 1 to 5). Experiments with tumors implanted into transparent chambers suggested that the growth of solid tumors was accompanied by new capillary growth.[6] Furthermore, it was demonstrated that tumor cells separated from the vascular bed of a host by a Millipore filter induced the growth of new capillaries.[7-9] These results suggested that tumor-derived diffusible factors were responsible for stimulating blood vessel growth in the host. Subsequently, a number of tumor-derived angiogenesis factors were identified and isolated.[10-13] However, attempts to purify these factors were unsuccessful. The difficulties could be ascribed partly to the relatively low levels of the highly potent angiogenesis factors that could be found in tumors as well as the lack of suitable and reproducible bioassays. In the past 2 years, these obstacles have been mostly overcome. As a result, a number of angiogenesis factors have been purified from both tumors and normal tissue, their primary amino acid sequences determined, and their genes cloned. This chapter is a review of the recent work on angiogenesis factors.

II. BIOLOGY OF ANGIOGENESIS

A. Capillary Growth and Development

New capillaries arise as offshoots or sprouts from established vessels, mostly venules or other capillaries. This phenomenon of capillary sprouting is accomplished by a series of sequential steps.[14] In the initial phase of capillary sprouting, the basement membrane of endothelial cells in the parent blood vessel is degraded. Degradation of basement membrane is probably accomplished by the action of proteolytic enzymes endogenous to endothelial cells that are induced by tumor angiogenesis factors, most notably plasminogen activator and collagenase.[15] It is also possible that tumors act more directly to degrade basement membrane, by action of tumor-derived heparan sulfate endoglycosidases which degrade heparan sulfate, the major glycosaminoglycan constituent of basement membrane.[16] Once basement membrane is degraded, endothelial cells bud out from the preexisting vessel and migrate directionally into the perivascular space. Cells that have migrated out of the parent blood vessel proliferate so as to elongate the budding sprout. Subsequently, the endothelial cells form tubes[17,18] containing a lumen through which blood flows. As a final step, new basement membrane is synthesized and a mature new capillary is established. The steps in the formation of a new capillary can be summarized as basement membrane degradation, endothelial cell migration and proliferation, tube formation, and differentiation into a mature blood vessel.

B. Angiogenesis Assays

Angiogenesis factors are assayed both in vitro and in vivo. The in vitro assays are based on steps in capillary growth that can be recapitulated in vitro. These include: (1) induction of endothelial cell proteases that degrade basement membrane, (2) stimulation of endothelial cell motility, and (3) stimulation of endothelial cell proliferation. These in vitro assays depend on the availability of cultured endothelial cells. Umbilical vein, aortic, corneal, and capillary endothelial cell have all been shown to be responsive to angiogenesis factors.[19-24] Protease assays involve the induction of plasminogen activator and latent collagenase in endothelial cells.[15] Endothelial cell migration is assayed either by measuring endothelial cell motility (chemokinesis) using the phagokinetic track assay[25,26] or measuring directional endothelial cell migration in a concentration gradient (chemotaxis) using a Boyden migration chamber.[27] Endothelial cell proliferation is assayed either by counting cell number or by monitoring ^{3}H-thymidine uptake into DNA.

The most commonly used bioassays for measuring angiogenesis in vivo are (1) the developing chick chorioallantoic membrane (CAM), and (2) the corneal implant. The CAM bioassay is carried out either in an egg in which a window is prepared by removing a piece of eggshell or by totally removing the shell and culturing the egg in a petri dish.[28-30] Test substances are applied to either filters or plastic coverslips. Alternatively, they are incorporated into methylcellulose. These materials are subsequently placed onto an 8- to 10-day-old CAM. Neovascularization at the site of implant is monitored visually for 1 to 2 days and after fixation CAMs are evaluated histiligically. Great care and proper controls must be used in evaluating CAM results; the CAM is undergoing rapid intrinsic neovascularization and also is highly sensitive to nonspecific inflammatory substances such as eggshell fragments.

The most reliable bioassay for angiogenesis uses the avascular cornea of either the rabbit or rat eye.[30,31] Test substances are incorporated into sustained-release polymer pellets which are implanted into a pocket prepared in the cornea surgically. In a positive angiogenesis response, the directional growth of new capillaries from the limbal blood vessels towards the implant is seen after 5 to 6 days. In controls, the cornea remains avascular. The bioassay can be quantitated by measuring the rate of blood vessel growth and the density of the new blood vessels. The major disadvantage of the cornea bioassay is the limitation on the number of samples that can be feasibly tested.

Examples of endothelial cell proliferation, chorioallantoic membrane, and cornea bioassays are shown in Figure 1.

III. ANGIOGENESIS FACTORS

A. Fibroblast Growth Factors — a Family of Heparin-Binding Endothelial Cell Mitogens
1. Purification

One strategy in angiogenesis factor purification has been to isolate growth factors that stimulate endothelial cell motility and proliferation in vitro. The rationale is that growth factors that stimulate endothelial cell migration and proliferation, key components in capillary growth and development in vitro, might be angiogenic in vivo. Among the first endothelial cell growth growth factors to be identified were basic fibroblast growth factor (FGF) isolated from brain[32] and endothelial cell growth factor (ECGF) isolated from hypothalamus.[33] Basic FGF was initially reported to be a cationic 13,000 mol wt polypeptide[34] that stimulated the proliferation both of endothelial cells seeded at clonal densities in vitro[21] and of blood vessels in the rabbit cornea in vivo.[35] The initial claims that basic FGF had been purified to homogeneity and was structurally related to myelin basic protein were subsequently found to be erroneous.[36] ECGF was initially reported to be an anionic factor existing in two molecular weight forms of 70,000 and 17,000 to 25,000 that stimulated the proliferation of human umbilical vein endothelial (HUVE) cells.[37] Other sources of ECGFs were the retina (anionic, mol wt of about 50,000), the eye (anionic, mol wt 17,500), and cartilage (cationic, mol wt 16,400).[38-40] Despite much effort, these ECGFs could not be purified to homogeneity until 1984.

A major advance in purifying ECGFs came as a result of the observation that an ECGF

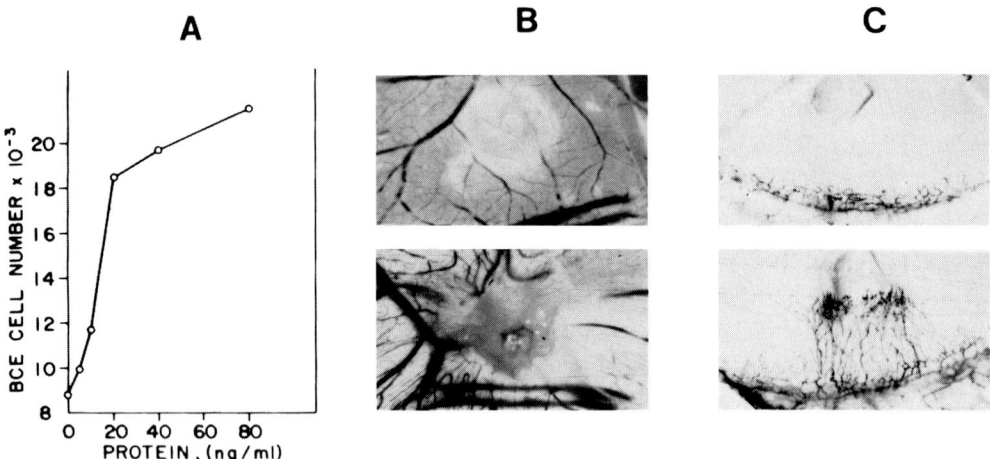

FIGURE 1.   Bioassays for angiogenesis factors. (A) Stimulation of bovine capillary endothelial (BCE) cells. BCE cells are plated sparsely (10,000 cells per 16-mm-diameter microtiter well in a 24-well microtiter plate) and on the following day the medium (DMEM 10% calf serum) is removed and increasing concentrations of chondrosarcoma-derived basic FGF is added in fresh medium. After 3 days in culture, cells are removed by trypsinization and counted in a Coulter® model Zf electron particle counter. (B) Chorioallantoic membrane assay (CAM). Bottom panel: chondrosarcoma-derived basic FGF (600 units, 120 ng) is incorporated along with 10 μg of bovine serum albumin and 2 units of heparin into 10 μℓ 0.5% methylcellulose. The methylcellulose pellets are implanted onto the CAMs of 9.5-day-old chick embryos through a window previously made in the eggshell on day 4. After a 24-hr incubation period the CAM is removed, fixed, and photographed. Top panel: albumin/heparin/methylcellulose control without basic FGF. (C) Rat cornea. Chondrosarcoma-derived FGF (50 units, 10 ng) is incorporated into an ethylene-vinyl acetate copolyme pellet. The pellet is implanted into a pocket made in the rat cornea as previously described.[30] After 6 days, ink is injected into the corneas of anesthetized animals for better visualization of the blood vessels, and the corneas are removed, fixed, and photographed. Top: a control in which the pellet is boiled prior to implantation, a treatment that inactivates basic FGF. Bottom: active basic FGF.

derived from rat chondrosarcoma had a marked affinity for heparin.[41] A rapid two-step procedure combining cation exchange chromatography and heparin-Sepharose chromatography was used to purify a chondrosarcoma-derived growth factor (ChDGF) to homogeneity. ChDGF is a cationic 18,000 mol wt endothelial cell growth factor that stimulates the proliferation of capillary endothelial cells at about 1 ng/mℓ.[41] ChDGF stimulates noninflammatory angiogenesis in the chick CAM with as little as 60 ng per egg and in the rat cornea with as little as 10 ng per eye.[30]

Subsequently, it was found that many, if not all, ECGFs had a marked affinity for heparin. The evidence for heparin affinity was threefold. First, the ECGFs bound to heparin so tightly that 1 to 1.5 M NaCl was required for their elution from columns of heparin-Sepharose. These salt concentrations are comparable to those needed to elute well known heparin-binding proteins such as anti-thrombin III.[41,42] Second, binding of the ECGFs for heparin was highly specific. They did not adhere at all to other highly anionic glycosaminoglycans, e.g., chondroitin sulfate and hyaluronic acid.[43] Third, binding to heparin was independent of the pI of the endothelial cell growth factors. Anionic ECGFs isolated from hypothalamus and retina with pIs of approximately five adhered to highly anionic heparin. In fact, they adhered much more tightly than did platelet-derived growth factor (PDGF), a polypeptide with a pI of ten that is not an endothelial cell mitogen.[42,44]

Heparin affinity chromatography has facilitated the purification of several ECGFs, including those found in brain,[36,45,46,48,49] hypothalamus,[45] eye,[47] and cartilage.[43] With the availability of homogeneous growth factor, it was possible for the first time in 1985 to obtain primary amino acid sequences. The first reported primary amino acid sequences were those

Table 1
CLASSIFICATION OF HEPARIN-BINDING GROWTH FACTORS

Growth Factors	Ref.
Acidic FGF	
(pIs of 5—7, elution at 1 *M* NaCl)	
Brain-derived acidic fibroblast growth factor	45,49,56
Endothelial cell growth factor	46,48
Eye-derived growth factor-2	47
Retina-derived growth factor	44
Astroglial growth factor-1	57
Hypothalamus growth factor-alpha	42
Heparin-binding growth factor-alpha	45
Basic FGF	
(pIs of 8—10, elution at 1.5 *M* NaCl)	
Brain-derived basic fibroblast growth factor	36,45,53
Eye-derived growth factor-1	47
Astroglial growth factor-2	57
Hypothalamus growth factor-beta	42
Heparin-binding growth factor-beta	45
Cartilage-derived growth factor	43
Chondrosarcoma-derived growth factor	41
Hepatoma-derived growth factor	66

of bovine brain and pituitary basic FGF, identical polypeptides of 146 amino acids,[50] and of bovine brain acidic FGF, a polypeptide of 140 amino acids.[51-53] Basic and acidic FGF were found to be structurally related, having a 53% absolute sequence homology.[53]

2. Classification and Structure

Analysis of the various heparin-binding ECGFs by heparin-affinity column elution profiles, protein sequences, immunological cross-reactivity, and receptor binding has greatly clarified the relationship of these polypeptides to one another.[54,55] There are not as many different ECGFs as previously thought. In fact, most ECGFs can be subdivided into two classes, one containing growth factors structurally related to acidic FGF and the other to basic FGF. The growth factors within a given class are either identical or represent multiple-molecular-weight forms of the same polypeptide (Table 1).

The class of acidic FGF consists of anionic polypeptides that elute from heparin-Sepharose columns with approximately 1.0 *M* NaCl. They have isoelectric points of 5 to 7 and molecular weights of 15,000 to 20,000. The acidic FGF class of heparin-binding growth factors has been found mainly in neural tissue and includes brain-derived acidic FGF,[45,49,56] ECGF,[48] eye-derived growth factor II,[47] acidic retina-derived growth factor,[44] astroglial growth factor-1,[57] and bone-derived growth factor-1.1.[58]

Acidic FGF has been found in several molecular-weight forms. Beta-ECGF is a larger-molecular-weight form of acidic FGF extended at the amino-terminal by 14 amino acids and blocked at its amino-terminus.[59] Thus, beta-ECGF is a 154-amino-acid polypeptide that may be a precursor to acidic FGF. Alpha-ECGF is a truncated form of acidic FGF, missing the 20 N-terminal amino acids of beta-ECGF.

The other class of heparin-binding endothelial cell mitogens consists of cationic polypeptides that elute from heparin-Sepharose with around 1.5 *M* NaCl. They have isoelectric points of 8 to 10, molecular weights between 16,000 and 19,000, and appear to be identical to, or multiple-molecular-weight forms of, basic FGF. The cationic class of heparin-binding growth factor appear to be far more ubiquitous than the anionic class. Polypeptides of this class have been isolated from sources such as pituitary,[50,60] brain,[36,45,53,60] hypothalamus,[42,45] eye,[47] cartilage,[43] bone,[58] corpus luteum,[61] adrenal cortex,[62] kidney,[63] placenta,[64] macrophages,[65] chondrosarcoma,[41] hepatoma cells,[66] endothelial cells,[67] astroglial cells,[57] and developing brain.[68]

Originally described as a 146-amino-acid polypeptide,[50] basic FGF appears to exist in both larger and smaller molecular-weight forms. An amino-terminal extension of at least two amino acids was found in hepatoma cell-derived basic FGF.[66] A 154-amino-acid molecular-weight form of basic FGF blocked at the amino-terminus has been recently described.[69] Truncated forms of basic FGF lacking the first 15 amino acids of the 146-amino-acid form have been isolated from corpus luteum, adrenal cortex, and kidney.[61-63] Truncation of basic FGF to smaller-molecular-weight forms is due in part to the action of acid proteinase cleavage at the amino-terminal end.[70]

In summary, it appears that both basic and acidic FGF are 154-amino-acid, N-terminally blocked polypeptides with greater than 50% structural homology. In various tissues, shorter-molecular-weight forms of basic FGF containing 146 and 131 amino acids can be found. Similarly, shorter-molecular-weight forms of acidic FGF containing 140 and 134 amino acids can be found. Whether the truncated forms represent cleavage products that result from posttranslational modifications or are artifacts of purification techniques is not known. The biological significance of having two classes of structurally related ECGFs is also yet unknown.

3. Fibroblast Growth Factor Genes

The availability of acidic and basic FGF primary amino acid sequences has resulted in the cloning of acidic and basic FGF genes. A human complimentary DNA (cDNA) clone encoding ECGF was isolated from a human brain stem cDNA library and its nucleotide sequence determined.[71] Southern blot analysis has suggested that there is a single ECGF gene that maps to human chromosome 5. The size of the human brain stem ECGF mRNA transcript is 4.8 kilobases. The predicted amino acid sequence of the open reading frame after the initiating methionine[71] is the same as the primary amino acid sequence of beta-ECGF, the 154-amino-acid precursor form of acidic FGF.[59]

A bovine cDNA clone encoding basic FGF was isolated from a bovine pituitary cDNA libary and the nucleotide sequence determined.[72] Basic FGF mRNA transcripts of 7 kilobases were found in hypothalamus and heptatoma cells. An additional 3.7-kilobase basic FGF mRNA transcript was found only in the hepatoma cells. The predicted amino acid sequence of the open reading frame corresponded to that of basic FGF but appeared to have a nine-amino-acid extension on the amino-terminal side. Thus, basic FGF without the initiating methionine is predicted to be a 154-amino-acid polypeptide[72] in agreement with structural analysis.[69]

In summary, gene and protein structure analysis suggest that both ECGF (acidic FGF) and basic FGF are 154-amino-acid polypeptides with a high degree of homology.

4. Fibroblast Growth Factor Receptors

Covalent cross-linking of purified ^{125}I-ECGF (acidic FGF) and ^{125}I-basic FGF have resulted in the identification of several FGF receptors on the surface of different cell types. A 150,000 mol wt receptor for ECGF has been identified on murine lung capillary endothelial cells.[73] Two receptors for basic FGF with molecular weights of 145,000 and 125,000 have been identified on the surface of BHK-21 cells.[74] The BHK-21 cells have 120,000 binding sites per cell. EDGF-I, the retinal form of basic FGF, has a receptor on the surface of bovine epithelial lens cells with a molecular weight of 130,000.[75] These cells have 20,000 binding sites per cell. A single receptor for both acidic and basic FGF with a molecular weight of

165,000 has been identified on the surface of Swiss 3T3 cells and myoblasts.[76] There are 60,000 binding sites on the 3T3 cells. However, at present the amount of data concerning FGF receptors is relatively sparse. It is not clear yet whether there are distinct receptors for acidic and basic FGF, nor has it been demonstrated that FGF receptors are tyrosine phosphorylated.[74]

5. Angiogenesis Activity

Heparin-binding growth factors are potent endothelial cell mitogens in vitro and, therefore, are good candidates for being angiogenesis factors in vivo as well. Basic FGF induces capillary endothelial cells to migrate into three-dimensional collagen matrices to form capillary-like tubes,[77] thus mimicking in vitro some of the events that occur during neovascularization in vivo. Both basic and acidic FGF are angiogenic in vivo in CAM,[30,50,51,78] cornea,[30,78] and wound healing models.[79-81] As little as 60 ng of chondrosarcoma-derived basic FGF induces angiogenesis in the chick CAM in an 18-hr period.[30] Blood vessel growth is directional towards the site of implantation. A marked hyperplasia of connective tissue cells at the site of implantation can also be found. Bovine brain basic FGF at 500 ng per egg[50] and brain acidic FGF at 160 ng per egg[78] have also shown to be angiogenic in the CAM.

As little as 10 ng of chondrosarcoma-derived basic FGF and 80 ng of brain acidic FGF will induce angiogenesis in the cornea in a 6-day period.[30,78] In neither the CAM nor the cornea bioassay is there evidence of inflammation.

Cartilage-derived basic FGF is active in a wound healing model at a dose of 500 ng.[79] In this model system, polyvinyl sponges are implanted into rats and, subsequently, basic FGF is injected into the sponges. After 3 days, histological examination of the sponges shows that FGF induces the formation of granulation tissue that is very vascular and characterized by highly dilated blood vessels.

It can be concluded from experiments carried out both in vitro and in vivo that heparin-binding ECGFs are angiogenesis factors as well. Their angiogenic potential is probably related to their ability to stimulate the migration and proliferation of endothelial cells.

B. Angiogenin

Angiogenin was first purified from the conditioned medium of a human adenocarcinoma cell line by a combination of cation exchange and reverse-phase high-performance liquid chromatography.[82-84] This purification procedure yielded 0.5 μg of angiogenin per liter of conditioned medium. Angiogenin is a single-chain polypeptide of 123 amino acids with a molecular weight of 14,400 and an isoelectric point of 9.5. Angiogenin has a 35% absolute sequence homology to a family of pancreatic ribonucleases. The major active site residues of ribonuclease in angiogenin are conserved, as are three of the four disulfide bonds. Angiogenin is inactive toward the more conventional substrates of ribonuclease such as wheat germ RNA, poly(C), poly(U), and RNA-DNA hybrids. However, angiogenin does cleave both 28S and 18S ribosomal RNA to relatively large products that are 100 to 500 nucleotides in length.[85] Whether this relatively specific hydrolytic activity has physiological significance is not known at this time. Human cDNA coding for angiogenin was isolated from a human liver cDNA library.[84] The nucleotide sequence of the angiogenin gene was determined and the predicted amino acid sequence was in complete agreement with that determined by amino acid sequence analysis. Angiogenin is a potent stimulator of angiogenesis in the chick CAM in the range of 0.5 to 290 ng per egg in the chick CAM and 50 ng per eye in the rabbit cornea.[82] Angiogenin differs both structurally and biologically from the heparin-binding growth factors in that (1) it lacks sequence homology to either acidic or basic FGF; (2) it does not bind to heparin-Sepharose; and (3) it has a gene sequence coding for a signal peptide of 22 to 24 amino acids, unlike both acidic and basic FGF which apparently have no signal peptide domains.[71,72] This structural difference is consistent with the observation that angiogenin is secreted by cells in culture[82] while basic FGF is not.[66] Angiogenin further differs from heparin-binding growth factors in that it does not appear to be a growth factor for endothelial cells. In all, these results suggest that angiogenin and the family of heparin-binding endothelial cell growth factors stimulate angiogenesis via different mechanisms which have yet to be elucidated.

C. Transforming Growth Factors

Transforming growth factors (TGF) are polypeptides that alter the phenotype of normal cells to that of transformed cells. These polypeptides are also capable of inducing angiogenesis in vivo. Two structurally distinct TGFs, TGF-alpha[86,87] and -beta,[88,89] have been purified and their structures have been determined by protein sequencing and cDNA cloning. TGF-alpha is a 50-amino-acid polypeptide synthesized by various transformed cells. It binds to the EGF receptor and has a 35% homology with EGF. Both TGF-alpha and EGF stimulate microvascular endothelial cell proliferation at 1 to 5 ng/mℓ.[90] When these polypeptides are injected subcutaneously into the hamster cheek pouch, they both stimulate capillary proliferation and an increase in the labeling index of endothelial cells. However, TGF-alpha is a far more potent angiogenic factor, being angiogenic at a dose of 0.3 to 1 μg, whereas EGF is angiogenic only at a dose of 10 μg.[90]

TGF-beta is a 25,000-molecular-weight homodimer found in tumors and normal tissue cells such as placenta, kidney, and platelets. When injected subcutaneously into newborn mice, TGF-beta, at a dose of 1 μg, stimulates angiogenesis and collagen production by fibroblasts to form a highly vascular granulation tissue at the site of injection in 2 to 3 days.[91] Neither EGF nor PDGF had the same effect in the same bioassay. Paradoxically, TGF-beta inhibits the growth of aortic endothelial cells in vitro.[92,93] Thus it is difficult at this point to reconcile the angiogenic activity of TGF-beta in vivo with its inhibitory effects on endothelial cell growth in vitro.

D. Other Angiogenesis Factors

A number of other factors, not as well characterized in terms of structure and function as those factors described above, have been shown to be angiogenic.

1. Chemotactic Factors

Angiogenesis factors that stimulate the locomotion but not the proliferation of endothelial cells have been isolated from wound fluid.[94] These factors appear to be polypeptides with molecular weights in the range of 2,000 to 14,000, but have yet to be purified. The material induces angiogenesis in the cornea at a dose of 150 ng. The wound fluid-derived angiogenesis factors stimulate the migration of capillary endothelial cells in a Boyden chamber assay but do not stimulate the proliferation of these cells. It is possible that these factors are responsible for the endothelial migration component of angiogenesis and that mitogens in wound fluid, perhaps synthesized by macrophages, are responsible for capillary growth and elongation.

2. Lipids

Several angiogenesis factors which are lipid rather than peptide in nature have been described. 3T3 cells that have undergone adipose differentiation in vitro secrete factors that stimulate angiogenesis in the CAM.[95] These factors stimulate the motility of aortic and capillary endothelial cells in a Boyden chamber assay but not the proliferation of these cells. Characterization of the angiogenesis/chemotactic factors, including chemical analysis and the use of inhibitors of prostaglandin synthesis, suggests that they may be a mixture of prostaglandins E_1 and E_2 and as yet uncharacterized polar lipids that constitute the major angiogenic activity.[96] Separate factors that stimulate endothelial cell proliferation but not

Table 2
BIOLOGICAL ACTIVITIES OF ANGIOGENIC FACTORS

Name	Angiognesis	EC proliferation	EC motility
Acidic FGF	Yes	Yes	Yes
Basic FGF	Yes	Yes	Yes
Angiogenin	Yes	No	N.D.[a]
TGF-alpha	Yes	Yes	N.D.
TGF-beta	Yes	Inhibition	N.D.
Wound fluid	Yes	No	Yes
Adipocyte lipids	Yes	No	Yes
Prostaglandins	Yes	No	N.D.

[a]   N.D. = not determined.

endothelial cell motility are also produced by differentiated 3T3 cells.[96,97] These endothelial cell mitogens apparently are not lipids but their identity has not yet been determined.

Prostaglandins, in particular PGE_1 and PGE_2, have been shown to directly stimulate angiogenesis.[98-100] PGE_1 at 1 µg will stimulate angiogenesis in the cornea, and PGE_2 and 0.2 to 20 ng will stimulate angiogenesis in the CAM. Other prostaglandins, e.g., the A or F series, are not angiogenic. Prostaglandins levels are elevated in tumors, activated macrophages, wounds, and inflammatory exudates.[98] Since these cells and fluids are associated with neovascularization, it may well be that certain prostaglandins play a role in angiogenesis.

A lipid angiogenic factor extractable in chloroform-methanol has been isolated from omentum.[101] When this material is injected directly into the cornea, intense vascularization occurs. The omentum-derived angiogenesis factor has not yet been purified.

3. Tumor-Derived Low-Molecular-Weight Angiogenesis Factors

Low-molecular-weight compounds ranging from 200 to 1000 that stimulate endothelial cell proliferation and angiognesis in the CAM and cornea have been isolated from Walker 256 tumors.[11-13] Similar factors have been found in synovial fluid.[102] These factors appear to be neither peptide, protein, nucleic acid, or prostaglandin and their purification has not been accomplished as of yet.

IV. CONCLUSIONS AND PERSPECTIVES

A. Mechanisms of Angiogenesis

It is clear that there are a number of different factors that are capable of stimulating angiogenesis in vivo. Some of these, such as acidic and basic FGF, angiogenin, TGF-alpha, and TGF-beta have been purified and sequenced and their genes cloned. Others have yet to be fully purified. Although all of these factors are angiogenic in vivo, their effects on endothelial cell motility and proliferation in vitro vary (Table 2). For example, while acidic and basic FGF are mitogenic for endothelial cells, angiogenin appears not to be mitogenic for these cells, and TGF-beta is actually an inhibitor of endothelial cell proliferation. Some angiogenic factors such as acidic and basic FGF stimulate both endothelial cell motility and proliferation. However, others such as the wound-derived angiogenesis factor and the 3T3-adipocyte lipid angiogenesis factors are chemotactic but not mitogenic for endothelial cells. These results suggest that different angiogenesis factors might act via different mechanisms. Some suggested mechanisms are as follows. First, some angiogenesis factors might act by directly stimulating endothelial cell motility and proliferation. Second, some angiogenesis factors might stimulate endothelial cell motility alone but induce the total angiogenic response by positioning endothelial cells so they can be stimulated to proliferate by mitogenic factors. Third, some angiogenesis factors might not affect endothelial cells directly, but rather act indirectly by mobilizing secondary cells to stimulate angiogenesis. For example, these factors might stimulate macrophages to produce basic FGF[65] and other angiogenesis factors.[103-105] Finally, some factors might be angiogenic because they liberate angiogenesis factors that are stored in tissues.

B. Regulation of Angiogenesis

The ubiquity of angiogenesis factors such as basic FGF and TGF-beta suggests that they must be tightly regulated under normal conditions. For example, considerable quantities of angiogenic ECGF is found in normal tissues such as brain, yet the vascular endothelial cell turnover time in most normal tissue is measured in years. Certain cells and tissues such as lymphocytes and ovary display brief angiogenesis but then return to the quiescent state. In contrast, the vascular endothelial cells within a tumor are maximally proliferating. Their turnover time is measured in days. There may be many ways to regulate angiogenesis activity. Possible regulatory mechanisms include (1) controlled expression of angiogenesis genes; (2) processing and posttranslational modification of angiogenesis factors; (3) sequestering of angiogenesis factors so that they are not available; (4) controlled release of angiogenesis factors from cells and tissue; and (5) and interaction of angiogenesis stimulators with specific angiogenesis inhibitors. Abnormal angiogenesis, such as occurs in tumors and diabetic retinopathy, might be the result of the breakdown of some of these postulated regulatory mechanisms.

C. Future Directions

The rapid progress of the last 2 years in purifying angiogenesis factors and cloning their genes promises to lead angiogenesis research into important new directions such as; (1) identification of which cells are producers of angiogensis factors and which are targets; (2) purification of angiogenesis factor receptors; (3) elucidation of the mechanisms by which angiogenesis factors act; (4) modulation of angiogenesis activity in vivo; (5) administration of angiogenesis factors in vivo in order to repair damaged tissue such as occurs in myocardial infarctions, wounds and bone fractures; and (6) suppression of pathological angiogenesis such as occurs in tumors, diabetic retinopathy, and rheumatoid arthritis.

REFERENCES

1. **Folkman, J.,** Tumor angiogenesis in *Cancer, Vol. 3, A Comprehensive Treatise, Biology of Tumors: Cellular Biology and Growth,* Becker, F. F., Ed., Plenum Press, New York, 1975, 355.
2. **Folkman, J.,** Angiogenesis, in *Biology of Endothelial Cells* Jaffe, E. A., Ed., Martinus Nijhoff, Boston, 1984, 412.
3. **Ausprunk, D. H.,** Tumor angiogenesis, in *Chemical Messengers of the Inflammatory Process,* Houck, J. C., Ed., North-Holland, New York, 1979, 317.
4. **Fenselau, A.,** Tumor and related angiogenesis, in *Growth and Maturation Factors,* Guroff, G., Ed., John Wiley & Sons, New York, 1984, 87.
5. **Vallee, B. L., Riordan, J. F., Lobb, R. R., Fett, J.W., Crossley, G., Buhler, R., Budzik, G., Breddam, K., Bethune, J. L., and Alderman, E. M.,** Tumor-derived angiogenesis factors from Walker 256 carcinoma: an experimental investigation and review, *Experentia,* 41, 1, 1985.
6. **Algire, G. H. and Chalkley, H. W.,** Vascular reactions of normal and malignant tissues in vivo. I. Vascular reactions of mice to wounds and to normal and neoplastic transplants, *J. Natl. Cancer Inst.,* 6, 73, 1945.
7. **Greenblatt, M. and Shubik, P.,** Tumor angiogenesis; transfilter diffusion studies in the hamster by the transparent chamber technique, *J. Natl. Cancer Inst.,* 41, 111, 1968.

8. **Ehrman, R. L. and Knoth, M.,** Choriocarcinoma: transfilter stimulation of vasoproliferation in the hamster cheek pouch — studied by light and electron microscopy, *J. Natl. Cancer Inst.,* 41, 1329, 1968.
9. **Gitterman, C. A. and Luell, S.,** Transfilter induction of vascular proliferation on the chorioallantoic membranes of embryonated eggs, *Proc. Am. Assoc. Cancer Res.,* 10, 29, 1969.
10. **Tuan, D., Smith, S., Folkman, J., and Merler, E.,** Isolation of the non-histone proteins of rat Walker Carcinoma 256: their association with tumor angiogenesis, *Biochemistry,* 12, 3159, 1973.
11. **Phillips, P., Steward, J. K., and Kumar, S.,** Tumor angiogenesis factor (TAF) in human and animal tumors, *Int. J.Cancer,* 17, 549, 1976.
12. **McAuslan, B. R. and Hoffman, H.,** Endothelium stimulating factor from Walker carcinoma cells, *Exp. Cell Res.,* 119, 181, 1979.
13. **Fenselau, A., Watt, S., and Mello, R. J.,** Tumor angiogenic factor: purification from the Walker 256 rat tumor, *J. Biol. Chem.,* 256, 9605, 1981.
14. **Ausprunk, D. H. and Folkman, J.,** Migration and proliferation of endothelial cells in preformed and newly formed blood vessels during tumor angiogenesis, *Microvasc. Res.,* 14, 53, 1977.
15. **Gross, J. L., Moscatelli, D., and Rifkin, D. B.,** Increased capillary endothelial cell protease activity in response to angiogenic stimuli in vitro, *Proc. Natl. Acad. Sci.,* 80, 2623, 1983.
16. **Vladofsky, I., Fuks, Z., Bar-Ner, M., Ariav, Y., and Schirrmacher, V.,** Lymphoma cell mediated degradation of sulfated proteoglycans in the subendothelial cell extracellular matrix: relationship to tumor metastasis, *Cancer Res.,* 43, 2704, 1983.
17. **Folkman, J. and Haudenschild, C. C.,** Angiogenesis in vitro, *Nature (London),* 288, 551, 1980.
18. **Maciag,T., Hoover, G. A., van der Spek, J., Stemerman, M. B., and Weinstein, R.,** Growth and differentiation of human umbilical-vein endothelial cells in culture, in *Growth of Cells in Hormonally Defined Media,* Book, A., Sato, G. H., Pardee, A. B., and Sibarsku, D. A., Eds., Cold Spring Harbor Laboratory, Cold Spring Harbor, New York, 1982, 525.
19. **Gimbrone, M. A., Jr.,** Culture of vascular endothelium, *Prog. Hemostasis Thromb.,* 3, 1, 1976.
20. **Duthu, G. S. and Smith, J. R.,** In vitro proliferation and lifespan of bovine aorta endothelial cells: effect of culture conditions and fibroblast growth factor, *J. Cell. Phys.,* 103, 385, 1980.
21. **Gospodarowicz, D., Moran, J., Braun, D., and Birdwell, C.,** Clonal growth of bovine vascular endothelial cells; FGF as a survival agent, *Proc. Natl. Acad. Sci. U.S.A.,* 73, 4120, 1976.
22. **Gospodarowicz, D., Mescher, A. L., and Birdwell, C.,** Stimulation of corneal endothelial cell proliferation in vitro by fibroblast and epidermal growth factors, *Exp. Eye Res.,* 25, 75, 1977.
23. **Maciag, T., Hoover, G. A., Stemerman, M. B., and Weinstein, R.,** Serial propogation of endothelial cells in vitro, *J. Cell Biol.,* 91, 420, 1981.
24. **Folkman, J., Haudenschild, C. C., and Zetter, B. R.,** Long-term culture of capillary endothelial cells, *Proc. Natl. Acad. Sci. U.S.A.,* 76, 5217, 1979.
25. **Zetter, B. R.,** Migration of capillary endothelial cells is stimulated by tumor-derived factors, *Nature (London),* 285, 41, 1980.
26. **Azizkhan, J. C., Sullivan, R., Azizkhan, R., Zetter, B., and Klagsbrun, M.,** The stimulation of capillary endothelial cell migration by chondrosarcoma-derived growth factors, *Cancer Res.,* 43, 3281, 1983.
27. **Glaser, B. M.,D'Amore, P. A., Seppa, H., Seppa, S., and Schiffmann, E.,** Adult tissues contain chemoattractants for vascular endothelial cells, *Nature (London),* 288, 483, 1980.
28. **Klagsbrun, M., Knighton, D., and Folkman, J.,** Tumor angiogenesis in cells grown in tissue culture, *Cancer Res.,* 36, 110, 1976.
29. **Ausprunk, D. H., Knighton, D. R., and Folkman, J.,** Differentiation of vascular endothelium in the chick chorioallantois: a structural and autoradiographic study, *Dev. Biol.,* 38, 237, 1974.
30. **Shing, Y., Folkman, J., Haudenschild, C., Lund, D., Crum, R., and Klagsbrun, M.,** Angiogenesis is stimulated by a tumor-derived endothelial cell growth factor, *J. Cell. Biochem.,* 29, 275, 1985.
31. **Gimbrone, M. A., Jr., Leapman, S. B., Cotran, R. S., and Folkman, J.,** Tumor dormancy in vivo by prevention of neovascularization, *J. Exp. Med.,* 136, 261, 1972.
32. **Gospodarowicz, D., Bialecki, H., and Greenburg, G.,** Purification of the fibroblast growth factor activity from bovine brain, *J. Biol. Chem.,* 253, 3736, 1978.
33. **Maciag, T., Cerundolo, J., Ilsey, S., Kelley, P. R., and Forand, R.,** An endothelial cell growth factor from bovine hypothalamus: identification and partial purification, *Proc. Natl. Acad. Sci. U.S.A.,* 76, 5674, 1979.
34. **Gospodarowicz, D.,** Purification of a fibroblast growth factor from bovine pituitary, *J. Biol. Chem.,* 250, 2515, 1975.
35. **Gospodarowicz, D., Bialecki, H., and Thakral, T. K.,** The angiogenic activity of the fibroblast and epidermal growth factor, *Exp. Eye Res.,* 28, 501, 1979.
36. **Gospodarowicz, D., Cheng, D., Lui, G.-M., Baird, A., and Bohlen, P.,** Isolation of brain fibroblast growth factor by heparin-sepharose affinity chromatography: identity with pituitary fibroblast growth factor, *Proc. Natl. Acad. Sci. U.S.A.,* 81, 6963, 1984.

37. **Maciag, T., Hoover, G. A., and Weinstein, R.,** High and low molecular weight forms of endothelial cell growth factor, *J. Biol. Chem.,* 257, 5333, 1982.
38. **D'Amore, P. A., Glaser, B. M., Brunson, S. K., and Fenselau, A.,** Angiogenic activity from bovine retina, partial purification and characterization, *Proc. Natl. Acad. Sci. U.S.A.,* 78, 3068, 1981.
39. **Barritault, D., Arruti, C., and Courtois, Y.,** Is there a ubiquitous growth factor in the eye?, *Differentiation,* 18, 29, 1981.
40. **Klagsbrun, M. and Smith, S.,** Purification of a cartilage-derived growth factor, *J. Biol. Chem.,* 255, 10869, 1980.
41. **Shing, Y., Folkman, J., Sullivan, R., Butterfield, C., Murray, J., and Klagsbrun, M.,** Heparin affinity: purification of a tumor-derived capillary endothelial cell growth factor, *Science,* 223, 1296, 1984.
42. **Klagsbrun, M. and Shing, Y.,** Heparin affinity of anionic and cationic capillary endothelial cell growth factors: analysis of hypothalamus-derived growth factors and fibroblast growth factors, *Proc. Natl. Acad. Sci. U.S.A.,* 82, 805, 1985.
43. **Sullivan, R. and Klagsbrun, M.,** Purification of cartilage-derived growth factor by heparin affinity chromatography, *J. Biol. Chem.,* 260, 2399, 1985.
44. **D'Amore, P. A. and Klagsbrun, M.,** Endothelial cell mitogens derived from retina and hypothalamus: biochemical and biological similarities, *J. Cell Biol.,* 99, 1545, 1984.
45. **Lobb, R. R. and Fett, J. W.,** Purification of two distinct growth factors from bovine neural tissue by heparin affinity chromatography, *Biochemistry,* 23, 6295, 1984.
46. **Maciag, T., Mehlman, T., Friesel, R., and Schreiber, A. B.,** Heparin binds endothelial cell growth factor, the principal cell mitogen in bovine brain, *Science,* 225, 932, 1984.
47. **Courty, J., Loret, C., Moenner, M., Chevallier, B., Lagente, O., Courtois, Y., and Barritault, D.,** Bovine retina contains three growth factor activities with different affinity for heparin: eye-derived growth factor I, II and III, *Biohimie,* 67, 265, 1985.
48. **Burgess, W. H., Mehlman, T., Friesel, R., Johnson, W. V., and Maciag, T.,** Multiple forms of endothelial cell growth factor; rapid isolation and biological and chemical characterization, *J. Biol. Chem.,* 260, 11389, 1985.
49. **Conn, G. and Hatcher, V. B.,** The isolation and purification of two anionic endothelial cell growth factors from human brain, *Biochem. Biophys. Res. Commun.,* 124, 262, 1984.
50. **Esch, F., Baird, A., Ling, N., Ueno, N., Hill, F., Denoroy, L., Klepper, R., Gospodarowicz, D., Bohlen, P., and Guillemin, R.,** Primary structure of bovine pituitary basic fibroblast growth factor (FGF) and comparison with the amino-terminal sequence of bovine acidic FGF, *Proc. Natl. Acad. Sci. U.S.A.,* 82, 6507, 1985.
51. **Thomas, K. A., Rios-Candelore, M., Gimenez-Gallego, G., DiSalvo, J., Bennett, J. C., Rodkey, J., and Fitzpatrick, S.,** Pure brain-derived acidic fibroblast growth factor is a potent angiogenic vascular endothelial cell mitogen with sequence homology to Interleuken I, *Proc. Natl. Acad. Sci. U.S.A.,* 82, 6409, 1985.
52. **Giminez-Gallego, G., Rodkey, J., Bennett, C., Rios-Candelore, M., Disalvo, J., and Thomas, K.,** Bovine-derived acidic fibroblast growth factor: complete amino acid sequence and homologies, *Science,* 230, 1385, 1985.
53. **Esch, F., Ueno, N., Baird, A., Hill, F., Denoroy, L., Ling, N., Gospodarowicz, D., and Guillemin, R.,** Primary structure of bovine brain acidic fibroblast growth factor (FGF), *Biochem. Biophys. Res. Commun.,* 133, 554, 1985.
54. **Lobb, R. R., Sasse, J., Shing, Y., D'Amore, P. A., Sullivan, R., Jacobs, J., and Klagsbrun, M.,** Purification and characterization of heparin-binding growth factors, *J. Biol. Chem.,* 261, 1924, 1986.
55. **Schreiber, A. B., Kenney, J., Kowalski, J., Thomas, K. A., Gimenez-Gallego, G., Rios-Candelore, M., Di Salvo, J., Barritault, D., Courty, J., Courtois, Y., Moenner, M., Loret, C., Burgess, W. H., Mehlman, T., Friesel, R., Johnson, W., and Maciag, T.,** A unique family of endothelial cell polypeptide mitogens: the antigenic and receptor cross-reactivity of bovine endothelial cell growth factor, brain-derived acidic fibroblast growth factor, and eye-derived growth factor II, *J. Cell. Biol.,* 101, 1623, 1985.
56. **Thomas, K. A., Rios-Candelore, M., and Fitzpatrick, S.,** Purification and characterization of acidic fibroblast growth factor from bovine brain, *Proc. Natl. Acad. Sci. U.S.A.,* 81, 357, 1984.
57. **Pettman, B., Weibel, M., Sensenbrenner, M., and Labourdette, G.,** Purification of two astroglial growth factors from brain, *FEBS Lett.,* 189, 102, 1985.
58. **Hauschka, P., Iafrati, M. D., Doleman, S. E., and Klagsbrun, M.,** Growth factors in bone matrix: isolation of multiple types by affinity chromatography on Heparin-Sepharose, *J. Biol. Chem.,* 261, 12665, 1986.
59. **Burgess, W. H., Mehlman, T., Marshak, D. R., Fraser, B. A., and Maciag, T.,** Structural evidence that endothelial cell growth factor-beta is the precursor of both endothelial cell growth factor-alpha and acidic fibroblast growth factor, *Proc. Natl. Acad. Sci. U.S.A.,* 83, 7216, 1986.
60. **Bohlen, P., Baird, A., Esch, F., Ling, N., and Gospodarowicz, D.,** Isolation and partial molecular characterization of pituitary fibroblast growth factor, *Proc. Natl. Acad. Sci. U.S.A.,* 81, 5364, 1984.

61. **Gospodarowicz, D., Cheng, D., Lui, G-M., Baird, A., Esch, F., and Bohlen, P.,** Corpus luteum angiogenic factor is related to fibroblast growth factor, *Endocrinology,* 117, 2283, 1985.
62. **Gospodarowicz, D., Baird, A., Cheng, D., Lui, G-M., Esch, F., and Bohlen, P.,** Isolation of fibroblast growth factor from bovine adrenal gland: physicochemical and biological characterization, *Endocrinology,* 118, 82, 1986.
63. **Baird, A., Esch, F., Bohlen, P., Ling, N., and Gospodarowicz, D.,** Isolation and partial characterization of an endotheial cell growth factor from the bovine kidney: honology with basic fibroblast growth factor, *Regul. Peptides,* 12, 201, 1985.
64. **Moscatelli, D., Presta, M., and Rifkin, D. B.,** Purification of a factor from human placenta that stimulates capillary endothelial cell protease production, DNA synthesis and migration, *Proc. Natl. Acad. Sci. U.S.A.,* 83, 2091, 1986.
65. **Baird, A., Mormede, P., and Bohlen, P.,** Immunoreactive fibroblast growth factor in cells of peritoneal exudate suggests its identity with macrophage-derived growth factor, *Biochem. Biophys. Res. Commun.,* 126, 358, 1985.
66. **Klagsbrun, M., Sasse, J., Sullivan, R., and Smith, J. A.,** Human tumor cells synthesize an endothelial cell growth factor that is structurally related to basic fibroblast growth factor, *Proc. Natl. Acad. Sci. U.S.A.,* 83, 2448, 1986.
67. **Vlodavsky, I., Folkman, J., Sullivan, R., Fridman, R., Ishai-Michaeli, R., Sasse, J., and Klagsbrun, M.,** Endothelial cell-derived basic fibroblast growth factor: synthesis and deposition into subendothelial extracellular matrix, *Proc. Natl. Acad. Sci. U.S.A.,* 84, 2292, 1987.
68. **Risau, W.,** Developing brain produces an angiogenic factor, *Proc. Natl. Acad. Sci. U.S.A.,* 83, 3855, 1986.
69. **Ueno, N., Baird, A., Esch, F., Ling, N., and Guillemin, R.,** Isolation of an amino-terminal extended form of basic fibroblast growth factor, *Biochem. Biophys. Res. Commun.,* 138, 580, 1986.
70. **Klagsbrun, M., Smith, S., Sullivan, R., Shing, Y., Davidson, S., and Sasse, J.,** Processing of brain and tumor-derived fibroblast growth factors by acid-activated proteases, *Proc. Natl. Acad. Sci. U.S.A.,* 84, 1839, 1987.
71. **Jaye, M., Howe, R., Burgess, W., Ricca, G. A., Chiu, I-M., Ravera, M. W., O'Brian, S. J., Modi, S. W., Maciag, T., and Drohan, W. N.,** Human endothelial cell growth factor: cloning, nucleotide sequence, and chromosome localization, *Science,* 233, 541, 1986.
72. **Abraham, J. A., Mergia, A., Whang, J. L., Tumolo, A., Friedman, J., Hjerrild, K. A., Gospodarowicz, D., and Fiddes, J. C.,** Nucleotide sequence of a bovine clone encoding the angiogenic protein, basic fibroblast growth factor, *Science,* 233, 545, 1986.
73. **Friesel, R., Burgess, W. H., Mehlman, T., and Maciag, T.,** The characterization of the receptor for endothelial cell growth factor (ECGF) by covalent ligand attachment, *J. Biol. Chem.,* 261, 7581, 1986.
74. **Neufeld, G. and Gospodarowicz, D.,** The identification and partial characterization of the fibroblast growth factor receptor of baby hamster kidney cells, *J. Biol. Chem.,* 260, 13860, 1985.
75. **Moenner, M., Chevalier, B., Badet, J., and Barritault, D.,** Evidence and characterization of the receptor to eye-derived growth factor I, the retinal form of basic fibroblast growth factor, on bovine epithelial lens cells, *Proc. Natl. Acad. Sci. U.S.A.,* 83, 5024, 1986.
76. **Olwin, B. B. and Hauschka, S. D.,** Identification of the fibroblast growth factor receptor of swiss 3T3 cells and mouse skeletal muscle myoblasts, *Biochemistry,* 25, 3487, 1986.
77. **Montesano, R., Vassalli, J.-D., Baird, A., Guillemin, R., and Orci, L.,** Basic fibroblast growth factor induces angiogenesis in vitro, *Proc. Natl. Acad. Sci. U.S.A.,* 83, 7297, 1986.
78. **Lobb, R. R., Alderman, E. M., and Fett, J. W.,** Induction of angiogenesis by bovine brain-derived class I heparin-binding growth factor, *Biochemistry,* 24, 4969, 1985.
79. **Davidson, J., Klagsbrun, M., Hill, K., Buckley, A., Sullivan, R., Brewer, P., and Woodward, S.,** Accelerated wound repair, cell proliferation, and collagen accumulation are produced by a cartilage-derived growth factor, *J. Cell Biol.,* 100, 1219, 1985.
80. **Buntrock, P., Buntrock, M., Marx, I., Kranz, D., Jentzch, K. D., and Heder, G.,** Stimulation of wound healing using brain extract with fibroblast growth factor activity, *Exp. Pathol.,* 26, 247, 1984.
81. **Hunt, T. K.,** Can repair processes be stimulated by modulators (cell growth factors, angiogenic factors, etc.) without adversely affecting normal processes?, *J. Trauma,* 24, S39, 1984.
82. **Fett, J. W., Strydom, D. J., Lobb, R. F., Alderman, E. M., Bethune, J. L., Riordan, J. F., and Vallee, B. L.,** Isolation and characterization of angiogenin, an angiogenic protein from human carcinoma cells, *Biochemistry,* 24, 5480, 1985.
83. **Strydom, D. J., Fett, J. W., Lobb, R. R., Alderman, E. M., Bethune, J. L., Riordan, J. F., and Vallee, B. L.,** Amino acid sequence of human tumor-derived angiogenin, *Biochemistry,* 24, 5486, 1985.
84. **Kurachi, K., Davie, E. W., Strydom, D. J., Riordan, J. F., and Vallee, B. L.,** Sequence of the cDNA and gene for angiogenin, a human angiogenesis factor, *Biochemistry,* 24, 5494, 1985.
85. **Shapiro, R., Riordan, J. F., and Vallee, B. L.,** Characteristic ribonucleolytic activity of human angiogenin, *Biochemistry,* 25, 3527, 1986.

86. **Marquadt, H., Hunkapiller, M. W., Hood, L. E., and Todaro, G. J.,** Rat transforming growth factor type I: structure and relationship to epidermal growth factor, *Science,* 223, 1079, 1984.

87. **Derynck, R., Roberts, A. B., Eaton, D. H., Winkler, M. E., and Goeddel, D. V.,** Human transforming growth factor-alpha: precursor sequence, gene structure and heterologous expression, in *Cancer Cells, 3. Growth Factors and Transformation,* Feramisco, J., Ozanne, B., and Stiles, C., Eds., Cold Spring Harbor Laboratory, Cold Spring Harbor, New York, 1985, 79.

88. **Derynck, R., Jarrett, J. A., Chen, E. Y., Eaton, D. H., Bell, J. R., Assoian, R. K., Roberts, A. B., Sporn, M. B., and Goeddel, D.,** Human transforming growth factor-beta complimentary DNA sequence and expression in normal and transfected cells, *Nature (London),* 316, 701, 1985.

89. **Childs, C. B., Proper, J. A., Tucker, R. F., and Moses, H. L.,** Serum contains a platelet-derived transforming growth factor, *Proc. Natl. Acad. Sci. U.S.A.,* 79, 5312, 1982.

90. **Schreiber, A. B., Winkler, M. E., and Derynck, R.,** Transforming growth factor-alpha: a more potent angiogenic mediator than epidermal growth factor, *Science,* 232, 1250, 1986.

91. **Roberts, A. B., Sporn, M. B., Assoian, R. K., Smith, J. M., Roche, N. S., Wakefield, L. M., Heine, U. I., Liotta, L. A., Falanga, V., Kehrl, J. H., and Fauci, A. S.,** Transforming growth factor type beta: rapid induction of fibrosis and angiogenesis in vivo and stimulation of collagen formation in vitro, *Proc. Natl. Acad. Sci. U.S.A.,* 83, 4167, 1986.

92. **Baird, A. and Durkin, T.,** Inhibition of endothelial cell proliferation by type beta-transforming growth factor: interactions with acidic and basic fibroblast growth factors, *Biochem. Biophys. Res. Commun.,* 138, 476, 1986.

93. **Heimark, R. L., Twardzik, D. R., and Schwartz, S. M.,** Inhibition of endothelial regeneration by type-beta transforming growth factor from platelets, *Science,* 233, 1078, 1986.

94. **Banda, M., Knighton, D. R., Hunt, T. K., and Werb, Z.,** Isolation of a nonmitogenic factor from wound fluid, *Proc. Natl. Acad. Sci. U.S.A.,* 79, 7773, 1982.

95. **Castellot, J. J., Jr., Karnovsky, M. J., and Spiegelman, B. M.,** Differentiation-dependent stimulation of neovascularization and endothelial cell chemotaxis by 3T3 adipocytes, *Proc. Natl. Acad. Sci. U.S.A.,* 79, 5597, 1982.

96. **Dobson, D. E., Castellot, J. J., and Spiegelman, B. M.,** Angiogenesis stimulated by 3T3-adipocytes is mediated by prostanoid lipids, *J. Cell Biol.,* 101, 109a, 1985.

97. **Castellot, J. J., Karnovsky, M. J., and Spiegelman, B. M.,** Potent stimulation of vascular endothelial cell growth by differentiated 3T3 adipocytes, *Proc. Natl. Acad. Sci. U.S.A.,* 77, 6007, 1980.

98. **Form, D. M. and Auerbach, R.,** PGE2 and angiogenesis, *Proc. Soc. Exp. Biol. Med.,* 172, 214, 1983.

99. **Ziche, M., Jones, J., and Gullino, P.,** Role of prostaglandin E1 and copper in angiogenesis, *J. Natl. Cancer Inst.,* 69, 475, 1982.

100. **Ben Ezra, D.,** Neovasculogenic ability of prostaglandins, growth factors and synthetic chemoattractants, *Am. J. Opthalmol.,* 86, 455, 1978.

101. **Goldsmith, H. S., Griffith, A. L., Kupferman, A., and Catsimpoolas, N.,** Lipid angiogenic factor from omentum, *J. Am. Med. Assoc.,* 252, 2034, 1984.

102. **Brown, R. A., Weiss, J. B., Tomlinson, I. W., Phillips, P., and Kumar, S.,** Angiogenic factor from synovial fluid resembling that from tumors, *Lancet,* 8170, 682, 1980.

103. **Polverini, P. J., Cotran, R. S., Gimbrone, Jr., M. A., and Unanue, E. R.,** Activated macrophages induce vascular proliferation, *Nature (London),* 269, 804, 1977.

104. **Thakral, K. K., Goodson, W. H., III, and Hunt, T. K.,** Stimulation of wound blood vessel growth by wound macrophages, *J. Surg. Res.,* 26, 430, 1979.

105. **Knighton, D. R., Hunt, T. K., Scheuenstahl, H., Halliday, B. J., Werb, Z., and Banda, M. J.,** Oxygen tension regulates the expression of angiogenesis factor by macrophages, *Science,* 221, 1283, 1983.

Chapter 16

GROWTH FACTOR PRODUCTION BY ENDOTHELIAL CELLS

Paul E. DiCorleto and Paul L. Fox

TABLE OF CONTENTS

I. INTRODUCTION

The endothelium is no longer thought to function solely as a passive barrier between the blood and tissues. The diversity of biologically active molecules produced by endothelial cells (EC) has suggested to many investigators that the EC is an active participant in such physiological processes as vessel development and remodeling, wound healing, formation of collateral circulation, vessel contraction and relaxation, and immune system responses. The endothelium of each organ may be considered to be a mini endocrine gland that responds to external stimuli by the production of paracrine hormones and growth factors which act on neighboring smooth muscle cells (SMC), glial cells, pericytes or other organ-specific cell types. Cultured EC have been found to also secrete factors such as colony-stimulating activity,[1,2] interleukin-1,[3] interferon,[4] and neutrophil/monocyte chemoattractants,[5-7] which modulate immune system cells and/or immature and circulating leukocytes. This chapter will focus on the capacity of EC to produce polypeptide growth factors (mitogens) which specifically stimulate the proliferation of connective tissue cells such as fibroblasts and vascular SMC.

The EC was the first proliferative, diploid cell shown to secrete mitogenic activity for fibroblasts (including 3T3 cells) and SMC. In 1978, three groups of investigators, Fass and co-workers,[8] Gajdusek and Schwartz,[9] and DiCorleto and Ross,[9] independently observed in EC lysates, EC-conditioned media, and EC-SMC cocultures, respectively, that SMC could be stimulated to multiply by an EC product. These initial studies were performed with EC from porcine and bovine aorta and from human umbilical vein, but growth-promoting activity for connective tissue cells has since been identified in EC-conditioned media from every source tested, including monkey aorta,[10] rat heart,[11] and human kidney,[12] as well as dog jugular vein and adult human aorta.[62] Growth inhibitors for SMC have also been found in EC-conditioned media,[13,14] but when SMC were cocultured with EC under nonproliferative conditions, i.e., in cell-free plasma-derived serum which lacks platelet mitogens, the EC-derived growth-promoting activity appeared to dominate and the SMC multiplied as if in the presence of maximal concentrations of exogenous mitogen.[9,15] As discussed in the following sections, however, growth factor production is a highly regulated process, and conditions may exist in vitro and in vivo in which secreted growth-inhibitory activity dominates. This chapter will concentrate on the nature of SMC growth-promoting activity that is secreted by EC, the regulation of growth-factor production by EC, and the possible role of this mitogenic activity in the growth control of EC and SMC in physiological and pathological processes.

II. CHARACTERISTICS OF ENDOTHELIAL CELL-DERIVED GROWTH FACTORS

Cultured bovine aortic EC constitutively secrete, for many weeks under serum-free conditions, mitogenic activity for SMC.[16] Thus sufficient amounts of crude EC-derived growth factor(s) can be obtained to allow biological characterization and biochemical fractionation. The growth-promoting activity in EC-conditioned media can be separated into at least two distinct mitogens by cation exchange chromatography.[17] The more basic of the two growth factors has been found to be indistinguishable from and possibly identical to platelet-derived growth factor (PDGF). This growth factor, designated PDGFc, binds to the PDGF receptor on fibroblasts and SMC,[18] is recognized by antibody to homogeneous human PDGF,[17,18] and exhibits identical stability and chromatographic properties to PDGF.[17] In addition, the cellular or proto-oncogene, c-*sis*, which codes for the B-chain of human PDGF, is expressed by cultured EC as measured by Northern blot analysis using v-*sis* or c-*sis* cDNA probes.[19-21] A *sis*-homologous cDNA clone has been recently isolated from human EC and sequenced.[20,21] Controversy exists, however, as to whether the *sis* gene product is in fact secreted, and whether the PDGFc in EC-conditioned media is the product of the *sis* gene.[22,23] As discussed in greater detail in a later section, the PDGFc found in EC-conditioned media may be a homodimer of the A-chain of PDGF, as is the case for several transformed cell lines,[24] rather than a B-chain homodimer or an A-B heterodimer.

The level of PDGFc found in the conditioned media of cultured EC is in the nanograms per milliliter per day range,[11] which is higher than that produced by most cultured neoplastic cells including those transformed by simian sarcoma virus, the source of the v-*sis* gene.[25-28] This substantial level of PDGFc, however, was found to represent only approximately 25% of the growth-promoting activity found in the conditioned media of bovine aortic EC and less than 15% of the mitogen produced by human umbilical vein EC or human saphenous vein EC.[17,18] A second major SMC mitogen secreted by EC was separable from PDGFc by carboxymethyl-Sephadex chromatography.[17] Elution occurred at a lower salt concentration than that required for PDGFc. This novel growth factor was also distinguished from PDGFc by a lack of binding to the PDGF receptor and a lack of recognition by antibody to PDGF. The non-PDGF mitogen did, however, exhibit stability properties similar to PDGF, i.e. the growth-promoting activity was sensitive to reducing agents but stable to boiling. Little else is known about this unique growth factor, although it appears to be distinct from epidermal growth factor and basic fibroblast growth factor (FGF). Interleukin-1 is produced by cultured EC,[3] but this lymphokine is virtually inactive in our human foreskin fibroblast mitogen assay, and it is, therefore, unlikely that the EC growth factor is interleukin-1. Without a specific radioreceptor or radioimmunological assay for the non-PDGF mitogen, experiments to probe its regulation of production, as discussed below for PDGFc, are difficult to perform. The majority of published studies on EC mitogens have concentrated on either PDGFc activity or c-*sis* gene expression, and, therefore, much of the discussion in later sections of this chapter will deal with this specific mitogen.

EC and other cells contain intracellular mitogens (for fibroblasts and SMC) that can be released by cell lysis. Gajdusek and Schwartz[29] compared the intracellular mitogenic activity in EC lysates to the secreted growth-promoting activity. The lysate mitogen, unlike PDGFc, was found to be heat (56°C) sensitive and relatively stable to trypsin treatment.[29] Recent studies by others indicate that the EC intracellular mitogen is related to basic FGF.[61] This identification was made by demonstrating similarities between EC lysate and basic FGF with respect to (1) elution from heparin-Sepharose, (2) molecular weight (18,000), (3) recognition by antibody to FGF, and (4) mitogenic activity for EC. The significance of a growth factor such as FGF which is only released to the environment during cell lysis remains to be determined, but one could speculate on a role for FGF as an autocrine repair factor to facilitate regeneration of a damaged EC monolayer.

III. STIMULATORS OF PDGFc PRODUCTION BY ENDOTHELIAL CELLS

Cultured EC constitutively secreted PDGFc into culture media in the presence or absence of serum or plasma-derived serum.[30] The level of production was not related to the growth state of the EC since sparse cultures in the log phase of growth (> 90% cycling cells) produced the same amount of PDGFc per cell as confluent, quiescent cultures (< 5% cycling cells).[30] Sparse, quiescent EC were found to produce more PDGFc than confluent, quiescent EC, but overall protein synthesis was also higher in the sparse cultures.[30] Growth-factor production by EC has been observed by multiple laboratories using various types of culture media. One culture condition which has a marked effect on growth-factor production is pH. Alkaline media (pH 8) favored growth-factor production, whereas cultures maintained at pH 6.5 or lower exhibit little if any PDGFc production.[30] Growth factor production is therefore quite sensitive to changes in pH near physiological levels.

Agents which cause slow lethal damage to bovine aortic EC, including high concentrations of phorbol esters, e.g., phorbol myristic acid (PMA) or methyl PMA, or endotoxin, stimulated production of PDGFc. We calculated that a mortally wounded cell produced as much PDGFc as did a healthy cell in 19 days. This burst of growth-promoting activity was not simply due to the release of an intracellular storage pool, since very low levels of PDGFc were detected following freeze/thaw or hypotonic lysis of EC. In addition, agents, e.g., high concentrations of cholesterol or H_2O_2[62] which caused very rapid cell death, as opposed to the 6- to 10-day process that was initiated by phorbol esters or endotoxin, did not stimulate PDGFc production. The stimulatory activity of the phorbol esters was independent of the tumor-promoting activity of these molecules, since a nonpromoting analog (methyl PMA) was as lethal and stimulatory as the true tumor promoters. It should also be noted that the concentration of phorbol ester required to cause EC death and increase PDGFc production was several orders of magnitude greater than that reported to activate protein kinase C.[31] The action of phorbol esters on EC growth-factor production correlated quantitatively with the cytotoxic potency of the drug. This suggests a possible role for injury-induced mitogenic activity in vessel wall pathology as discussed in a later section of this chapter.

Both phorbol esters and endotoxin have also been shown to stimulate growth-factor production by specific types of EC in the absence of cytotoxic response. Human umbilical vein EC, unlike bovine aortic EC, are not lethally injured by endotoxin.[32] Low doses of endotoxin were found to cause a twofold increase in c-*sis* hybridizable mRNA in human umbilical vein EC; however, the level of PDGFc in the EC media was not quantitated.[33] Nontoxic concentrations of PMA also stimulated c-*sis* mRNA in cultured human renal microvascular EC.[12] In this case, increased amounts of PDGFc were detected in the EC media. This latter observation represents the only reported example of coordinate regulation in EC of c-*sis* mRNA levels and PDGFc production.

Two components of the coagulation system, α-thrombin[34] and Factor Xa,[35] induced a severalfold increase in the level of PDGFc in the media of cultured EC. The thrombin effect on human umbilical vein EC was observable in hours,[34] but augmented production persisted for days when cultures of bovine aortic EC were treated with multiple changes of media containing thrombin.[62] Thrombin stimulation of PDGFc production required the serine esterase activity of the enzyme, but was independent of *de novo* cellular protein synthesis over a 3-hr time period.[34] Activated coagulation Factor X (Factor Xa) acted in a manner similar to thrombin — stimulating PDGFc production by EC in a process requiring active enzyme but not cellular protein synthesis.[35] The Factor Xa effect, however, was not mediated through thrombin since the thrombin inhibitor hirudin did not prevent the stimulation of PDGFc production by Factor Xa. An altered level of v-*sis* hybridizable mRNA was not observed when bovine aortic EC were treated with Factor Xa; however, only a single 16-hr time point was examined.[35]

Daniel et al.[12] have reported that kidney microvascular EC respond to thrombin by increasing c-*sis* mRNA by three- to fivefold and by increasing PDGFc in the media. The peak in *sis* message occurred 4 hr following stimulation and returned to control levels by 8 hr. In contrast to the studies of Harlan et al.[34] with human umbilical vein EC, Daniel et al. observed full c-*sis* stimulating activity with diisopropyl fluorophosphate-treated (inactive) thrombin. This difference may reflect a differential response of these EC types or a discrepancy between the measurement of PDGFc and *sis* message. Though Daniel et al. measured PDGFc in EC media following thrombin treatment, no data were presented for the effect of inactive thrombin on the production of PDGFc activity.

IV. INHIBITORS OF PDGFc PRODUCTION BY ENDOTHELIAL CELLS

The growth of human EC in culture is regulated in part by a specific EC growth factor (ECGF) that has been isolated from bovine brain.[36] Jaye et al.[37] have reported that withdrawal of ECGF from human umbilical vein EC resulted in rearrangement of the monolayer into a network of tubular structures resembling microvessels. Formation of the organized network was accompanied by a decrease in the expression of the c-*sis* transcript to 23% of the value for untreated control cells. The suppression of c-*sis* was specific since the fibronectin transcript did not decrease, and in fact was elevated, under the same conditions. The morphological and functional changes in the cells were both reversed upon addition of ECGF. The authors speculated that the high expression of c-*sis* seen in proliferative monolayers may be an important component of neovascularization in vivo, where EC production of PDGF-like molecules may stimulate migration and proliferation of vascular fibroblasts and SMC. However, since the authors measured mRNA expression only and not the secretion of biologically active protein, the role of differentiation in growth-factor production remains speculative.

The composition of the extracellular matrix has been shown to influence morphological and functional properties of cultured EC including cell shape, adhesivity, cytoskeletal structure, migration, and proliferation (see Madri and Pratt[38] for review). We have recently initiated studies on the effect of matrix material on EC production of PDGFc. Bovine aortic EC grown to confluence on Vitrogen (solubilized types I and III bovine collagen, Flow Laboratories) produced in 5 days only one third the amount of PDGFc compared to control cultures on untreated tissue culture plastic.[39] This difference did not persist during a second incubation of 6 days, suggesting that the substratum may be modified by EC to resemble that of underlying cells on untreated plastic. Treatment of culture dishes with fibronectin did not alter PDGFc production by EC even during short intervals.[39]

EC situated on an atherosclerotic lesion may have an altered cellular lipid content reflecting the pathological lipid composition of the plasma or the diseased artery. With this idea in mind, we investigated the effect of lipoproteins and cellular cholesterol content on EC production of PDGFc. EC uptake of native low density lipoprotein (LDL) is limited by down regulation of the LDL receptor, but LDL made electronegative, e.g., by acetylation (acetyl-LDL), is endocytosed by a distinct EC surface receptor (scavenger receptor) resulting in cellular cholesterol content up to 60% greater than control cells.[40,41] We have found that incubating confluent cultures of bovine aortic EC with acetyl-LDL (150 μg of cholesterol per milliliter), but not native LDL, resulted in a decrease in production of PDGFc to 0 to 25% of the control, with half-maximal inhibition at 25 to 75 μg of cholesterol per milliliter.[11] The observed inhibition was not due to cell damage since there was no change in cell morphology or number, and furthermore, total cellular and protein synthesis, as measured by incorporation of [^{3}H]leucine into trichloroacetic acid-precipitable material, was unaffected. This last result also suggested that the inhibition was specific with respect to the proteins affected.

We have begun to investigate the mechanism for the acetyl-LDL-mediated inhibition of PDGFc production. EC isolated from bovine aorta, human umbilical vein, and rat heart all produced PDGFc at similar rates, but acetyl-LDL was shown to inhibit production in only the first two cell types.[11] We have examined these cells for the presence of an active scavenger receptor by measuring uptake and degradation of ^{125}I-labeled acetyl-LDL, and found that bovine aorta and human umbilical vein EC, but not the rat heart EC, possessed this receptor. These results suggested that receptor-mediated uptake of acetyl-LDL was required for the observed inhibition.

The requirement for lysosomal hydrolysis of acetyl-LDL was examined by incubating bovine aortic EC with several inhibitors of lysosome activity, including chlorquine, NH$_4$Cl, and monensin.[42] At inhibitor levels that completely prevented degradation of ^{125}I-labeled acetyl-LDL, there was no effect on the inhibitory ability of this lipoprotein, demonstrating that hydrolysis of the lipoprotein was not required for the suppression of PDGFc production.

LDL made cationic by reaction with dimethylpropane-diamine (DMPA-LDL) also inhibited EC production of PDGFc.[11] Both DMPA- and acetyl-LDL caused elevated cellular cholesterol content, suggesting that this lipid may be responsible for the inhibition. However, incubating bovine EC with sonicated cholesterol/albumin complexes did not inhibit PDGFc production despite the fact that cellular cholesterol content was higher than in acetyl-LDL treated cells.[42] Since the cholesterol was shown to be part of a metabolically active intra-cellular pool (by increased cellular esterification of cholesterol), we concluded that choles-terol is not the molecular species responsible for the inhibition of growth factor production.

Several observations suggested that an oxidized lipid may be involved in the regulation of PDGFc production.[42] First of all, LDL oxidized in vitro by dialysis against $FeSO_4$ inhibited PDGFc production in a manner dependent on lipoprotein dose and oxidation level. Secondly, the inhibition of PDGFc production by acetyl-LDL also was found to be dependent on lipid peroxidation. Acetyl-LDL, however, was inhibitory at much lower oxidation levels than LDL, probably reflecting its increased propensity for receptor-mediated endocytosis by the scavenger receptor. Finally, a lipid extract from oxidized LDL, but not native LDL, inhibited PDGFc production. Although these results indicate an involvement of lipid peroxidation, we cannot conclude that an oxidized lipid is the actual inhibitory species since oxidation may cause increased uptake of LDL, acetyl-LDL, or lipid extracts of LDL, thus increasing the cellular concentration of a putative, nonoxidized inhibitor.

In an attempt to identify the site of acetyl-LDL inhibition of PDGFc production, we performed Northern blot analysis using a v-*sis* probe on RNA extracted from control and treated cultures. The autoradiographs showed that PDGF B-chain mRNA levels were not altered by oxidized or acetylated LDL under conditions where PDGFc production was nearly completely inhibited.[42] This result suggests either that regulation by these lipoproteins is at a posttranscriptional level or, as discussed below, that actively secreted EC PDGFc is not encoded by the B-chain transcript.

We have summarized the published studies on the mechanism of regulation of PDGFc production by EC in Table 1. Comparisons of results from different laboratories have been hampered by several factors. First, few investigators have had all the tools available to quantitate (1) PDGFc protein levels in EC-conditioned media by radioreceptor or radioim-munoassay, (2) PDGF B-chain (c-*sis*) message levels in the same EC cultures, and (3) PDGF A-chain message levels. As a result of partial information from individual laboratories (e.g., just *sis* message or just PDGFc concentration), apparent inconsistencies in the literature have grown. Secondly, currently available antibodies to PDGF have been unsuitable for clean immunoprecipitation studies on the synthesis and processing of metabolically-labeled PDGFc by EC. Posttranslational processing may be very important in the regulation of growth-factor production by EC. The product of the v-*sis* gene (B-chain homodimer) has been shown by Robbins et al.[22] to pass through multiple processing steps during synthesis in virally trans-formed cells. A third deficiency in the EC growth factor field is the lack of amino acid sequence data on EC-derived PDGFc. Is this molecule an A-chain homodimer, similar to that produced by cultured osteosarcoma cells,[24] rather than an A-B heterodimer (human platelet PDGF)[43] or B-chain homodimer (pig platelet PDGF[44])? Until this question is an-swered, quantitation of c-*sis* message levels should be considered an independent pursuit from the measurement of PDGFc in the media. The resolution of these questions awaits purification of sufficient PDGFc from EC-conditioned media to allow N-terminal sequencing and the generation of high-affinity antibodies to native PDGFc that will be useful for immunoprecipitation studies.

V. PDGFc AND AUTOCRINE GROWTH CONTROL IN ENDOTHELIAL CELLS

Cultured EC are unusual in their ability to proliferate equally well in plasma- or serum-supplemented media, indicating a lack of requirement for platelet-derived mitogens in EC

Table 1
REGULATION OF GROWTH FACTOR PRODUCTION BY ENDOTHELIAL CELLS

Regulatory factor	Source of EC	Parameter quantitated	Expression level (%)	Ref.
Stimulators				
Endotoxin	Bovine aorta	PDGFc	200	30
Phorbol esters (cytotoxic doses)	Bovine aorta	PDGFc	350	30
Heparin	Bovine aorta	Mitogenic activity	300—500	60
Factor Xa	Bovine aorta	PDGFc	1000	35
Thrombin	Human umbilical vein	PDGFc	500—1000	34
Thrombin	Human kidney	c-*sis* mRNA; PDGFc	300—500; 210	12
PMA	Human kidney	c-*sis* mRNA; PDGFc	1500; 240	12
Endotoxin	Human umbilical vein	c-*sis* mRNA	230	33
Inhibitors				
"Tube formation"	Human umbilical vein	c-*sis* mRNA	23	37
acetyl-LDL, oxidized LDL DMPA-LDL	Human umbilical vein, bovine, porcine aorta	PDGFc[a]	0—25	11,42
Collagen	Bovine aorta	PDGFc	34	39

[a] c-*sis* mRNA also quantitated but unaltered by regulatory factor.

multiplication.[45] Addition of exogenous PDGF to media containing plasma-derived serum does not cause enhanced EC growth. Either serum components absent in plasma are not required for EC proliferation or the EC are capable of producing and responding to factors which substitute for serum components. One candidate for an EC autocrine mitogen is PDGFc. Several transformed cell lines have been documented to exhibit autocrine growth control in response to self-generated PDGFc.[46,47] In addition in vitro transformation of PDGFc-producing cells has been shown to correlate with the presence of PDGF receptors on the cell surface.[48,49]

We tested the hypothesis that cultured EC do not require platelet-derived mitogens to grow because they are responding to autologously produced PDGFc. One requirement for autocrine growth is the presence of specific receptors for the putative autocrine mitogen. Bovine aortic and rat heart EC were found to express little or no cell surface receptors for PDGF.[50] Treatment of EC with acetic acid, under conditions which released prebound PDGF from control cells without affecting subsequent receptor binding, did not reveal a population of filled or cryptic receptors. In addition, when rat EC were grown in the presence of sufficient PDGF antibody to completely neutralize PDGFc in the media, EC proliferation was unimpaired. An equivalent dose of antibody completely inhibited the mitogenic response of human fibroblasts that had been preincubated for 1 hr at 37°C with a dose of PDGF comparable to that produced by EC. No intracellular PDGF receptors were detected in rat heart or bovine aortic EC.[50] In total, these results force us to conclude that PDGFc is not an autocrine mitogen for cultured EC. The possibility remains, however, that one or more other EC-derived growth factors, either in EC-conditioned media or in EC lysates, is responsible for autocrine growth control of EC.

VI. PHYSIOLOGICAL AND PATHOLOGICAL IMPLICATIONS OF ENDOTHELIAL CELL GROWTH FACTOR PRODUCTION

In vitro evidence strongly supports the concept that EC have the capacity to secrete potent growth factors and chemoattractants for neighboring connective tissue cells. This EC function may be of great physiological importance in such processes as vessel neogenesis both during embryogenesis and during the development of collateral circulation in the adult animal. The EC, which leads the way in angiogenesis, may secrete PDGFc to attract and stimulate nearby SMC and pericytes in order to complete the formation of a blood vessel.[51] Wound repair may also involve the temporal, well regulated secretion of polypeptide mitogens for local fibroblasts. A role for EC-derived growth factors in these processes has not yet been demonstrated, but with such methods as *in situ* hybridization or highly sensitive immunocytochemistry, we may soon know whether growth factor genes are expressed in EC during vessel neogenesis or wound healing.

Uncontrolled secretion of paracrine growth factors could be easily visualized as a triggering mechanism for the development of proliferative lesions in tissue. Atherosclerotic plaque development involves hyperplasia of intimal SMC in response to an as yet unidentified mitogen.[52] Inappropriate expression of growth factor genes by EC in vivo may in fact be a critical step in the recruitment and mitogenic stimulation of SMC in the subendothelial space. The only in vivo data reported for EC growth factor production has been the observation that low but detectable levels of *sis* message were present in freshly isolated EC from human umbilical vein or bovine aorta; the message levels, however, were 1 to 2 orders of magnitude lower than in cultured cells.[19] A similar experiment has not been performed with a probe for the A-chain of PDGF. Local stimulation of EC by some type of activation process, not unlike the perturbation caused by culturing EC on plastic, may lead to focal SMC proliferation and atherosclerotic plaque development. The concept of EC injury as an initiating event in atherogenesis may be relevant, not so much as a function of exposure of a nonthrombogenic surface for platelet attachment, but rather due to the stimulation of growth factor production. Increased expression of EC growth factor genes may result from the action of either thrombin-like factors which stimulate EC without causing cell death or cytotoxic agents which induce a bolus of growth factor release as the EC dies.

An autocrine mechanism has been hypothesized for the uncontrolled growth of transformed cells. Cell lines with PDGF receptors exhibit phenotypic transformation when transfected with the *sis* gene.[53] An alternative trigger for chronic mitogenic stimulation of target cells could be a paracrine growth factor (PDGFc) that is continuously secreted by a dysfunctional neighboring cell type (EC). Multiple events are probably required for a normal cell to exhibit neoplastic growth. The possibility exists that a paracrine mitogen could substitute for one or more of these events. Dysfunctional or activated EC could affect the growth of a neighboring tumor in the absence of EC transformation. Combined immunocytochemical and *in situ* hybridization techniques will be required in future studies to determine whether EC paracrine action plays a role in some neoplasias.

EC growth factors may also have a role in the neointimal hyperplasia that is often responsible for the long-term failure of artificial vascular prostheses.[54,55] Evidence that EC may be involved in graft failure is derived from observations in several model systems in which there is a spatial correlation between areas of graft endothelialization and intimal thickening. This is most often observed at graft/vessel anastomoses in humans and dogs, where EC ingrowth is limited to a 1- to 2-cm ring which is generally accompanied by subendothelial SMC accumuation.[56] In a second example, perforated[57] or highly porous[58] grafts have been shown to be vascularized by transmural capillaries which promote endothelialization along the entire graft length; in these grafts a subendothelial thickened intima was observed in mid-graft regions as well as at the anastomoses. EC situated on grafts may be activated by the foreign environment of their substratum to produce growth factors, thus accounting for these observations.

VII. CONCLUSIONS

The endothelium appears to be capable of responding to humoral and neuronal factors by secreting secondary paracrine signals that modulate the function of neighboring cell types. The best known example of this property is the release of endothelium-derived relaxing factor by EC in response to vasorelaxants in the blood.[59] A similar situation may exist for growth factor production by EC in vivo. Developmental signals, either endocrine or neural, may trigger growth factor gene expression in the EC during vessel neogenesis. Pathological stimuli, such as undefined initiating factors in atherogenesis, may cause these same developmental genes to be expressed in the adult. Many unanswered questions remain as to the true role of EC growth factors in vivo and the regulation of expression of these growth factor genes both in vivo and in vitro. In the next few years tools should be developed and techniques refined such that critical studies both in cell culture and in whole animals can be performed to resolve questions on the regulation of EC growth factor production.

ACKNOWLEDGMENT

The studies reported from our laboratory were supported in part by grants from the American Heart Association, Northeast Ohio Affiliate, and the NIH (HL-29582). P.E.D. is the recipient of a Research Career Development Award (HL-1561) from the NIH.

REFERENCES

1. **Quesenberry, P. J. and Gimbrone, M. A., Jr.,** Vascular endothelium as a regulator of granulopoiesis: production of colony-stimulating activity by cultured human endothelial cells, *Blood,* 56, 1060, 1980.
2. **Gerson, S. L., Friedman, H. M., and Cines, D. B.,** Viral infection of vascular endothelial cells alters production of colony-stimulating activity, *J. Clin. Invest.,* 76, 1382, 1985.
3. **Miossec, P., Cavender, D., and Ziff, M.,** Production of interleukin 1 by human endothelial cells, *J. Immunol.,* 136, 2486, 1986.
4. **Einhorn, S., Eldor, A., Vlodavsky, I., Fuks, Z., and Panet, A.,** Production and characterization of interferon from endothelial cells, *J. Cell. Physiol.,* 122, 200, 1985.
5. **Mercandetti, A. J., Lane, T. A., and Colmerauer, M. E. M.,** Cultured human endothelial cells elaborate neutrophil chemoattractants, *J. Lab. Clin. Med.,* 104, 370, 1984.
6. **Berliner, J. A., Territo, M., Almada, L., Carter, A., Shafonsky, E., and Fogelman, A. M.,** Monocyte chemotactic factor produced by large vessel endothelial cells in vitro, *Arteriosclerosis,* 6, 254, 1986.
7. **Quinn, M. T., Parthasarathy, S., and Steinberg, D.,** Endothelial cell-derived chemotactic activity for mouse peritoneal macrophages and the effects of modified forms of low density lipoprotein, *Proc. Natl. Acad. Sci. U.S.A.,* 82, 5949, 1985.
8. **Fass, D. N., Downing, M. R., Meyers, P., Bowie, E. J. W., and Witte, L. D.,** A mitogenic factor from porcine arterial endothelial cells, Council on Atherosclerosis, 32nd Annual Meeting, American Heart Association, 11, 1978.
9. **Gajdusek, C., DiCorleto, P., Ross, R., and Schwartz, S. M.,** An endothelial cell-derived growth factor, *J. Cell Biol.,* 85, 467, 1980.
10. **Wang, C.-H., Largis, E. E., and Schaffer, S. A.,** The effects of endothelial cell-conditioned media on the proliferation of aortic smooth muscle cells and 3T3 cells in culture, *Artery,* 9, 358, 1981.
11. **Fox, P. L. and DiCorleto, P. E.,** Modified low density lipoproteins suppress production of a platelet-derived growth factor-like protein by cultured endothelial cells, *Proc. Natl. Acad. Sci. U.S.A.,* 83, 4774, 1986.
12. **Daniel, T. O., Gibbs, V. C., Milfay, D. F., Garovoy, M. R., and Williams, L. T.,** Thrombin stimulates c-*sis* gene expression in microvascular endothelial cells, *J. Biol. Chem.,* 261, 9579, 1986.
13. **Willems, Ch., Astaldi, G. C. B., de Groot, Ph. G., Janssen, M. C., Gonsalvez, M. D., Zeijlemaker, W. P., Van Mourik, J. A., and Van Aken, W. G.,** Media conditioned by cultured human vascular endothelial cells inhibit the growth of vascular smooth muscle cells, *Exp. Cell Res.,* 139, 191, 1982.

14. **Castellot, J. J., Jr., Addonizio, M. L., Rosenberg, R., and Karnovsky, M. J.,** Cultured endothelial cells produce a heparinlike inhibitor of smooth muscle cell growth, *J. Cell Biol.,* 90, 372, 1981.

15. **Davies, P. F., Truskey, G. A., Warren, H. B., O'Connor, S. E., and Eisenhaure, B.H.,** Metabolic cooperation between vascular endothelial cells and smooth muscle cells in co-culture: changes in low density lipoprotein metabolism, *J. Cell Biol.,* 101, 871, 1985.

16. **DiCorleto, P. E., Gajdusek, C. M., Schwartz, S. M., and Ross, R.,** Biochemical properties of the endothelium-derived growth factor: comparison to other growth factors, *J. Cell. Physiol.,* 114, 339, 1983.

17. **DiCorleto, P. E.,** Cultured endothelial cells produce multiple growth factors for connective tissue cells, *Exp. Cell Res.,* 153, 167, 1984.

18. **DiCorleto, P. E. and Bowen-Pope, D. F.,** Cultured endothelial cells produce a platelet-derived growth factor-like protein, *Proc. Natl. Acad. Sci. U.S.A.,* 80, 1919, 1983.

19. **Barrett, T. B., Gajdusek, C. M., Schwartz, S. M., McDougall, J. K., and Benditt, E. P.,** Expression of the *sis* gene by endothelial cells in culture and *in vivo, Proc. Natl. Acad. Sci. U.S.A.,* 81, 6772, 1984.

20. **Collins, T., Ginsburg, D., Boss, J. M., Orkin, S. H., and Pober, J. S.,** Cultured human endothelial cells express platelet-derived growth factor B chain: cDNA cloning and structural analysis, *Nature (London),* 316, 748, 1985.

21. **Tong, B. D., Levine, S. E., Jaye, M., Ricca, G., Drohan, W., Maciag, T., and Deuel, T. F.,** Isolation and sequencing of a cDNA clone homologous to the v-*sis* oncogene from human endothelial cells, *Mol. Cell. Biochem.,* 6, 3018, 1986.

22. **Robbins, K. C., Leal, F., Pierce, J. H., and Aaronson, S. A.,** The v-*sis*/PDGF-2 transforming gene product localizes to cell membranes but is not a secretory protein, *EMBO J.,* 4, 1783, 1985.

23. **Betsholtz, C., Johnsson, A., Heldin, C.-H., Westermark, B., Lind, P., Urdea, M. S., Eddy, R., Shows, T. B., Philpott, K., Mellor, A. L., Knott, T. J., and Scott, J.,** cDNA sequence and chromosomal localization of human platelet-derived growth factor A-chain and its expression in tumour cell lines, *Nature (London),* 320, 695, 1986.

24. **Heldin, C.-H., Johnsson, A., Wennergren, S., Wernstedt, C., Betsholtz, C., and Westermark, B.,** A human osteosarcoma cell line secretes a growth factor structurally related to a homodimer of PDGF A-chains, *Nature (London),* 319, 511, 1986.

25. **Heldin, C.-H., Westermark, B., and Wasteson, A.,** Chemical and biological properties of a growth factor from human-cultured osteosarcoma cells: resemblance with platelet-derived growth factor, *J. Cell. Physiol.,* 105, 235, 1980.

26. **Deuel, T. F., Huang, J. S., Huang, S. S., Stroobant, P., and Waterfield, M. D.,** Expression of a platelet-derived growth factor-like protein in simian sarcoma virus transformed cells, *Science,* 221, 1348, 1983.

27. **Owen, A. J., Pantazis, P., and Antoniades, H. N.,** Simian sarcoma virus-transformed cells secrete a mitogen identical to platelet-derived growth factor, *Science,* 225, 54, 1984.

28. **Bowen-Pope, D. F., Vogel, A., and Ross, R.,** Production of platelet-derived growth factor-like molecules and reduced expression of platelet-derived growth factor receptors accompany transformation by a wide spectrum of agents, *Proc. Natl. Acad. Sci. U.S.A.,* 81, 2396, 1984.

29. **Gajdusek, C. M. and Schwartz, S. M.,** Comparison of intracellular and extracellular mitogenic activity, *J. Cell. Physiol.,* 121, 316, 1984.

30. **Fox, P. L. and DiCorleto, P. E.,** Regulation of production of a platelet-derived growth factor-like protein by cultured bovine aortic endothelial cells, *J. Cell. Physiol.,* 121, 298, 1984.

31. **Nishizuka, Y.,** Studies and perspectives of protein kinase C, *Science,* 233, 305, 1986.

32. **Harlan, J. M., Harker, L. A., Reidy, M. A., Gajdusek, C. M., Schwartz, S. M., and Striker, G. E.,** Lipopolysaccharide-mediated bovine endothelial cell injury *in vitro, Lab. Invest.,* 48, 269, 1983.

33. **Libby, P., Ordovas, J. M., and Janicka, M. W.,** Regulation of c-*sis* oncogene expression in human vascular endothelial cells by endotoxin, *Arteriosclerosis,* 6, 524a, 1986.

34. **Harlan, J. M., Thompson, P. J., Ross, R. R., and Bowen-Pope, D. F.,** α-Thrombin induces release of platelet-derived growth factor-like molecule(s) by cultured human endothelial cells, *J. Cell Biol.,* 103, 1129, 1986.

35. **Gajdusek, C., Carbon, S., Ross, R., Nawroth, P., and Stern, D.,** Activation of coagulation releases endothelial cell mitogens, *J. Cell Biol.,* 103, 419, 1986.

36. **Jaye, M., Howk, R., Burgess, W., Ricca, G. A., Chiu, I.-M., Ravera, M. W., O'Brien, S. J., Modi, W. S., Maciag, T., and Drohan, W. N.,** Human endothelial cell growth factor: Cloning, nucleotide sequence, and chromosome localization, *Science,* 233, 541, 1986.

37. **Jaye, M., McConathy, E., Drohan, W., Tong, B., Deuel, T., and Maciag, T.,** Modulation of the *sis* gene transcript during endothelial cell differentiation in vitro, *Science,* 228, 882, 1985.

38. **Madri, J. A. and Pratt, B. M.,** Endothelial cell-matrix interactions: in vitro models of angiogenesis, *J. Histochem. Cytochem.,* 34, 85, 1986.

39. **DiCorleto, P. E. and Chisolm, G. M., III,** Participation of the endothelium in the development of the atherosclerotic plaque, *Prog. Lipid, Res.,* in press.

40. **Stein, O. and Stein, Y.,** Bovine aortic endothelial cells display macrophage-like properties towards ace-tylated ^{125}I-labelled low density lipoprotein, *Biochim. Biophys. Acta,* 620, 631, 1980.

41. **Baker, D. P., Van Lenten, B. J., Fogelman, A. M., Edwards, P. A., Kean, C., and Berliner, J. A.,** LDL, scavenger, and β-VLDL receptors on aortic endothelial cells, *Arteriosclerosis,* 4, 248, 1984.

42. **Fox, P. L., Chisolm, G. M., and DiCorleto, P. E.,** Lipoprotein-mediated inhibition of endothelial cell production of PDGF-like protein depends on free radical lipid peroxidation, *J. Biol. Chem.,* 262, 6042, 1987.

43. **Antoniades, H. N. and Hunkapiller, M. W.,** Human platelet-derived growth factor (PDGF): amino-terminal amino acid sequence, *Science* 220, 963, 1983.

44. **Stoobant, P. and Waterfield, M. D.,** Purification and properties of porcine platelet-derived growth factor, *EMBO J.,* 3, 2963, 1984.

45. **Schwartz, S. M., Gajdusek, C. M., and Selden, S. C., III,** Vascular wall growth control: the role of the endothelium, *Arteriosclerosis,* 1, 107, 1981.

46. **Huang, J. S., Huang, S. S., and Deuel, T. F.,** Transforming protein of simian sarcoma virus stimulates autocrine growth of SSV-transformed cells through PDGF cell-surface receptors, *Cell,* 39, 79, 1984.

47. **Betsholtz, C., Westermark, B., Ek, B., Heldin, C.-H.,** Coexpression of a PDGF-like growth factor and PDGF receptors in a human osteosarcoma cell line: implications for autocrine receptor activation, *Cell,* 39, 447, 1984.

48. **Johnsson, A., Betsholtz, C., Heldin, C.-H., and Westermark, B.,** Antibodies against platelet-derived growth factor inhibit acute transformation by simian sarcoma virus, *Nature (London),* 317, 438, 1985.

49. **Leal, F., Williams, L. T., Robbins, K. C., and Aaronson, S. A.,** Evidence that the v-*sis* gene product transforms by interaction with the receptor for platelet-derived growth factor, *Science,* 230, 327, 1985.

50. **Kazlauskas, A. and DiCorleto, P. E.,** Cultured endothelial cells do not respond to a platelet-derived growth-factor-like protein in an autocrine manner, *Biochim. Biophys. Acta* 846, 405, 1985.

51. **Folkman, J.,** Regulation of angiogenesis: a new function of heparin, *Biochem. Pharmacol.,* 34, 905, 1985.

52. **Ross, R.,** The pathogenesis of atherosclerosis — an update, *N. Engl. J. Med.,* 314, 488, 1986.

53. **Leal, F., Williams, L. T., Robbins, K. C., and Aaronson, S. A.,** Evidence that the v-*sis* gene product transforms by interaction with the receptor for platelet-derived growth factor, *Science,* 230, 327, 1985.

54. **Carson, S. N., Hunter, G., French, S., Lord, P., and Wong, H. N.,** Occurrence of occlusive intimal changes in an expanded polytetrafluoroethylene graft, *J. Cardiovasc. Surg.,* 21, 503, 1980.

55. **DeWeese, J. A.,** Anastomotic neointimal fibrous hyperplasia: pathogenesis and prevention, in *Critical Problems in Vascular Surgery 2,* Veith, F. J., Ed., Appleton-Century-Crofts, New York, 1982, 151.

56. **Madras, P. N., Ward, C. A., and Singh, P. I.,** Anastomotic hyperplasia, *Surgery* 90, 922, 1981.

57. **Kusaba, A., Fischer, R., III, Matulewski, T. J., and Matsumoto, T.,** Experimental study of the influence of porosity on development of neointima in Gore-Tex® grafts: a method to increase long-term patency rate, *Am. Surg.,* 47, 347, 1981.

58. **Clowes, A. W., Kirkman, T. R., and Reidy, M. A.,** Mechanisms of arterial graft healing. Rapid transmural capillary ingrowth provides a source of intimal endothelium and smooth muscle in porous PTFE prostheses, *Am. J. Pathol.,* 123, 220, 1986.

59. **Peach, M. J., Singer, H. A., and Loeb, A. L.,** Mechanisms of endothelium-dependent vascular smooth muscle relaxation, *Biochem. Pharmacol.,* 34, 1867, 1985.

60. **Gajdusek, C. M.,** Release of endothelial cell-derived growth factor (ECDGF) by heparin, *J. Cell. Physiol.,* 121, 13, 1984.

61. **Vlodavsky, I. and Klagsbrun, M.,** personal communication.

62. **DiCorleto, P. E. and Fox, P. L.,** unpublished observations.

Chapter 17

ENDOTHELIAL HETEROGENEITY: INFLUENCE OF VESSEL SIZE, ORGAN LOCALIZATON, AND SPECIES SPECIFICITY ON THE PROPERTIES OF CULTURED ENDOTHELIAL CELLS

Bruce R. Zetter

TABLE OF CONTENTS

I. INTRODUCTION

The field of endothelial cell biology has seen rapid development in recent years. This progress has come as a direct result of improvements in our ability to grow endothelial cells in vitro. Although the first successful long-term endothelial cell cultures were derived from large vessels such as the bovine aorta[1] or human umbilical vein,[2-4] endothelial cells can now be cultured from blood vessels derived from different organs as well as from different species. The results emerging from the descriptions of these varying cultures allow us to compare these cells with regard to their morphologic, physiologic, and metabolic properties.

Although all vascular endothelial cells share certain common functions, it has become clear that not all endothelial cells are identical. They differ with regard to antigenic determinants, cell surface molecules, adhesive properties, metabolic properties, permeability functions, and production of vasoactive mediators. In the text that follows, this author will attempt to describe some of the differences between the different types of endothelial cells that have been studied to date. The differences described come from studies conducted on vascular beds in vivo as well as from studies on cultured endothelial cells in vitro. It should be noted, however, that differences in culture conditions or experimental protocols may create artifactual differences between cell types that do not reflect genuine in vivo differences. Situations where different conditions may contribute to apparent differences between cells will be pointed out, but in many cases it is not possible to know whether those culture differences are sufficient to result in the phenotypic differences observed. Rather than discuss the properties of each type of endothelial cell that has been cultured, the chapter will be organized around specific cellular parameters in which significant differences have been noted between endothelial cells of varying origin.

II. ANTIGENIC DIFFERENCES

A. Antibodies Against Nonendothelial Cells

Some of the most compelling evidence for endothelial cell heterogeneity has come from th study of the antigens expressed by different endothelia. Although many different poly- and monoclonal antisera have been raised directly against endothelial cells, some of the antiendothelial cell antibodies were originally raised against other cell types or mixed cell preparations and later found to react with endothelial cells. An example of an antibody prepared against a nonendothelial cell that reacts with an endothelial antigen is the HCl antibody originally prepared against hairy cell leukemia cells. This antibody labels a gly-coplipd found on the endothelium of all tissues tested excluding brain endothelium and splenic sinusoidal cells.[5] This antigen is also found on basal epidermal cells.

Immunization against melanoma cells has also resulted in antibodies that recognize specific endothelial cell antigens.[6] Three monoclonal antibodies have been isolated, all of which recognize endothelial cells in a variety of tissues. Only one of these antibodies, however, also recognizes an antigen on hepatic sinusoidal endothelial cells.

B. Antibodies Against Brain Blood Vessels

Several investigators have successfully made antibodies against components of the brain vasculature. Ghandour et al., for example, prepared a monoclonal antibody (MESA-1) against cultures of dissociated mouse cerebellum.[7] Immunofluorescence and immunocyto-chemistry of the mouse cerebellum revealed that the antigen was localized specifically on the luminal surface of the cerebellar endothelium and was not present on fibroblasts or astrocytes. Consequently, the investigators tested other murine tissues and found the antigen present on endothelial cells of capillaries, veins, and arteries in the brain, heart, lung, and kidneys. When the mouse liver was sectioned and stained, however, the antigen was found on endothelial cells of the hepatic veins but not on the cells of the hepatic sinusoidal small vessels. These results reveal a differential expression of a cell surface antigen that is present on virtually all endothelial cells with the exception of the hepatic sinusoids.

C. Antibodies to Basement Membrane Components

Blood vessels can be distinguished not only by antigens that are integral parts of the endothelial cells themselves but also by components of the subendothelial basement membrane. An example is found in the work of Barsky et al.,[8] who found laminin and type IV collagen in the subendothelial basement membranes of virtually all blood vessels with the exception of the lymphatic circulation. Lymphatic capillaries lack both collagen IV and laminin, whereas larger lymphatic vessels display both antigens. The authors suggest that these different staining patterns can be used to distinguish between the lymphatic and vascular capillaries in frozen sections from a variety of tissues.

D. Antibodies to Specific Endothelial Cells

Antigenic differences between endothelial cells from different organs have been detected in an elegant series of experiments by Auerbach and colleagues. Using mono- and polyclonal antibodies, they have shown that capillary endothelial cells express an array of cell surface antigens that vary from organ to organ. In their initial studies, these investigators identified antibodies that could distinguish between the capillary endothelial cells of mouse brain and those of mouse ovary.[9] More recently, this group has isolated a panel of murine microvascular endothelial cells from lung, liver, brain, placenta, bladder, mammary gland, epididymal fat pad, kidney, and ovary as well as larger vessel endothelial cells from mouse aorta and thoracic duct.[10] The propagation of these diverse murine vascular cell lines will allow a careful comparison of the antigenic and physiologic properties of endothelial cells from diverse organs.

Further evidence for organ-specific endothelial cell antigens comes from the work of Michalak et al.,[11] who prepared monoclonal antibodies to isolated rat brain microvessels. From 156 hybridomas that produced antibodies against brain blood vessels, two were found that reacted with endothelial cells in the brain but not those from other organs. One of these, termed EBM by the authors, was reported to label endothelial cells in frozen sections of brain microvessels but not endothelial cells in brain arteries or veins or in the endothelium of the choroid plexus. The antigen was found in the endothelial cytoplasm as well as on the luminal cell surface. The only other cells labeled by this antigen were brush border cells in the proximal tubules of the kidney and epithelial cells of the hepatic bile duct. The second antibody, 6E1, stained the basement membrane surrounding intermediate-sized cerebral vessels with a diameter of 15 to 20 μm. No staining was associated with brain capillaries or with larger arteries and veins. These results are important because they suggest antigenic specificity is associated not only with organ-specific cytoplasmic or cell surface determinants but also with basement membrane components that vary with the size of the blood vessel.

In a related series of experiments,[12] Pardridge et al. prepared antibodies against isolated plasma membranes from bovine brain microvessels. These antibodies reacted with a 46,000-molecular weight protein and was found by immunofluorescence to be localized on capillaries in the brain cortex but not on capillaries in the heart, liver, or kidney. Cultured bovine capillary endothelial cells also expressed the 46K antigen. Together, the work from these various groups clearly demonstrate that brain endothelial cells express antigens in vivo and in vitro that are specific to the brain and are not found in the capillaries of many other tissues. The function of these antigens, particularly their relationship to the blood-brain barrier properties of brain microvessels, is not yet known.

E. Antibodies to Angiotensin-Converting Enzyme

The production of angiotensin-converting enzyme (ACE) is often used to identify endothelial cells in vitro. Although ACE production is not restricted to endothelial cells, it is produced by many different types of endothelium. In their recent survey, Gumkowski et al.[10] found the enzyme on all endothelial cells tested from brain, placenta, mammary gland, lung, kidney, bladder, adrenal, liver, heart, ovary, and thoracic duct. The intensity of the staining was not uniform, however, with brightest staining seen with cells from the heart and liver. Additional work with this enzyme further indicates that the ACE molecules produced in different organs may not be structurally identical. Monoclonal antibodies prepared against rat-lung ACE were found to react preferentially with lung ACE relative to kidney ACE in a quantitative enzyme-linked immunoassay (ELISA), and these results were confirmed by immunofluorescent staining of lung and kidney sections.[13] These results suggest that caution should be exercised when using antibodies to ACE as a method to confirm that cultured cells are of endothelial origin because antibodies prepared against ACE from one organ may not reliably label the enzyme in cells from another location.

F. Antibodies to von Willebrand Factor

The von Willebrand factor or factor VIII-related antigen (vWF, FVIII-RAg) has long been a favored endothelial cell marker since this antigen is found only in endothelial cells, platelets, and megakaryocytes. Although vWF is present in nearly all endothelial cells, it is not present in the renal glomerular capillaries[14] or in lymphatic endothelium.[15] Another antibody with a similar distribution is OKM5 which reacts with an antigen found in monocytes and platelets in addition to endothelium in all organs including splenic sinusoidal endothelial cells but excluding renal glomerular capillaries and the renal cortical vasculature.[16]

G. Other Organ-Specific Antigens

The distribution of organ-specific endothelial antigens is not easily predicted. Monoclonal antibody EN 3, for example, was isolated after immunization of mice with isolated human umbilical vein endothelial cells.[17] This antibody recognizes umbilical vein endothelium, capillary endothelium in the esophagus and tonsils, and tumor-induced blood vessels. Other endothelial cells, including those of the human umbilical artery and most other larger arteries and veins, showed reduced staining with this antibody. Similar results were found with another antiendothelial antibody, PAL-E.[18] This antibody stains the endothelial layer of capillaries, venules, and small veins in frozen sections of human and some animal tissues but fails to react with the endothelium of arterioles, arteries, or larger veins. These antibodies appear to describe a class of antigen present in microvessels and in the umbilical vein but absent from most large arteries and veins.

H. Species-Specific Endothelial Cell Antigens

In addition to organ-specific differences and differences related to vessel size, there have also been reports of antigenic differences between endothelial cells from different species. Hamburger et al.[19] have developed three monoclonal antibodies against antigens on human umbilical vein endothelial cells that differ in their ability to recognize bovine endothelial cells. Although all three antibodies reacted with the human endothelial cells, only antibody 12 C6 reacted with a cell surface antigen on bovine aortic and bovine pulmonary artery endothelial cells. Antibodies 14 E5 and 10 B9 both failed to react with the bovine cells lines. These results suggest that antigenic differences exist between human endothelial cells and bovine endothelial cells.

Interspecies differences between macrovascular endothelial cells have also been reported concerning the differential ability of human endothelial cells and guinea pig endothelial cells in acting as antigen-presenting cells in immune activation. Whereas human endothelial cells can be induced to express Ia antigens and subsequently to stimulate a mixed leukocyte reaction, guinea pig endothelial cells do not readily present Ia restricted antigens and cannot function alone as antigen presenting cells to primed T lymphocytes.[20] Although such results suggest a significant difference between human and guinea pig endothelium, it must be noted that the human cells were derived from the umbilical vein, whereas the guinea pig cells were of aortic origin. Further work will be necessary to compare cells from the same organ bed in the two species.

I. Conclusion

The above experiments clearly demonstrate endothelial heterogeneity with regard to antigen expression. Differences are seen between endothelium of different species, different organs and even different sizes or types of vessels within a given tissue. Some of these differences are summarized in Table 1.

III. LECTIN-BINDING DETERMINANTS

A. *Dolichos biflorus* Agglutinin

Lectins are carbohydrate-binding molecules that, when labeled, can be used to recognize cell surface determinants in vivo as well as in vitro. Ponder and Wilkinson[21] studied the ability of the *Dolichos biflours* agglutinin (DBA) to react with endothelial cell membranes in frozen sections prepared from different mouse organs. Their results reveal differences in DBA binding to endothelial cells from different organs as well as from different vessels within the same organ. Uniform binding to endothelium was seen in several sites including skin, skeletal, cardiac, and smooth muscle, adipose tissue, mammary gland, salivary gland, GI tract, pancreas, gall bladder, gonads, ureters, bladder, trachea, thyroid, and parathyroid. In contrast, no DBA reactive endothelial cells were found in the lymph nodes or the adrenal gland and very little binding was observed in the liver. In the brain cortex, no DBA binding sites were found on brain microvessels, although binding was found on choroidal and meningeal vessels. In the lung, DBA reactive sites vary with vessel size and type. Binding sites were found in highest number on the pulmonary arteries with fewer found on the pulmonary veins and no detectable binding on the pulmonary microvasculature. These intriguing results demonstrate a variable distribution for the N-acetyl galactosamine-containing molecule or molecules recognized by the DBA lectin.

B. *Ulex europeus* Agglutinin

Species differences have also been reported with regard to lectin-binding sites on endothelial cells. In the case of the *Ulex europeus* agglutinin (UEA) which binds to terminal fucose residues, bright staining was found on endothelial cells from human kidney but not from a variety of other animals.[22] In another study, the *Ulex* lectin stained endothelial cells in a variety of tissues from human patients diagnosed with fucosidosis. In marked contrast, frozen sections from dogs with canine fucosidosis did not bind the lectin.[23] The *Ulex europeus* lectin has also been shown to differentiate between vascular endothelium and lymphatic sinusoids in normal as well as benign and malignant tumors. On the basis of these findings, Walker[24] has suggested that peroxidase-labeled UEA lectin can be used to distinguish between vascular and lymphatic vessels in solid tumors.

C. Peanut Agglutinin and Soybean Agglutinin

Gumkowski et al.[10] tested a panel of fluorescent lectins for their ability to bind to murine capillary endothelial cells isolated from a variety of organs. In their studies, all endothelial cells stained with concanavalin A, *Ricinus communis* agglutinin, and wheat germ agglutinin. With peanut agglutinin (PNA), however, an organ-specific distribution was found. Liver and brain endothelial cells uniformly bound PNA, whereas only a few endothelial cells in the lung or thoracic duct showed any binding. The authors suggest that cell sorting after

Table 1
ANTIBODIES THAT DEMONSTRATE ENDOTHELIAL CELL HETEROGENEITY

Antigen	Endothelial cells recognized	Endothelial cells not recognized	Nonendothelial cells recognized	Ref.
MESA-1	Large and small vessels in many tissues	Hepatic sinusoids	None	7
FVIII-RAg	Large and small vessels in many tissues	Renal glomeular capillaries	Megakaryocytes, platelets	14
OKM5	Vascular and splenic endothelium	Renal capillaries	Monocytes, platelets	16
HCl	Vascular endothelial cells	Brain capillaries, splenic sinusoids	Hairy cell leukemia basal epidermal cells	5
MBE-1	Brain capillaries	Lung or ovary	Brain parenchymal cells	9
EBM	Brain, kidney, and bile duct capillaries	Brain large vessels, choroid plexus	Renal tubular epithelial cells	11
6E1	Basement membranes of medium-size brain blood vessels	All other vessels tested	—	11
46K	Bovine brain capillaries	Liver, heart, or kidney capillaries	—	12
EN 3	Human umbilical vein, some human capillaries	Other human large arteries or veins	—	17
PAL-E	Capillaries, venules, and small veins	Arterioles, arteries, and large veins; lymphatic sinusoids	—	18
ACE	Rat lung ACE	Rat kidney ACE	—	13
H4-7/33	Vasa vasorum hepatic large vessels	Coronary artery hepatic sinusoids	—	6
14 E5	Human umbilical vein	Bovine aorta or pulmonary artery	Human macrophages	19
10 B9	Human umbilical vein	Bovine aorta or pulmonary artery	Human monocyte-like cell lines	19
Ii	Large vessels in mouse kidney, heart, or lungs	Microvessels	Many epithelium	99
12 C6	Human umbilical vein, bovine artery and aorta		Human fibroblasts monkey kidney epithelial cells	19

Table 2
DIFFERENTIAL LECTIN BINDING TO VASCULAR ENDOTHELIAL CELLS

Lectin	Positive binding	Negative binding	Ref.
Dolichos biflorus agglutinin	Endothelial cells in many tissues	Lymph nodes, adrenal; most liver vessels; brain and pulmonary microvessels	21
Ulex europeus agglutinin	Most human vascular endothelial cells	Human lymphatic sinusoids; most nonhuman endothelium	10,22,23,24
Peanut agglutinin	Liver and brain endothelium	Most lung and thoracic duct endothelium	10
Soybean agglutinin	Most large blood vessels	Most microvessels	10

PNA labeling is a useful method for separating hepatic endothelial cells from hepatocytes in primary isolates or mixed cultures. Soybean agglutinin also discriminated among various endothelial cell types. In the case of SBA, cells derived from large blood vessels were brightly stained by the lectin, whereas all microvascular endothelial cells stained weakly or not at all. No UEA binding was observed with any of the murine cell lines.

The results summarized in Table 2 clearly demonstrate that endothelial cells from different species, organs, and sizes of vessel display differences in the amounts of cell surface carbohydrates that mediate the binding of specific lectins. While these results strengthen the conclusion that endothelial cells from different sites are heterogeneous with regard to cell surface composition, very few of the lectin-binding molecules are currently known. Since lectin-affinity chromatography is a powerful tool for purifying lectin-binding molecules, it is likely that the application of this technique could result in the purification of distinct cell surface molecules from endothelial cells in different vascular beds.

IV. CELLULAR ADHESION TO ENDOTHELIAL CELLS

A variety of cell types, including neutrophils, monocytes, lymphocytes, and tumor cells have been shown to bind to endothelial cells in vivo and in vitro.[5] These adhesive interactions are considered necessary for the cellular extravasation through vascular walls that plays an essential role in inflammation, immune reactions, and tumor metastasis. Because many of these processes appear to be localized in particular tissues and organs, many investigators have asked whether these adhesive specificities can be observed in vitro using cryostat sections or cultured endothelial monolayers derived from different tissues.

A. Lymphocyte Adhesion to Endothelial Cells

In vivo lymphocytes have the ability to recognize and emigrate through specific vascular beds in central and peripheral lymph nodes, Peyer's patches, and in chronic inflammatory sites. The endothelium in the venules at these sites maintain a characteristic columnar morphology and are generally referred to as "high endothelial venules" (HEV). In order to determine whether the preferential passage of lymphocytes through HEV reflected a specific adhesive interaction between these two cell types, Stamper and Woodruff designed an assay to study the adhesion of freshly isolated lymphocytes to cryostat sections prepared from specific tissues.[26] Their results clearly demonstrated that lymphocytes preferentially adhere to HEV relative to other endothelial cells or other cell types in lymphatic tissue.

Distinct lymphocyte populations migrate preferentially to distinct sites (Peyer's patches, peripheral lymph nodes, inflamed tissue) in vivo, and these preferences are also reflected in the in vitro binding assays.[27-29] More recently, some of the lymphocyte surface molecules mediating these specific adhesive events have been isolated and identified,[30-32] and in one case the gene for an adhesive receptor protein has been cloned.[33] These elegant experiments demonstrate a functional heterogeneity among lymphoid endothelia at different sites and give us an insight into the lymphocyte adhesion molecules involved. The nature of the specific molecules on the endothelial cells is not as well understood, although the adhesive interactions have been suggested to involve endothelial cell surface carbohydrates, specifically mannose phosphates[34] and sialic acid.[35] The complete characterization of the endothelial cell adhesins responsible for lymphocyte adhesion in specific target organs is essential for the further progress of this important line of research.

B. Neutrophil Adherence

Although considerable research has been directed toward understanding neutrophil adhesion to endothelial monolayers, much less is known about endothelial heterogeneity in this regard since most of the experiments published to date have employed a single cell type — the human umbilical vein endothelial cell. In vivo, the majority of adherent neutrophils are found in postcapillary venules, although the mechanism for this predilection is still unknown.[36] In vitro, neutrophil adherence to vascular endothelial cells is enhanced by pretreatment of the endothelial cell monolayer with agents such as interleukin 1,[37] lipopolysacharide (LPS),[38] or tumor necrosis factor.[39] In a recent study, Pohlman et al.[39] found that LPS induction of endothelial adherence for neutrophils was found with human umbilical vein endothelial cells, but not with human dermal fibroblasts or bovine aortic endothelial cells. These results indicate that it will be important to test other relevant cell types such as endothelial cells from human postcapillary venules in order to determine whether results obtained with human umbilical vein endothelial cells can be extrapolated to other vessel beds.

C. Tumor Cell Binding to Vascular Endothelial cells

Considerable interest has recently been expressed in the role of cell adhesion in organ-specific tumor metastasis.[40,41] As in the case of lymphocyte adhesion, it has been shown that tumor cells adhere preferentially to cryostat sections prepared from the organs to which those tumors most readily metastasize.[42] The correlation between organ-specific adhesion and metastasis was strengthened by experiments demonstrating that when murine tumor cells were selected for their ability to adhere to pulmonary tissues in vitro, the selected cells demonstrated a sixfold increase in their ability to colonize lung tissue in vivo.[43]

Metastatic tumor cells generally travel to nearby sites via the lymphatic circulation and to distant sites via the bloodstream. To successfully colonize a lymph node or other organ, the tumor cell must become arrested in the vessel, extravasate through the vessel wall into the tissue parenchyma, and proliferate at the new site. Each of these events is a potential control point for metastatic localization. Because tumor cells most commonly penetrate capillaries or postcapillary venules, the cell adhesion molecules most likely to play a role in the metastatic process may be localized on or immediately beneath the microvascular endothelial cells in the target organ.

In 1964, Greene and Harvey[44] postulated that ''...organic retention may be effected by the adhesion of tumor cells to vascular endothelium in the manner of leukocytes in the vicinity of an inflammatory focus, the strength of the adhesive bond varying with different tumors and with the vascular endothelium at different sites.'' Twenty years later, Alby and Auerbach[45] demonstrated a strong correlation between organ-specific metastatic profiles and preferential adhesion of tumor cells to microvascular endothelial cells derived from the preferred organs. These investigators showed that an ovary colonizing mouse testicular teratoma preferentially adhered to mouse ovary endothelial cells. These results have more recently been extended to several other tumor/endothelial cell systems.[101] Other investigators

Table 3
CELLULAR ADHESION TO SPECIFIC VASCULAR
ENDOTHELIAL CELLS

Adherent cell	Preferential attachment to	Ref.
Lymphocytes	High endothelial venules in lymph nodes or Peyer's patches	26—29
Neutrophils	Activated venular endothelium; not aortic endothelial cells	37—39
Glioma cells	Brain endothelial cells	45
Teratoma cells	Ovarian endothelial cells	45
Lymphosarcoma cells	Hepatic endothelial cells	46

have also measured the in vitro adhesion of metastatic tumor cell lines to endothelial cells derived from particular organs. Roos et al.[46] have described a high-affinity binding between lymphoid tumor cells that normally colonize mouse liver and isolated rat liver endothelial cells.

While these results suggest a role for tumor-endothelial cell adhesion in modulating organ-specific metastasis, they in no way rule out the important contribution of other parameters such as organ-specific chemotactic factors[47] and organ-specific growth factors.[48] Cell adhesion factors in the subendothelial extracellular matrix may also play a role in determining the location of tumor metastases. The above results do, however, confirm the hypothesis that endothelial cells derived from different organs have different cell-surface determinants that can be discriminated by cell-cell interactions with leukocytes and tumor cells. These results are summarized in Table 3.

V. ENDOTHELIAL CELL METABOLISM

The uptake and metabolism of vasoactive compounds is an essential component of endothelial cell function. Endothelial cells also produce vasoactive substances that result in leukocyte adherence, platelet aggregation, and inflammatory cell accumulation. Studies on the production and metabolism of vasoactive substances by vascular endothelial cells have revealed many intriguing site-specific differences. According to Shepro and Dunham,[49] "It is an axiom in endothelial cell research that metabolism may vary not only between intimal and microvessel cells, but from organ to organ, and even within a given microvascular bed".

A. Biogenic Amines
Metabolism and degradation of biogenic amines takes place primarily in the lung. Removal of 5-hydroxytryptamine (5-HT) from the bloodstream, for example, occurs almost totally in the pulmonary microcirculation.[50] In vitro studies have shown that cultured pulmonary artery endothelial cells express a high-affinity receptor that mediates the internalization and subsequent metabolism of 5-HT.[51] Bovine aortic endothelial cells, in contrast, do not express a saturable 5-HT uptake system.[52]

Histamine metabolism is under the control of two enzymes, diamine oxidase (DAO) and histamine methyltransferase (HMT). Recent studies have shown that HMT activity is present in microvascular endothelial cells isolated from guinea pig perirenal fat pad, heart ventricles, and brain cortex, but the same preparations from rat had no detectable HMT activity.[53] DAO, however, was found in rat as well as guinea pig microvessels. In work designed to map the sites of histamine receptors on endothelial cells in vivo, Heltianu et al. have demonstrated that vascular histamine (H2) receptors are predominantly distributed in post-capillary venules, in comparison with arterioles or veins.[54] These results demonstrate vessel-as well as site-specific difference in the ability of microvascular endothelial cells to produce particular metabolic receptors and enzymes.

B. Glycosaminoglycans

There is increasing evidence that the types of glycosaminoglycan (GAG) synthesized and secreted into the extracellular matrix differ in various types of endothelium. Species differences were first noted by Merrilees and Scott,[55] who reported that rat aortic endothelial cells primarily produce nonsulfated GAGs whereas pig aortic endothelial cells produced mostly sulfated GAGs. Bar et al. have recently examined the synthesis, release, and localization of sulfated GAGs by vascular endothelial cells.[56] They found considerable differences in the patterns of GAG distribution between a variety of endothelial cells from vessels of differing size or from different vascular beds. Their findings suggest, for example, that higher amounts of cellular and pericellular dermatan sulfate are found in bovine adipose capillary endothelial cells than in cells derived from bovine pulmonary artery or vein, bovine aorta, or human umbilical vein. These findings suggest that the extracellular matrix elaborated by different types of endothelial cell will differ in composition and should not be considered identical. Such findings are consistent with the work of Sage et al., who showed that different types of endothelial cells synthesize different types of collagen.[57]

C. Insulin Binding

Pathological changes in the structure and function of small as well as large blood vessels are a frequent complication of diabetes.[58] There is consequently considerable interest in the interactions of endothelial cells with insulin. King and colleagues have compared insulin binding and uptake on a variety of cultured endothelial cells and perivascular cells (smooth muscle cells and pericytes).[59] They found that the concentration of insulin receptors was the same for each vascular cell type tested with the exception of aortic smooth muscle cells which had tenfold fewer receptors than the other cell types. Insulin stimulated glucose incorporation into glycogen in retinal capillary endothelial cells and pericytes as well as aortic smooth muscle cells, but had no effect on aortic endothelial cells. Bar and colleagues have also investigated insulin binding to endothelial cells.[60] They found a differential binding capacity for insulin in human arterial vs. venous endothelial cells in primary culture. Together, these studies suggest that the cellular response of endothelial cells to insulin may vary with vessel size or organ localization.

D. Lipoprotein Uptake

The receptor mediated uptake of low-density lipoprotein (LDL) by a variety of cell types has been studied in detail.[61] An alternative pathway for the metabolism of chemically modified lipoproteins such as acetylated-LDL (Ac-LDL) has also been described,[62] and has been termed the "scavenger cell pathway" due to its occurrence in monocytes and macrophages.[63]

In addition to monocytes and macrophages, endothelial cells have also been shown to metabolize modified LDL at an accelerated rate compared with other cell types.[64] Although a variety of endothelial cells from large and small blood vessels take up Ac-LDL in vitro[65] and in vivo,[66] brain capillary endothelial cells[67] and heart endothelial cells[68] reportedly lack the receptor for modified LDL. Pitas et al. have reported that uptake of acetoactylated LDL is preferentially carried out in vivo by the sinusoidal endothelial cells of the liver, spleen, and bone marrow.[69] These results were confirmed by in vitro studies demonstrating that isolated rat liver sinusoidal endothelial cells were able to internalize and degrade AcAc-LDL ten times more effectively than bovine aortic endothelial cells, although the uptake by the aortic cells was still quite significant. Together these results indicate that the ability to internalize and degrade Ac-LDL is found in most but not all vascular beds. Receptor activity is greater in most endothelial cells than in all nonendothelial cells except for monocytes and macrophages. Among endothelial cells, activity is greatest in sinusoids and lowest in the brain and heart.

E. Prostacyclin

Prostacyclin (PGI$_2$) is an arachidonic acid derivative that functions as a vasodilator and as an inhibitor of platelet activation. Many different types of endothelial cell synthesize prostacyclin from an arachidonic acid precursor;[70-72] however, prostacyclin is not the primary arachidonic acid derivative produced by human dermal microvascular endothelial cells.[73] Prostacyclin metabolism also varies in different endothelium. When labeled prostacyclin is incubated with rat aortic endothelial cells in vitro, the agent is metabolized to 6,15 keto-acids by the enzyme 15-hydroxyprostaglandin dehydrogenase;[74] yet when the same agent is incubated with canine pulmonary microvascular endothelial cells, no significant metabolic degradation is observed.[75]

VI. PROPERTIES OF SPECIALIZED ENDOTHELIUM

The ability to culture capillary endothelial cells from a variety of vascular beds has improved dramatically in the past few years. Although, it is difficult to compare the properties of cells that have been cultured under variable conditions, certain properties that have been associated with specific endothelial cell lines do seem to correlate with their in vivo functions. A few of these are listed below.

A. Cerebral Endothelial Cells

Microvascular endothelial cells from the brains of cows and rats are among the best studied small vessel endothelia. They are relatively easy to culture,[76] and maintain several of the differentiated properties of brain endothelial in vivo. These include the ability to synthesize the enzymes Y-glutamyl transpeptidase,[77] choline acetyltransferase,[78] and L-DOPA decarboxylase.[79] There may be some species specificity to these properties since endothelial cells from chick brain capillaries were reported not to contain Y-glutamyl tanspeptidase.[80] Cerebral endothelial cells in vitro are able to form tight junctions[81,82] that are reminiscent of the cerebral tight junctions found in vivo that have been postulated to play a role in maintaining the blood-brain barrier. Features similar to those of brain capillary endothelial cells are also found in endothelial cells cultured from bovine retina[83] and from the bovine inner ear.[84]

B. Adipose Endothelial Cells

Endothelial cells have been cultured from rat epididymal fat,[85,86] as well as from human adipose tissue.[87] The rat adipose-derived endothelial cells were found to proliferate when grown on substrata of interstitial collagen but to stop growing and to differentiate into capillary tubes when grown on basement membrane collagens.[86] Bovine adipose capillary endothelial cells bind, internalize, and degrade insulin[88,89] and secrete heparin-like cell surface glycosaminoglycans.[90]

C. Hepatic Sinusoidal Endothelial Cells

The sinusoidal endothelium of the liver, spleen, and bone marrow are characterized in vivo by gaps between adjacent cells and intracellular pores or fenestrae. Only in the past 2 to 3 years have sinusoidal endothelial cells been cultured and compared with other endothelium. Most of these studies have been conducted with endothelial cells isolated from rat or guinea pig liver. These studies have revealed significant differences when hepatic sinusoidal endothelial cells are compared with endothelial cells from hepatic veins or with endothelial cells from other microvascular beds.

In recent studies, it has been shown that formaldehyde-denatured albumin was rapidly internalized by hepatic sinusoidal endothelial cells (HSE)[91] but not by hepatic venous endothelia (HVE).[92] In areas where the two cell types were in contact, each maintained its characteristic level of denatured albumin internalization.

Caperna and colleagues have conducted a series of experiments to determine the properties of hepatic endothelial cells and their differences from hepatocytes and Kupffer cells. They found, for example, that isolated rat HSE synthesize high amounts of metallothionein, a binding protein for zinc, copper and cadmium, whereas Kuppfer cells failed to synthesize metallothionein and had a lesser ability to accumulate these heavy metals from the culture medium.[93] Both HSE and Kupffer cells isolated from neonatal porcine liver accumulate iron-dextran complexes at a faster rate than hepatocytes.[94] The accumulated iron is rapidly depleted from the Kupffer cells but remains within the HSE for several days.[95]

Shaw et al. have isolated guinea pig HSE by use of centrifugal elutriation and have cultured these cells for up to nine passages.[96] The cells were able to grow best when cultured on a substratum of guinea pig liver or skin collagen but failed to grow on uncoated plastic or on plastic coated with rat-tail collagen. The cells contain vWF, release angiotensin converting enzyme and prostacyclin, and maintain morphological structures known as sieve plates that are characteristic of sinusoidal endotheium in vivo. Uptake of modified low-density lipoprotein in mouse or rat liver is associated primarily with the sinusoidal endothelium and is not characteristic of Kupffer cells or hepatocytes.[97]

VII. SUMMARY

All vascular endothelial cells share common functions in vivo as a result of their position at the blood/tissue interface. Morphologists have long known, however, that endothelial cells in different vascular beds have varying shapes, sizes, and types of intercellular contacts. Recent progress in cellular immunology and in the culture of vascular endothelial cells have provided us with evidence that the cellular properties of the individual endothelial cells vary from organ to organ and from vessel to vessel. Perhaps the most abundant data has come from studies on the antigenic determinants that occupy the endothelial cell surface. The studies discussed in Section II clearly demonstrate that different endothelial cells express different cell survace antigens. Unfortunately, the functions of these antigenic molecules are generally unknown. In many cases the endothelial antigens, which are maintained in long-term cell culture, are also expressed on the parenchymal cells of the organ from which the endothelial cells were isolated. Such results suggest that blood vessels and the tissues in which they arise share common cell surface determinants that may be produced independently by different cell types within the same organ.

One possible function for these cell surface molecules is to serve as recognition molecules for tumor cells, immune cells, and inflammatory cells that extravasate from the blood stream to the tissue side of the blood vessels at particular locations. Variable cell surface molecules may also act as receptors for vasoactive substances or carrier proteins for molecules that are selectively internalized and transported across the endothelium in different tissues.[99] Finally the recent finding that HTLV-III, the causative agent for acquired immune deficiency syndrome (AIDS), selectively infects endothelial cells in the brain and not in other organs[100] suggests that endothelial cells in different locations may have different receptors for viruses and other infectious agents.

Endothelial cells from different sizes and types of blood vessels express well known differences in metabolic properties, especially in the production and response to vasoactive mediators.[49] It is likely that as the number and role of these mediators becomes better understood, the multiplicity of responses to them by different types of endothelium will become even better established.

It is most important to recognize that the field of endothelial cell culture is still in its infancy and that many types of endothelial cells from specific organ beds, especially microvascular cells, have not yet been successfully cultivated. As this manuscript is written, the first reports on cultures of sinusoidal endothelial cells from liver and lymphatics are just appearing, and the successful culture of splenic endothelium or bone marrow sinusoidal endothelium has barely yet begun. It is clear from the work cited here, however, that each vascular bed has its own distinct properties and its own set of responses to circulating cells, solutes, and biological response modifiers. It can no longer be considered sufficient to use aortic endothelial cells or umbilical vein endothelial cells as prototypical cells whose results are then extrapolated to other endothelial cell types. Those who work with endothelial cells in the future must take care to assure that they use the appropriate cell type to explore specific questions of vascular physiology and metabolism so that others may reasonably interpret their results.

REFERENCES

1. **Booyse, F. M., Sedlak, B. J., and Rafelson, M. E.,** Culture of arterial endothelial cells. Characterization and growth of bovine aortic cells, *Thromb. Diath. Haemorrh.*, 34, 825, 1975.
2. **Maruyama, Y.,** The human endothelial cell in tissue culture, *Z. Zellforsch.*, 60, 69, 1963.
3. **Jaffe, E. A., Nachman, R. L., Becker, C. G., and Minick, C. R.,** Culture of human endothelial cells derived from umbilical veins. Identification by morphologic and immunologic criteria, *J. Clin. Invest.*, 52, 2745, 1973.
4. **Gimbrone, M. A., Cotran, R. S., and Folkman, J.,** Human vascular endothelial cells in culture. Growth and DNA synthesis, *J. Cell Biol.*, 60, 673, 1974.
5. **Posnett, D. N., Chiorazzi, N., and Kunkel, H. G.,** A membrane antigen (HCl) selectively present on hairy cell leukemia cells, endothelial cells, and epidermal basal cells, *J. Immunol.*, 132, 2700, 1984.
6. **Hagemeier, H. H., Goerdt, S., and Sorg, C.,** Phenotypic heterogeneity of endothelial cells in normal and diferent pathological tissues detected by a panel of monoclonal antibodies, *Abstracts of the Fourth International Symposium Conference on the Biology of the Vascular Endothelial Cell*, 1986, 156.
7. **Ghangour, S., Langley, K., Gombos, G., Hirn, M., Hirsch, M. R., and Gordis, C.,** A surface marker for murine vascular endothelial cells defined by monoclonal antibody, *J. Histochem. Cytochem.*, 30, 165, 1982.
8. **Barsky, S. H., Togo, S., Baker, A., Liotta, L. A., and Siegel, G. P.,** Use of anti-basement membrane antibodies to distinguish blood vessel capillaries from lymphatic capillaries, *Am. J. Surg. Pathol.*, 7, 667, 1983.
9. **Auerbach, R., Alby, L., Morrissey, L. W., Tu, M., and Joseph, J.,** Expression of organ-specific antigens on capillary endothelial cells, *Microvasc. Res.*, 29, 401, 1985.
10. **Gumkowski, F., Kaminska, G., Kaminski, M., Morrissey, L.W., and Auerbach, R.,** Heterogenity of mouse vascular endothelium: in vitro studies of lymphtic, large blood vessel and microvascular endothelial cells, *Blood Vessels*, in press.
11. **Michalak, T., White, F. P., Gard, A. L., and Dutton, G. R.,** A monoclonal antibody to the endothelium of rat brain microvessels, *Brain Res.*, 379, 320, 1986.
12. **Pardridge, W. M., Yang, J., Eisenberg, J., and Mietus, L. J.,** Antibodies to blood-brain barrier bind selectively to a brain capillary endothelial lateral membranes and to a 46K protein, *J. Cereb. Blood Flow Metab.*, 6, 203, 1986.
13. **Moore, M. G., Chrzanowski, R. R., McCormick, J. R., Cieplinski, W., and Schwink, A.,** Production of monoclonal antibodies to rat lung angiotensin-converting enzyme, *Clin. Immunol. Immunopathol.*, 33, 301, 1984.
14. **Mukai, K., Rosai, J., and Burgdorf, W. H. C.,** Localization of Factor VIII-related antigen in vascular endothelial cells using an immunoperoxidase method, *Am. J. Surg. Pathol.*, 4, 273, 1980.
15. **Harach, H. R., Jasani, B., Williams, E. D.,** Factor VIII as a marker of endothelial cells in follicular carcinoma of the thyroid, *J. Clin. Pathol.*, 36, 1050, 1983.
16. **Knowles, D. M., II, Tolidjian, B., Marboe, C., D'Argati, V., Grimes, M., and Chess, L.,** Monoclonal anti-human monocyte antibodies OKM1 and OKM5 possess distinctive tissue distributions including differential reactivity with vascular endothelium, *J. Immunol.*, 132, 2170, 1984.

17. **Cui, Y. C., Tai, P.C., Gatter, K. C., Mason, D. Y., and Spry, C. J.**, A vascular endothelial cell antigen with restricted distribution in human foetal, adult and malignant tissues, *Immunology,* 49, 183, 1983.
18. **Schlingemann, R. D., Dingjan, G. M., Emeis, J. J., Bok, J., Warnaar, S. D., and Ruiter, D. J.**, Monoclonal antibody PAL-E specific for endothelium, *Lab. Invest.,* 52, 71, 1985.
19. **Hamburger, A. W., Reid, Y. A., Pelle, B. A., Breth, L. A., Beg, N., Ryan, U., and Cines, D. B.**, Isolation and characterization of monoclonal antibodies reactive with endothelial cells, *Tissue Cell,* 17, 451, 1985.
20. **Roska, A. K., Geppert, T. D., and Lipsky, P. E.**, Immunoregulation by vascular endothelial cells, *Immunobiology,* 168, 470, 1984.
21. **Ponder, B. A. J. and Wilkinson, M. M.**, Organ-related differences in binding of *dolichos biflorus* agglutinin to vascular endothelium, *Dev. Biol.,* 96, 535.
22. **Holthofer, H.**, Lectin binding sites in kidney. A comparative study of 14 animal species, *J. Histochem. Cytochem.,* 31, 531, 1983.
23. **Alroy, J., Ucci, A. A., and Warren, C. D.**, Human and canine fucosidosis: a comparative lectin histochemistry study, *Acta Neuropathol.,* 67, 265, 1985.
24. **Walker, R. A.**, *Ulex europeus* I — peoxidase as a marker of vascular endothelium: its application in routine histopathology, *J. Pathol.,* 146, 123, 1985.
25. **Harlan, J. M.**, Leukocyte-endothelial interactions, *Blood,* 65, 513, 1985.
26. **Stamper, H. B. and Woodruff, J. J.**, Lymphocyte homing into lymph nodes: in vitro demonstration of the selective affinity of recirculating lymphocytes for high-endothelial venules, *J. Exp. Med.,* 144, 828, 1976.
27. **Butcher, E., Scollay, R., and Weissman, I.**, Organ specificity of lymphocyte interaction with organ-specific determinants on high endothelial venules, *Eur. J. Immunol.,* 10, 556, 1980.
28. **Stevens, S. K., Weissman, I. L., and Butcher, E. C.**, Differences in the migration of B and T lymphocytes: organ specific localization *in vivo* and the role of lymphocyte-endothelial cell recognition, *J. Immunol.,* 128, 844, 1986.
29. **Jalkanen, S., Steere, A. C., Fox, R. I., and Butcher, E. C.**, A distinct endothelial cell recognition system that controls lymphocyte traffic into inflamed synovium, *Science,* 233, 556, 1986.
30. **Chin, Y. H., Carey, G. D., and Woodruff, J. J.**, Lymphocyte recognition of lymph node high endothelium. V. Isolation of adhesion molecules from lysates of rat lymphocytes, *J. Immunol.,* 131, 1368, 1983.
31. **Rasmussen, R. A., Chin, Y. H., Woodruff, J. J., and Easton, T. G.**, Lymphocyte recognition of lymph node high endothelium. VII. Cell surface proteins involved in adhesion defined by monoclonal antiHEBFLN (A.11) antibody, *J. Immunol.,* 135, 19, 1985.
32. **Gallatin, W. M., Weissman, I. L., and Butcher, E. C.**, A cell surface molecule involved in organ-specific homing of lymphocytes, *Nature (London),* 304, 30, 1983.
33. **St. John, T., Gallatin, W. M., Siegelman, M., Smith, H. T., Fried, V. A., and Weissman, I. L.**, Expression cloning of a lymphocyte homing receptor cDNA: ubiquitin is the reactive species, *Science,* 231, 845, 1986.
34. **Stoolman, L. M., Tenforde, T. S., and Rosen, S. D.**, Phosphomannosyl receptors may participate in the adhesive interaction between lymphocytes and high endothelial venules, *J. Cell Biol.,* 99, 1535, 1984.
35. **Rosen, S. D., Singer, M. S., Yednock, T. A., and Stoolman, L. M.**, Involvement of sialic acid on endothelial cells in organ-specific lymphocyte recirculation, *Science,* 228, 1005, 1985.
36. **Mayrovitz, H., Wiedeman, M., and Tuma, R.**, Factors influencing leukocyte adherence in microvessels, *Thromb. Haemost.,* 38, 823, 1977.
37. **Bevilaqua, M.P., Pober, J. S., Wheeler, M. Z., Mendrick, D., Cotran, R. S., and Gimbrone, M. A.**, Interleukin-1 acts on cultured human vascular endothelial cells to increase the adhesion of polymorphonuclear leudocytes, monocytes and related leukocyte cell lines, *J. Clin. Invest.,* 76, 2003, 1985.
38. **Schleimer, R. P. and Rutledge, B. K.**, Cultured human vascular endothelial cells acquire adhesiveness for neutrophils after stimulation with interleukin 1, endotoxin, and tumor-promoting phorbol diesters, *J. Immunol.,* 136, 649, 1986.
39. **Pohlman, T. H., Stanness, K. A., Beatty, P. G., Ochs, H. D., and Harlan, J. M.**, An endothelial cell surface factor(s) induced *in vitro* by lipopolysacharide, interleukin 1 and tumor necrosis Factor-alpha increases neutrophil adherence by a CDw18-dependent mechanism, *J. Immunol.,* 136, 4547, 1986.
40. **Roos, E.**, Cellular adhesion, invasion and metastasis, *Biochem. Biophys. Acta,* 738, 263, 1984.
41. **Netland, P. A. and Zetter, B. R.**, Tumor cell interactions with blood vessels during cancer metastasis, in *Progressive Stages of Neoplasia,* Kaiser, H., Ed., Martinus Nijhoff, Boston, 1986.
42. **Netland, P. A. and Zetter, B. R.**, Organ-specific adhesion of metastatic tumor cells, *Science,* 224, 1113, 1984.
43. **Netland, P.A. and Zetter, B. R.**, Metastatic potential of B16 melanoma cells after *in vitro* selection for organ-specific adherence, *J. Cell Biol.,* 101, 720, 1986.

44. **Greene, H. S. N. and Harvey, E. K.,** The relationship between the dissemination of tumor cells and the distibution of metastases, *Cancer Res.,* 24, 799, 1964.

45. **Alby, L., and Auerbach, R.,** Differential adhesion of tumor cells to capillary endothelial cells *in vitro, Proc. Natl. Acad. Sci. U.S.A.,* 81, 5739, 1984.

46. **Roos, E., Tulp, A., Middlekoop, O. P., and van de Pavert, I. V.,** Interactions between lymphoid tumor cells and isolated liver endothelial cells, *J. Natl. Cancer Inst.,* 72, 1173, 1984.

47. **Hujanen, E. S. and Terranova, V. P.,** Migration of tumor cells to organ-derived chemoattractants, *Cancer Res.,* 45, 3517, 1985.

48. **Nicolson, G. L., and Dulski, K. M.,** Organ specificity of metastatic colonization is related to organ-selective growth properties of malignant cells, *Int. J. Cancer,* 38, 289, 1986.

49. **Shepro, D. and Dunham, B.,** Endothelial cell metabolism of biogenic amines, *Annu. Rev. Physiol.,* 48, 335, 1986.

50. **Gillis, C. N. and Roth, J. A.,** Pulmonary disposition of circilating vasoactive hormones, *Biochem. Pharmacol.,* 25, 2547, 1976.

51. **Kjellstrom, T., Ahlman, H., Dahlstrom, A., Hanson, G. K., and Risberg, B.,** The uptake of 5-hydroxytryptamine in endothelial cells cultured from the pulmonary artery in rats, *Acta Physiol. Scand.,* 120, 243, 1984.

52. **Carson, M. P., Peterson, S. W., Hechtman, H. B., and Shepro, D.,** Serotonin uptake in ^{3}H-thymidine-selected cultures of bovine aortic endothelium, *Fed. Proc.,* 40, 610, 1981.

53. **Robinson-White, A. and Beaven, M. A.,** Presence of histamine and histamine-metabolizing enzyme in rat and guinea-pig microvascular endothelial cells, *J. Pharmacol. Exp. Ther.,* 223, 440, 1982.

54. **Heltianu, C., Simionescu, M., and Simionescu, N.,** Histamine receptors of the microvascular endothelium revealed *in situ* with a histamine-ferritin conjugate: characteristin high-affinity binding sites in venules, *J. Cell Biol.,* 93, 357, 1982.

55. **Merrilees, M. J. and Scott, L.,** Culture of rat and pig aortic endothelial cells — differences in their isolation, growth rate and glycosaminoglycan synthesis, *Atherosclerosis,* 38, 19, 1986.

56. **Bar, R. S., Dake, B. L., and Spanheimer, R. G.,** Sulfated glycosaminoglycans in cultured endothelial cells from capillaries and large vessels of human and bovine origin, *Atherosclerosis,* 56, 11, 1986.

57. **Sage, H., Prial, P., and Bornstein, P.,** Secretory phenotypes of endothelial cells in culture: comparison of aortic, venous, capillary and corneal endothelium, *Arteriosclerosis,* 1, 427, 1981.

58. **West, K. M.,** *Epidemiology of Diabetes and its Vascular Lesion,* Elsevier, New York, 1978.

59. **King, G. L., Buzney, S. M., Kahn, C. R., Hetu, N., Buchwald, S., MacDonald, S. G., and Rand, L. I.,** Differential responsiveness to insulin of endothelial and support cells from micro- and macrovessels, *J. Clin. Invest.,* 71, 974, 1983.

60. **Bar, R. S., Peacock, M. L., Spanheimer, R. G., Veenstra, R., and Hoak, J. C.,** Differential binding of insulin to human arterial and venous endothelial cells in primary culture, *Diabetes,* 29, 991, 1980.

61. **Goldstein, J. L. and Brown, M. S.,** A receptor mediated pathway for cholesterol homeostasis, *Science,* 232, 34, 1982.

62. **Mahley, R. W., Innerarity, T. L., Weisgraber, K. H., and Oh, S. Y.,** Altered metabolism (in vivo and in vitro) of plasma lipoproteins after selective chemical modification of lysine residues of the lipoprotein, *J. Clin. Invest.,* 64, 743, 1979.

63. **Brown, M. S., Basu, K., Falck, J. R., Ho, Y. K., and Goldstein, J. L.,** The scavenger cell pathway for lipoprotein degradation: specificity of the binding site that mediates the uptake of negatively-charged LDL by macrophages, *J. Supramol. Struct.,* 15, 67, 1980.

64. **Stein, O. and Stein, Y.,** Bovine aortic endothelial cells display macrophage-like properties towards acetylated ^{125}I-labeled low density lipoprotein, *Biochim. Biophys. Acta,* 620, 631, 1980.

65. **Voyta, J. C., Via, D. P., Butterfield, C. E., and Zetter, B. R.,** Identification and isolation of endothelial cells based on their increased uptake of acetylated-low density lipoprotein, *J. Cell Biol.,* 99, 2034, 1984.

66. **Netland, P. E., Zetter, B. R., Via, D. P., and Voyta, J. C.,** *In situ* labelling of vascular endothelium with fluorescent acetylated low density lipoprotein, *Histochem. J.,* 17, 1309, 1985.

67. **Gaffney, J., West, D., Arnold, F., Sattar, A., and Kumar, S.,** Difference in the uptake of modified low density lipoproteins by tissue cultured endothelial cells, *J. Cell Sci.,* 79, 317, 1985.

68. **Fox, P. L. and DiCorleto, P. E.,** Modified low density lipoproteins suppress production of a platelet-derived growth factor-like protein by cultured endothelial cells, *Proc. Natl. Acad. Sci. U.S.A.,* 83, 4774, 1986.

69. **Pitas, R. E., Boyles, J., Mahley, R. W., and Bissell, D. M.,** Uptake of chemically modified low density lipoproteins in vivo as mediated by specific endothelial cells, *J. Cell Biol.,* 100, 103, 1985.

70. **Weksler, B. B., Ley, C. W., and Jaffe, E.,** Stimulation of endothelial prostacyclin production by thrombin, trypsin and the ionophore A23187, *J. Clin. Invest.,* 62, 923, 1978.

71. **Ryan, J. W., Ryan, U. S., Habliston, D., and Martin, L.,** Synthesis of prostaglandins by pulmonary endothelial cells, *Trans. Assoc. Am. Physiol.,* 91, 343, 1978.

72. **Ali, A. E., Barrett, J. C., and Eling, T. E.,** Prostaglandin and thromboxane production by fibroblasts and vascular endothelial cells, *Prostaglandins,* 20, 667, 1980.

73. **Charo, I. F., Shak, S., Karasek, M. A., Davison, P. M., and Goldstein, I. M.,** Prostaglandin I_2 is not a major metabolite of arachidonic acid in cultured endothelial cells from human foreskin microvessels, *J. Clin. Invest.,* 74, 914, 1984.

74. **Sun, F. F. and Taylor, B. M.,** Metabolism of prostacyclin in rat, *Biochemistry,* 17, 4096, 1978.

75. **Dusting, G. J., Moncada, S., and Vane, J. R.,** Recirculation of prostacyclin (PGI_2) in the lung, *Br. J. Pharmacol.,* 64, 315, 1978.

76. **Bowman, P. D., Betz, A. L., Ar, D., Wolinsky, J. S., Penney, J. B., Shivers, R. R., and Goldstein, G. W.,** Primary culture of capillary endothelium from rat brain, *In Vitro,* 17, 353, 1981.

77. **DeBault, L. E. and Cancilla, P.,** Y-Glutamyl transpeptidase in isolated brain endothelial cells: Induction by glial cells in vitro, *Science,* 207, 653, 1979.

78. **Parnavelas, J. G., Kelly, W., and Burnstock, G.,** Ultrastructural localization of choline acetyltrasferase in vascular endothelial cells in rat brain, *Nature (London),* 316, 724, 1985.

79. **Wade, L. A. Katzman, R.,** *Am. J. Physiol.,* 228, 352, 1986.

80. **Stewart, P. A.,** Histochemical absence of Y-glutamyl transpeptidase in chick brain capillary endothelium, *Exp. Neurol.,* 67, 442, 1980.

81. **Betz, A. L. and Goldstein, G. W.,** Brain Capillaries — Structure and function, in *Handbook of Neurochemistry,* Vol. 7, Lajtha, A., Ed., Plenum Press, New York, 1984.

82. **Diglio, C. A., Grammas, P., Giacomellic F., and Wiener, J.,** Primary culture of rat cerebral microvascular endothelial cells — isolation, growth and characterization, *Lab. Invest.,* 46, 554, 1982.

83. **Buzney, S. M., Massicotte, S. J., Hetu, N., and Zetter, B. R.,** Retinal vascular endothelial cells and pericytes — differential growth characteristics in vitro, *Invest. Ophthalmol.,* 4, 470, 1983.

84. **Bowman, P. D., Rarey, K., Rogers, C., and Goldstein, G. W.,** Primary culture of capillary endothelial cells form the spiral ligament and stria vascularis of bovine inner ear, *Cell Tissue Res.,* 241, 479, 1985.

85. **Bjorntorp, P., Hansson, G. K., Jonasson, L., Pettersson, P., and Sypniewska, G.,** Isolation and characterization of endothelial cells from the epididymal fat pad of the rat, *J. Lipid Res.,* 24, 105.

86. **Madri, J. A. and Williams, S. K.,** Capillary endothelial cell cultures:phenotypic modulation by matrix components, *J. Cell Biol.,* 97, 153, 1983.

87. **Kern, P. A., Knedler, A., and Eckel, R. H.,** Isolation and culture of microvascular endothelium from human adipose tissue, *J. Clin. Invest.,* 71, 1822, 1983.

88. **Bar, R. S., Dolash, S., Dake, B. L., and Boes, M.,** Cultured capillary endothelial cells from bovine adipose tissue: a model for insulin binding and action in microvascular endothelium, *Metabolism,* 35, 317, 1986.

89. **Solenski, N. J. and Williams, S. K.,** Insulin binding and vesicular ingestion in capillary endothelium, *J. Cell Physiol.,* 124, 87, 1985.

90. **Marcum, J. A. and Rosenberg, R. D.,** Heparinlike molecules with anticoagulant activity are synthesized by cultured endothelial cells, *Biochem. Biophys. Res. Commun.,* 126, 365, 1985.

91. **Blomhoff, R., Smedsrod, B., Eskild, W., Granum, P. E., and Berg, T.,** Preparation of isolated liver endothelial cells and kupffer cells in high yield by means of an enterotoxin, *Exp. Cell Res.*

92. **Yokota, S.,** Functional differences between sinusoidal endothelial cells and interlobular or central vein endothelium in rat liver, *Anat. Rec.,* 212, 74, 1985.

93. **Caperna, T. J. and Failla, M. L.,** Cadmium metabolism by rat liver endothelial and Kupffer cells, *Biochem. J.,* 22, 631, 1984.

94. **Caperna, T. J., Failla, M. L., Steele, N. C., and Richards, M. P.,** Uptake and metabolism of iron-dextran by hepatocytes, Kupffer cells and endothelial cells in the neonatal pig liver, *J. Nutr.,* in press.

95. **Caperna, T. J., Failla, M. L., Richards, M. P., and Sttele, N. C.,** Uptake and distribution of iron dextran by neonatal swine liver cells, in *Swine in Biomedical Research,* Vol. 2, Tumbleson, M., Ed., 1986, 1059.

96. **Shaw, R. G., Johnson, A. L., Schulz, W. W., Zahlten, R. N., and Combes, B.,** Sinusoidal endothelial cells from normal guinea pig liver: Isolation, culture and characterization, *Hepatology,* 4, 591, 1984.

97. **Irving, M. G., Roll, F. J., Huang, S., and Bissell, D. M.,** Characterization and culture of sinusoidal endothelium from normal rat liver: lipoprotein uptake and collagen phenotype, *Gastroenterology,* 87, 1233, 1984.

98. **Momburg, F., Kock, N., Moller, P., Moldenhauer, G., Butcher, G. W., and Hammerling, G. J.,** Differential expression of Ia and Ia-associated invariant chain in mouse tissues after *in vivo* treatment with IFN-gamma, *J. Immunol.,* 136, 940, 1986.

99. **Ghitescu, L., Fixman, A., Simionescu, M., and Simionescu, N.,** Specific binding sites for albumin restricted to plasmalemmal vesicles of continuous capillary endothelium: receptor-mediated transcytosis, *J. Cell Biol.,* 102, 1304, 1986.

100. **Wiley, C. A., Schrier, R. D., Nelson, J., Lampert, P. W., and Oldstone, M. B.,** Cellular localization of human immunodeficiency virus infection within the brains of acquired immune deficiency syndrome patients, *Proc. Natl. Acad. Sci. U.S.A.,* 83, 7089, 1986.
101. **Auerbach, R.,** personal communication.

Chapter 18

CURRENT CONCEPTS ON THE ROLE OF THE ENDOTHELIAL CYTOSKELETON IN ENDOTHELIAL INTEGRITY, REPAIR, AND DYSFUNCTION

Avrum I. Gotlieb and Michael K. K. Wong

TABLE OF CONTENTS

I. INTRODUCTION

This chapter deals with the role of the endothelial cytoskeleton in large-vessel endothelial cells as it pertains to injury and repair and the pathogenesis of atherosclerosis. Studies on the endothelial cytoskeleton both from our own laboratory and from elsewhere support three current concepts. The first is that the endothelial cytoskeleton is a regulator of endothelial integrity. The second is that the endothelial cytoskeleton is important in the regulation of repair of the injured endothelium. The third is that the cytoskeleton is a target for some agents which promote endothelial injury. While much of the work to be discussed was performed initially in in vitro model systems, more recent studies are providing confirmation of some of the observations in in vivo models of injury and disease. As well, the impetus for carrying out many of the in vitro studies was based on ideas derived from in vivo work. Thus, in vitro and in vivo studies are proceeding hand in hand to elucidate this very important area of endothelial biology and pathobiology.

II. THE ENDOTHELIAL CYTOSKELETON

Microfilaments, microtubules, and intermediate filaments constitute the three major fibrous protein systems of the endothelial cell.[1] Although collectively referred to as the ''cell cytoskeleton'' these systems are distinct with respect to function and to biochemical, structural, and immunological properties. Interactions, however, have been described between these cytoskeletal proteins.[2-7] These systems are dynamic in nature, and are regulated with respect to cross-linking and polymerization by many ''associated proteins'' including actin-binding proteins[8,9] and microtubule-associated proteins.[5,10,11] Studies on the cytoskeleton have focused on understanding how these fibrous systems are organized and how they perform their many cellular functions.

A. Actin-Containing Microfilaments

Actin, a contractile protein, is present in endothelial cells,[12] as in most eucaryotic cells, in the filamentous form, F-actin, and in the monomeric form, G-actin.[13-15] It is likely that the shift in equilibrium between the monomeric and the polymeric forms of actin is associated with many actin-mediated cell functions.[14,15] For example, studies comparing nonmigrating and migrating endothelial cells showed that in the latter there is a shift in the ratio of G- to F-actin favoring G-actin while total actin remains unchanged.[16] Filamentous actin is organized into a diffuse network of short microfilaments[17] and into prominent microfilament bundles.[18-20] The diffuse network is present mainly in the cell cortex as well as in cellular processes such as lamellipodia.[21] The microfilament bundles include several different types (reviewed in Reference 22), including the stress fibers. The stress fiber bundles are composed of actin filaments in parallel alignment with nonuniform polarity,[22,23] and were first identified and studied in vitro in a variety of cell types. In addition to actin, the stress fibers contain myosin, tropomyosin, and alpha actinin,[24-28] and are thought to be contractile and under mechanical stress.[29-31] Stress fibers located on the ventral aspect of the cell may develop as the cell attempts to pull against the substratum at an adhesion site[31,32] into which they insert.[33,34] For example, stress fibers localized *in situ* in fish scale scleroblasts are distributed in such a way as to suggest that the fibers are helping the cells resist shear stress.[24] In well-spread nonmigrating cultured fibroblasts it is likely that at least some of the stress fibers function as long substrate adhesion complexes.[33] Studies have also shown that in some systems, microfilament bundles did not appear necessary for cell motility[35] and that the bundles became more prominent when cells became extremely flattened on the substratum and were nonmotile.[36] In contrast, data from other studies suggest that the stress fibers are associated with cell migration.[37-40] It is likely that there are differences in the number and

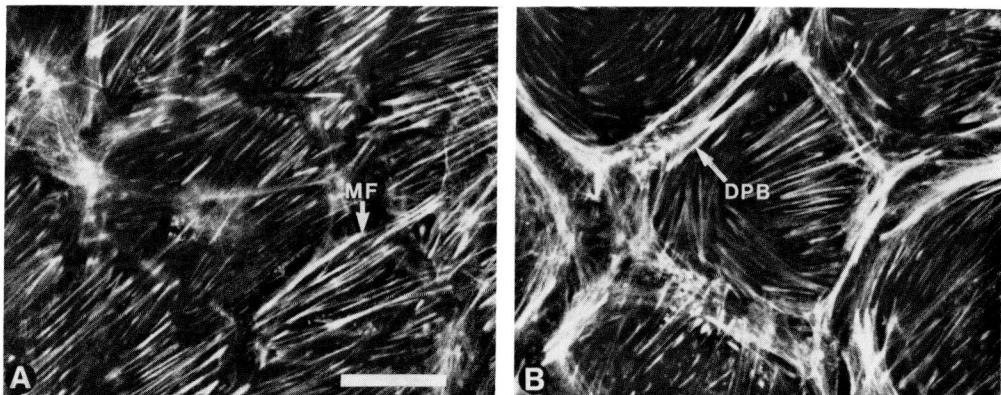

FIGURE 1.   Photomicrograph depicting the distribution of F-actin within porcine aortic endothelial cell cultures at various times after plating. (A) Subconfluent culture 3 days after plating showing many microfilament bundles (MF) in a cell located in the center of a group of cells; (B) confluent culture 5 days after plating. Note the presence of a dense peripheral band of microfilaments (DPB) around the circumference of the cell. Bar = 20 μm.

orientation of stress fibers necessary for a given cell type to migrate optimally on a given substratum.

The actin microfilament bundles have been identified in vessel wall endothelial cells from a variety of locations using transmission electron microscopy.[41-44] More recently, *in situ* localization of microfilaments has been achieved using fluorescence microscopy in both large-[45-48] and small-vessel endothelial cells.[49] The actin microfilaments are localized using antibodies to actin or by fluorescent phallotoxin probes. The toxin 7-nitrobenz-2-oxa-1,3-diazole-phalloidin (NBD-Ph), derived from a bi-cyclic heptapeptide phallotoxin isolated from the amanita family of mushrooms, has a very high affinity for F-actin.[50] Rhodamine phalloidin, more recently available, is useful due to its enhanced brightness and relative resistance to bleaching under ultraviolet light.[51] Studies with these probes have shown that actin microfilaments are located in two areas of the normal endothelial cell: at the periphery of the cell and centrally as stress fibers. *In situ* staining of these microfilaments has shown that they also contain myosin.[46]

Endothelial cell culture studies have allowed for the initial characterization of the F-actin-containing microfilaments. When the cells form a confluent contact-inhibited monolayer, we have shown that the periphery of the cell contains prominent microfilament bundles which we termed the "dense peripheral band" (DPB) as well as shorter central microfilament bundles which were randomly distributed[52] (Figure 1). The central bundles were reduced in number as the confluent monolayer became more tightly packed. The DPB did not appear until the monolayer became confluent, at which time a complete band was formed around each individual endothelial cell (Figure 1). In low-density culture, even in islands of endothelial cells where there is cell-to-cell contact, a DPB was not formed. Occasionally, there is a partial band in cells toward the center of an island of cells. Using double fluorescence and immunofluorescence microscopy to colocalize contractile proteins in the DPB, we found that within the DPB there was colocalization of actin with myosin, with tropomyosin, and with alpha-actinin (Figure 2). We have also shown that microtubules extended toward and into the DPB; however, there did not seem to be any preferential localization of microtubules within the band. Transmission electron microscopic examination of the DPB has shown that microfilaments emanated from the band and extended into junctions which had cytoplasmic plaques. Often the microfilaments of adjacent cells extended into their respective plaques and appeared to be in alignment with each other. Since we found actin microfilaments

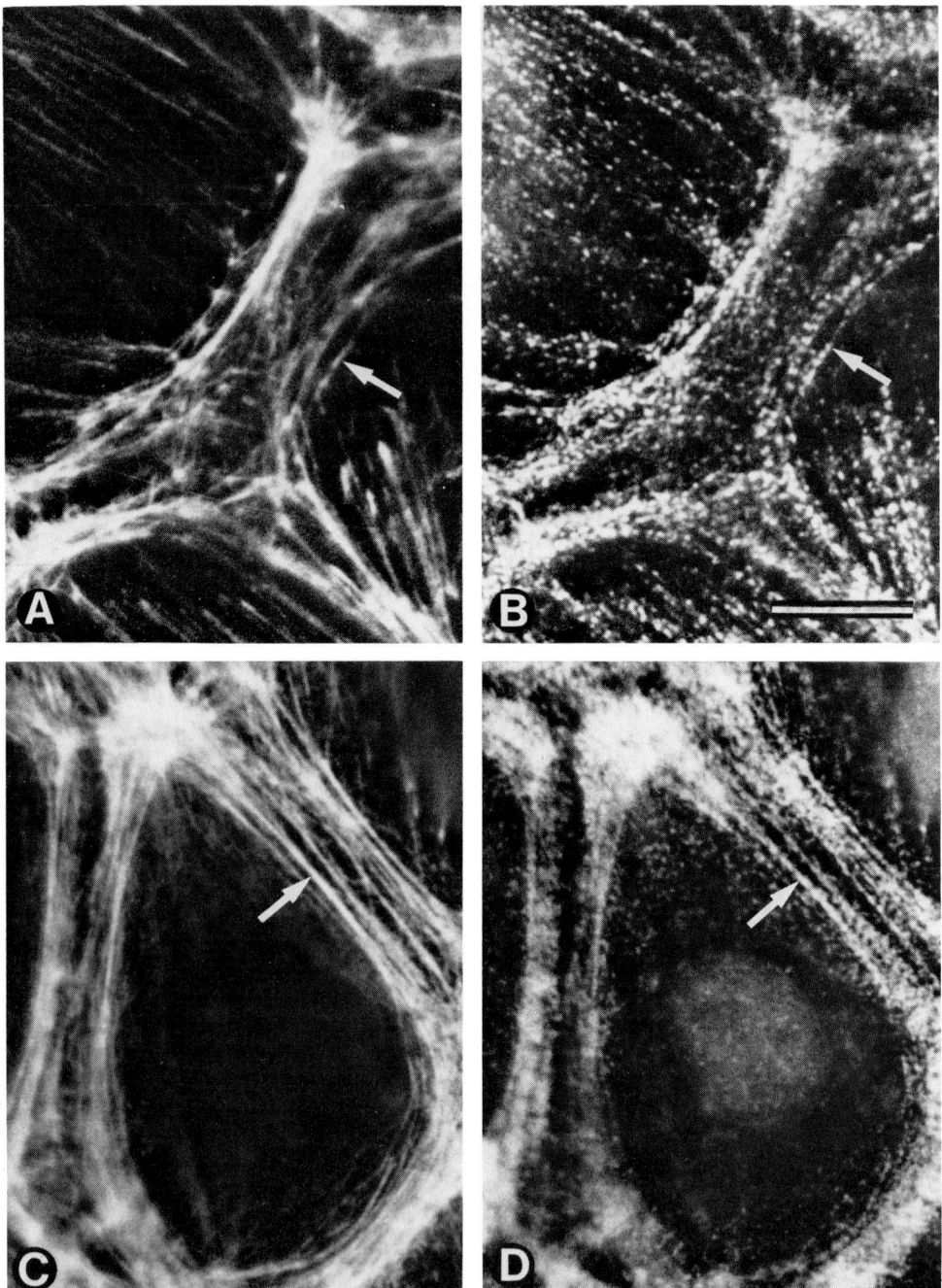

FIGURE 2.   Photomicrograph of a confluent endothelial monolayer showing the colocalization of F-actin with alpha-actinin (A, B) or myosin (C, D). Note that both alpha-actinin and myosin are associated with the microfilaments which make up the DPB. Arrows point to the same area in matched micrographs of the same cell. Bar = 10 μm.

extending from the DPB into vinculin plaques at cell-cell interfaces (Figure 3), we postulate that these plaques may be similar to adherent-type junctions.[52]

B. Microtubules and Centrosomes

Endothelial cells contain microtubules and centrosomes which consist of the paracentral paired centrioles and the amorphous material around them.[53] The centrosome acts as a microtubule organizing center,[54] and is thus thought to influence the spatial organization of microtubules.[55,56] In cell culture, microtubules are seen to emanate from the centrosomal area and are very thick toward the center of the cell, becoming thinner in density toward the periphery.[1] The distribution of microtubules is similar in both low-density and confluent endothelial cell cultures. Microtubules have been shown to play a role in many eucaryotic cell functions including mitosis, cell motility, particle transport, and morphogenesis. There are no data to suggest that microtubules in endothelial cells behave differently than those in other mammalian cells. The centrioles in porcine thoracic aortic endothelial cells but not in rabbit aorta have been shown to be preferentially located on the heart side of the nucleus.[57] This orientation is reestablished following reversal of a segment of the aorta. A similar orientation is observed in the porcine inferior vena cava suggesting that centriolar orientation is independent of blood flow. The function of this centrosomal polarity is not known.

C. Intermediate Filaments

Endothelial cells also contain the 8- to 10-nm intermediate filaments, and these have been well described elsewhere.[58,59] At the present time the physiological role of the filaments in endothelial function is not known; however, it is generally believed that intermediate filaments primarily have a mechanical role in cell function. They do not appear to be as dynamic as the microfilaments and microtubules. Antibodies to intermediate filaments microinjected into fibroblasts induce the filaments to coil around the nucleus; however, cell locomotion and cell morphology are unaltered,[60-62] suggesting that motility and shape are not regulated by these filaments. Further studies are needed to probe the structure-function relationship of these filaments in endothelial cells in normal and pathological states.

III. THE ENDOTHELIAL CYTOSKELETON IN INTEGRITY AND DYSFUNCTION

The actin microfilaments are very good candidates for intracellular structures which regulate the structural integrity of the endothelium. These filaments are arranged in two locations within the cell, central and peripheral. The data at hand suggest that the former are important in adherence of the cell to the subendothelial substratum while the latter are important in maintaining close contact between cells to prevent interendothelial gap formation.

A. Central Microfilaments

The central microfilaments within the endotheial cell cytoplasm are considered to be stress fibers. There is evidence from our own studies and those of others that these microfilaments are important in cell migration. This will be discussed below. With respect to endothelial integrity, the stress fibers appear to play an important role in cell-substratum adhesion. Gabbiani's group reported that the endothelial cells of newborn rats contain numerous actin stress fibers while stress fibers in adult rats are rare.[63] They suggested that stress fiber development was associated with endothelial cell turnover which is high in the new born and very low in the adult. White et al.[46] showed that thoracic aortic rat endothelial cells did contain stress fibers; however, they had a thread-like appearance. When they compared different areas of the aorta they found that in areas where increased hemodynamic stress is likely to be present (no direct measurements of shear stress were carried out, however) the

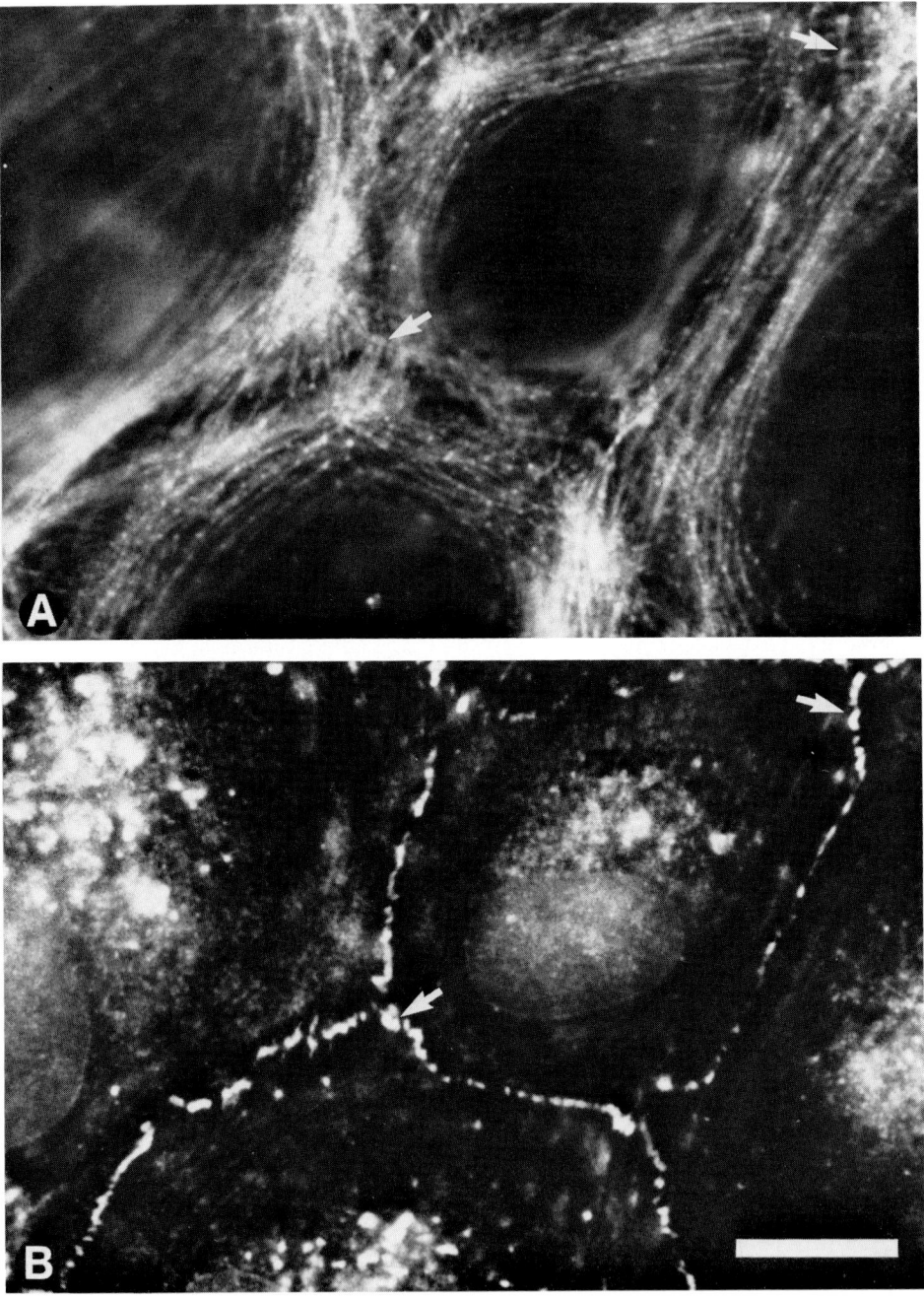

FIGURE 3.   Photomicrograph showing the association of vinculin with F-actin in a just-confluent culture of endothelial cells. (A) F-actin localization using rhodamine-phalloidin showing many short, thin microfilament bundles emanating from the DPB and radiating into the area where the cell abuts upon its neighbor; (B) indirect immunofluorescent labeling for vinculin in the same cell as in (A). Note the puntate distribution of vinculin at the cell-cell interface. Arrows point to the same location in each of the two micrographs. Note also that many of the vinculin plaques are associated with microfilament bundles from two DPBs. Bar = 10 μm.

number of cells with prominent stress fibers was increased. Similar findings were seen in the rabbit aorta,[64] and we find similar distributions in the porcine aorta.[65] Since there is a wide variation in the microfilament morphology throughout the vascular tree (Figure 4), more precise studies are needed to correlate F-actin organization and shear stress using improved techniques to visualize ostia and bifurcations and measure shear stress at these areas. Hypertension also induces an increase in stress fibers,[43] although White and Fujiwara[66] interpret their data to suggest that blood pressure is not the primary influence upon stress fiber distribution. Reported sex-related differences in the number of cells with stress fibers may be related to the higher blood pressures recorded in male rats.[46] Hemodynamic shear stress has been shown to alter endothelial cell shape, orientation, and repair.[67-69] In vitro studies have shown that fluid shear stress can effect stress fiber expression.[70,71] Since there are differences in the response of endothelial cells to flow in vitro when compared to in vivo reendothelialization studies, in vivo studies are necessary to substantiate in vitro findings.[71] The evidence to link stress fibers to the effects of hemodynamic shear stress is compelling; however what is indeed missing are in vivo studies which carefully explore the sequence of cytoskeletal changes in response to defined shear stresses. Since shear depends on a variety of factors including blood viscosity and velocity and vessel diameter, direct measurements of shear stress are essential.

Thus, the central microfilaments which are abluminal appear to be important in maintaining endothelial integrity by subserving an attachment function. They hold the endothelial cell to the wall. It has been shown that the endothelial cytoskeleton is affected by specific components of the subendothelial matrix and that the endothelial cells may be able to regulate repair by altering the matrix they deposit.[72-74] How these fibers are affected by the cell matrix and by the numerous metabolic activities of the endothelial cell is not known.

B. Peripheral Actin Microfilaments

There are several studies which show a correlation between the disruption of peripheral actin microfilaments and loss of endothelial integrity. These studies suggest that these microfilaments may be the target for some agents which promote endothelial dysfunction.

Shasby's group has carried out several studies examing endothelial injury and the association between changes in endothelial permeability and the cytoskeleton.[75] Using confluent monolayers of porcine pulmonary artery endothelial cells grown on micropore filters, they showed that reversible increases in albumin transfer were induced by oxidants.[76] The oxygen radical effect was associated with both changes in cell shape characterized by retraction and rounding and changes in actin microfilament distribution. They further presented data to show that calcium was an important mediator of endothelial permeability associated with cytoskeletal changes.[77] Wysolmerski and Lagunoff[78,79] showed that the pulmonary edema induced by the drug ethchlorovynol was associated with gaps between endothelial cells and loss of the dense peripheral band of actin microfilament bundles.

Thrombin[80,81] and phorbal 12-myristate 13-acetate (TPA)[82] are known to induce reversible shape changes in the endothelial cells of the confluent monolayer. We have shown that these changes are associated with reversible changes in the DPB.[83] In the presence of thrombin or TPA the confluent polygonal cells become elongated as the dense peripheral band is lost and the central microfilament bundles become more prominent. Other agents such as trypsin induce a reversible rounding of the endothelial cells; however, the dense peripheral band persists although the cells undergo retraction. Stolpen et al.[84] reported that recombinant tumor necrosis factor and immune interferon caused human endothelial cells to rearrange their actin cytoskeleton, elongate, overlap, and lose their stainable fibronectin matrix. The cytoskeletal changes are similar to those we described for the thrombin effect — a loss of the DPB and an increase in prominent central microfilament bundles. This reorganization

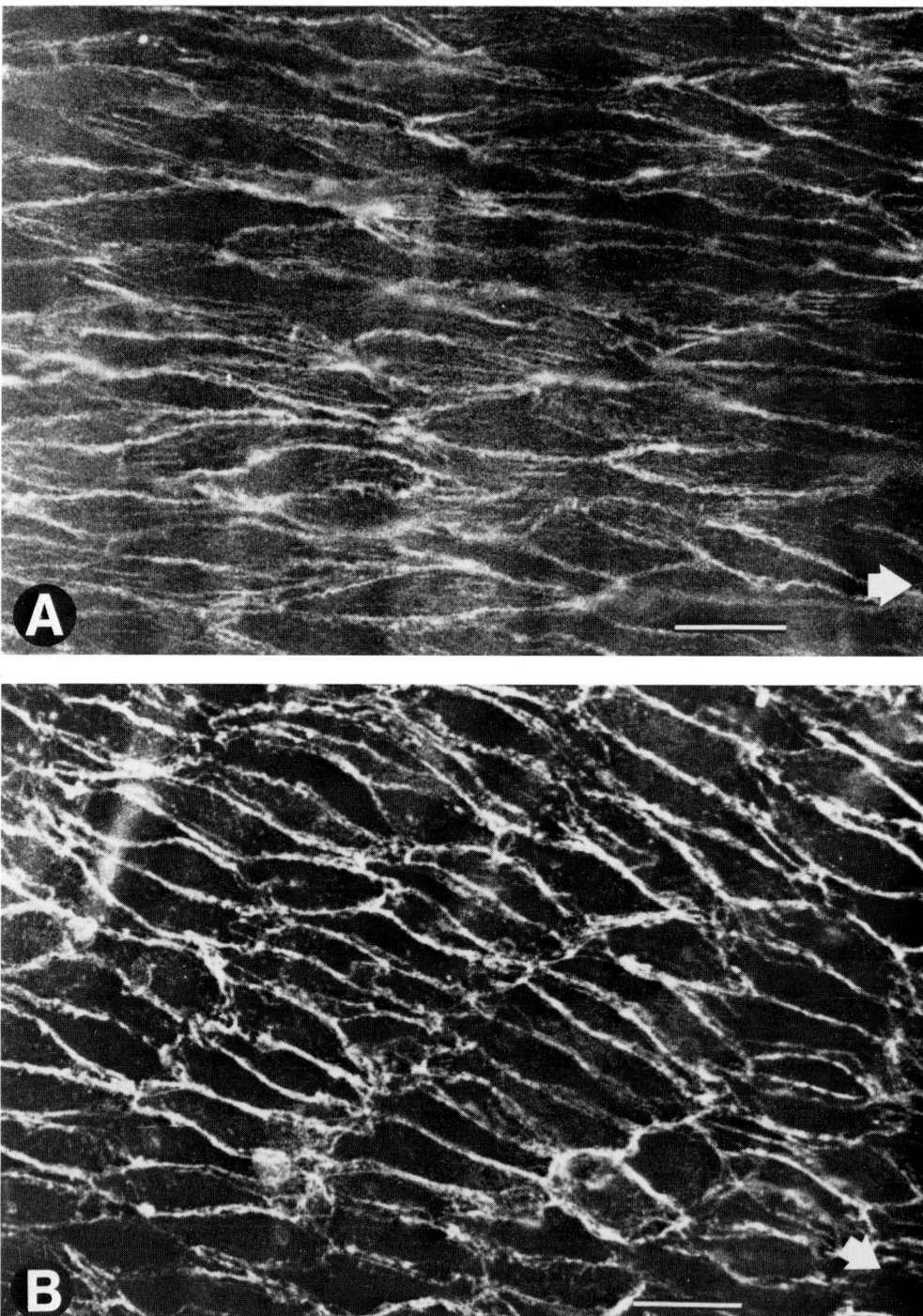

FIGURE 4. Photomicrographs showing the different distribution of F-actin in rabbit endothelial cells *in situ*. (A) Left common carotid artery 15 mm upstream of the innominate artery. Note the presence of both peripheral and centrally located microfilaments. (B) Left internal carotid artery 1 mm upstream from the common carotid artery. Note prominant peripheral microfilaments and paucity of central F-actin staining. (C) Common carotid artery at the left superior thyroid artery showing the distal lip of the branching site. Note the very prominent microfilaments in the endothelial cells immediately distal to the branch site. Star = lumen of the left superior thyroid artery. Arrows denote direction of blood flow. Bars = 20 μm.

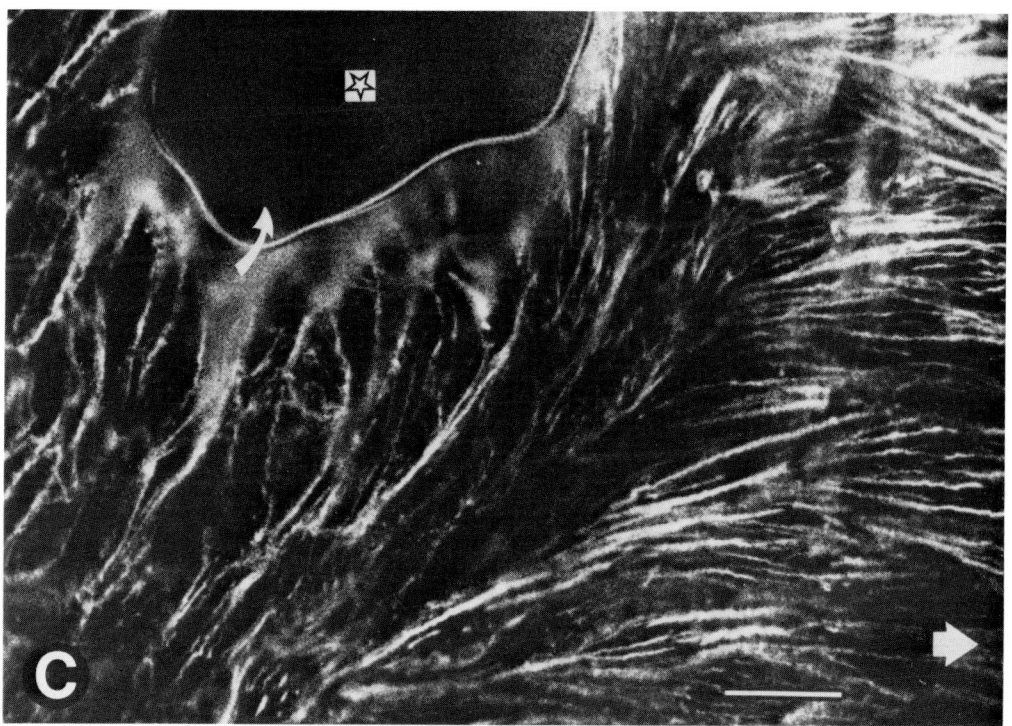

FIGURE 4C.

of the monolayer, which was specific for endothelial cells, may be the in vivo counterpart of immune changes seen in vascular endothelial cells characterized by increased permeability.

Vasoactive substances and mediators of inflammation have been shown to enhance endothelial permeability and form interendothelial gaps. It is hypothesized that these substances act by promoting active endothelial contraction mediated by microfilament bundles.[85,86] However, Welles et al.[87] showed that in vitro histamine causes a decrease in actin microfilament bundles in endothelial cells. Although further in vivo studies are needed to clarify the issue of endothelial contractility, another possibility that should be tested is that there may be no active contraction of the endothelial cell per se. Instead the interendothelial gaps are formed because cell-cell adhesion is disrupted due to a change in the junctional actin filament web concentrated along the junction.[44]

Based on the above studies, our hypothesis is that the DPB is important in maintaining the integrity of the endothelium. The DPB represents a specific organization of actin microfilament bundles located around the complete circumference of individual endothelial cells in a confluent monolayer. It differs from the diffuse submembranous microfilament mat or sheath found in most cell types since the mat is a diffuse array of very fine microfilaments located within the entire cortical cytoplasm.[17,88,89] It is, however, likely that there is an association between the two. The DPB is different than the central microfilament. Stress fibers are visualized in the ventral surface of the endothelial cell by phase contrast microscopy, while the DPB is not. The DPB also appears to be much more sensitive to the effects of cytochalasin B, as it is completely disrupted within 10 to 20 min of incubation with cytochalasin B, whereas many central microfilament bundles still persist at 24 hr.[52] It is also unlike the central F-actin bundles of microvilli, since the latter do not contain many of the contractile proteins present in the DPB.[90] The DPB may, however, share similarities

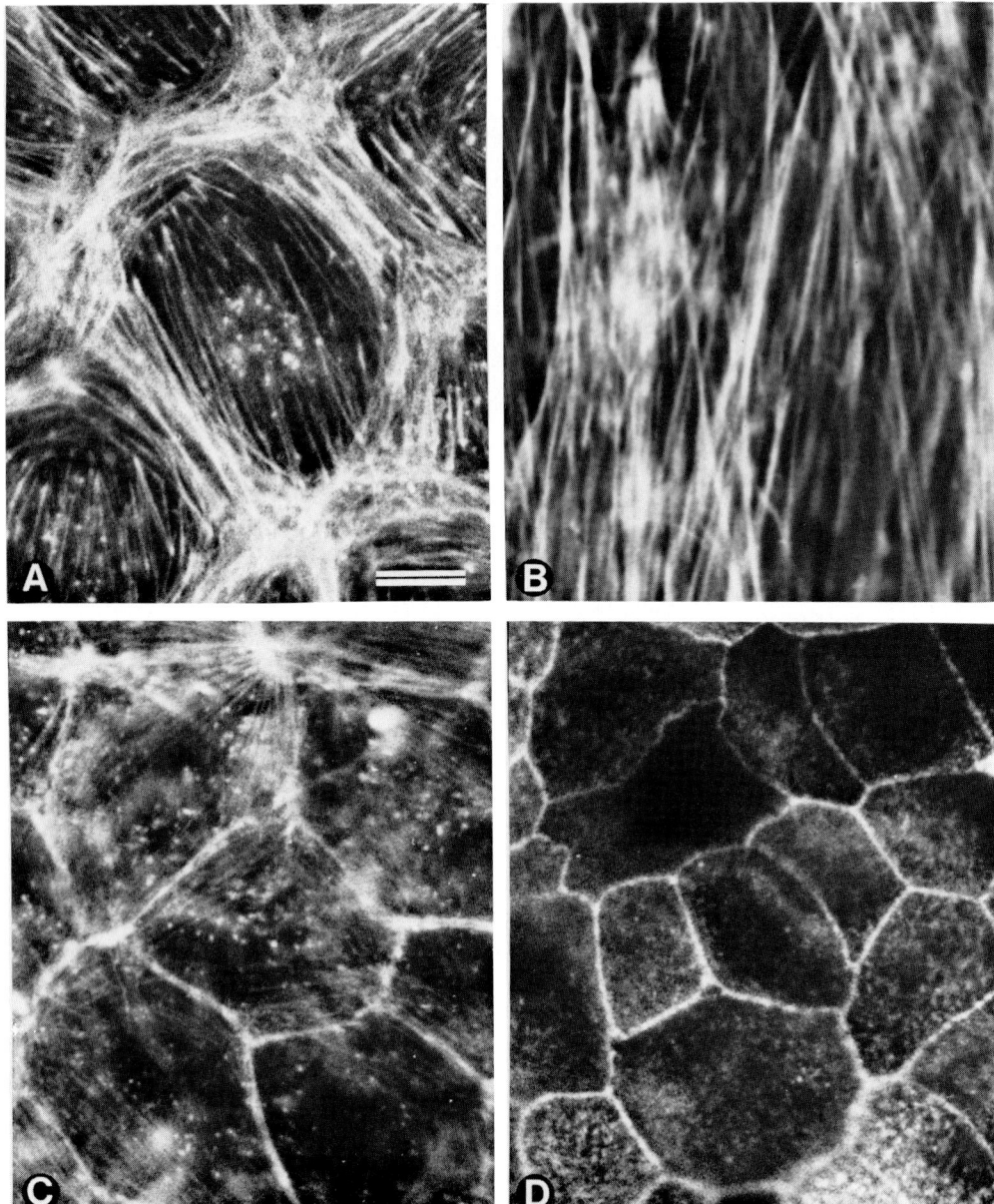

FIGURE 5.   Photomicrograph showing the localization of F-actin by rhodamine-phalloidin in different cell types at confluency. (A) Primary cultures of porcine aortic endothelial and, (B) smooth muscle cells; (C) PtK$_2$, rat kangaroo-derived epithelial cell line; (D) Madin-Darby canine kidney-derived cell line photographed at the top of the cell to show the apical circumferential band of microfilaments. Cells in (A), (C), and (D) show peripheral microfilament bundles. In contrast, smooth muscle cells (B) show prominant central microfilament bundles. The photograph is slightly out of focus to show the cell layers. Bar = 10 μm.

with the circumferential band of microfilaments seen in the apical portion of epithelial cells such as retinal pigmented epithelial cells[91,92] and the Madin-Darby bovine or canine kidney epithelial cell lines[93,94] (Figure 5). These cells form tight monolayers in which the apical bundles are associated with vinculin adherens junctions present at the interface between adjacent cells. This is similar to the close association between microfilaments making up the DPB and vinculin plaques at cell-cell interfaces. It is also of considerable interest that the microfilament-adherens junction complex has also been demonstrated to regulate permeability across a sheet of epithelial cells[94] and that the cellular organization of vinculin is sensitive to external soluble factors including platelet-derived growth factor.[98,99] Further structural, biochemical, and functional studies are needed to characterize the endothelial vinculin plaques to show that they are similar to the adherens junctions.[95,96] Relying on structural studies alone is not sufficient, as was shown in a very recent report on the identification of a new type of spot desmosome junction in epithelial cells which is also associated with microfilaments and not tonofilaments.[97] These plaques stained with actin, alpha-actinin, and vinculin and not with desmoplakin and were considered to be a new type of cell-cell adhesion junction.

In addition, the DPB may be formed in order to stabilize the endothelial cell against the physical forces exerted upon it by its neighbor as the cells pack into the two-dimensional space of the monolayer.[100] The colocalization of alpha-actinin, myosin, and tropomyosin in the microfilaments making up the DPB implies that this structure possesses the ability to be contractile. Anchored by vinculin at cell-cell junctions, the DPB may thus provide the tensile force to stabilize endothelial cell shape within the monolayer while maintaining cell-cell adhesion.

The DPB was described and characterized in polygonal shaped endothelial cells in confluent monolayer culture. How is this structure related to the peripheral F-actin observed in the elongated endothelial cells present in the vessel wall? Peripheral microfilament bundles have been described ultrastructurally in endothelial cells in vivo.[101,102] Localization of these microfilament bundles by light microscopic immunocytochemical methods show that the in vivo structure is not as well developed as the DPB seen in vitro. Although in vivo endothelial cell peripheral microfilaments also contain myosin like their in vitro counterparts, they are not made up of an aggregation of multiple distinct microfilament bundles, but rather of a few peripheral bundles. The reason for this difference is not known; however, it may be related to the differences in endothelial cell shape and size as well as to the absence of hemodynamic shear forces in our tissue culture conditions. These questions can be addressed in the future by carrying out in vivo studies in which a careful correlation is made between filamentous actin distribution and both cell shape and cell location within the vascular tree with respect to measured variations in shear stress and flow patterns. Modulation of the in vivo peripheral actin in pathological states has also been described, such as in endothelial regeneration. Peripheral actin, however, has been given little prominence by most authors as a cytoskeletal system important in the regulation of endothelial barrier integrity. Our studies suggest an essential role for peripheral actin microfilaments in normal endothelial cell function.

IV. THE ENDOTHELIAL CYTOSKELETON IN REPAIR

The ability of the endothelium to rapidly repair itself is very important in maintaining the thromboresistant and macromolecular barrier functions of the endothelium, at least in the acute situation prior to the vessel wall attempting to subserve these functions by other means.[103,104] Our studies have centered on the role of the microtubules and their associated centrosome and on the microfilaments in regulating repair.

A. Microfilaments in Endothelial Regeneration
We used the large wound experimental model to study the changes in the F-actin cytoskeleton during endothelial regeneration. As the sheet of endothelial cells migrates forward to cover the denuded area, we found that four zones could be defined at the wound edge

which have characteristic features.[38] The leading edge was made up of one row of endothelial cells. The endothelial cell contained prominent lamellipodia with F-actin in the lamellipodia and prominent central microfilaments distributed both parallel and perpendicular to the wound edge. Behind this leading edge there was a zone of elongated endothelial cells. The cells did not contain a DPB; however, prominent central microfilaments were observed extending parallel to the long axis of the cell which itself was perpendicular to the wound edge. Behind the elongated zone there was the transition zone of cells which had undergone some shape change but still had polygonal features. These endothelial cells contained a DPB and central microfilament bundles which were randomly distributed with respect to the wound edge. Behind this was the confluent monolayer which contained the prominent DPB and the central microfilaments arranged in a random fashion. Cells in the first two zones, the leading edge and elongated zone, showed prominent migration while the other zones did not. Thus there is a reorganization of both central and peripheral microfilaments in migrating cells. While central microfilament bundles are prominent in migrating cells, the presence of the DPB was associated with a marked reduction or inhibition of cell migration.

In order to study single cell wound repair, we developed an in vitro single-cell wound model using a micromanipulator to make a precise wound in a confluent monolayer.[105] We characterized the kinetics of wound repair using time-lapse cinemicrophotography and studied the dynamics of F-actin during this process (Figure 6). Immediately following the removal of a cell, that part of the DPB facing the wound showed some splaying and became more prominent. Microfilaments were seen to emanate from the band into the lamellipodia extending out to close the wound. Wounds incubated with cytochalasin B at concentrations which caused loss of the DPB with sparing of the central microfilament showed very little closure after a period of 6 hr, while complete closure occurred normally between 30 and 45 min. However, when the cytochalasin was washed out, the microfilaments of the DPB began to reappear immediately, followed thereafter by lamellipodia formation. Thus, reendothelization occurring by spreading is associated with an intact DPB, while repair involving cell translocation is associated with the breakdown of the DPB.

What is the function of these central microfilaments during repair? Gabbiani et al.[48] suggest that the stress fibers which appear in vivo in endothelial cells at the edge of a wound reflect cell adhesion and not motility. The event, however, which leads to the appearance of the stress fibers is denudation and subsequent cell migration. A comparison of our in vitro data with that in vivo shows the same orientation of stress fibers in migrating cells, suggesting that adhesion under hemodynamic stress is not necessarily the main factor in the development of prominent stress fibers. An important question that remains unanswered is whether the stress fibers are involved in isotonic contraction during cell migration.

B. Centrosomes and Microtubules in Endothelial Regeneration

Using the large wound in vitro model system,[37,106] we have shown that endothelial cells migrating into the wound redistributed their centrosomes to the front of the cell between the nucleus and the leading lamellipodia (Figure 7).[107] The centrosomal redistribution occurred rapidly following wound production and required the presence of intact microtubules.[108] If the wound was treated first with colcemid to break down the microtubules, redistribution did not occur. We have also shown that redistribution can occur independent of cell migration. Cells treated with cytochalasin B at concentrations which just inhibited migration were still able to redistribute their centrosomes. This redistribution, however, occurred more slowly than under normal conditions and thus suggests that the microfilaments may play some role in enhancing centrosome redistribution.

Since the distribution of various cytoskeletal components may change when cells are removed from their *in situ* environment and grown in vitro, there was a need to verify and in vitro observations with in vivo experiments.[47] Thus, it has been shown that centrosome

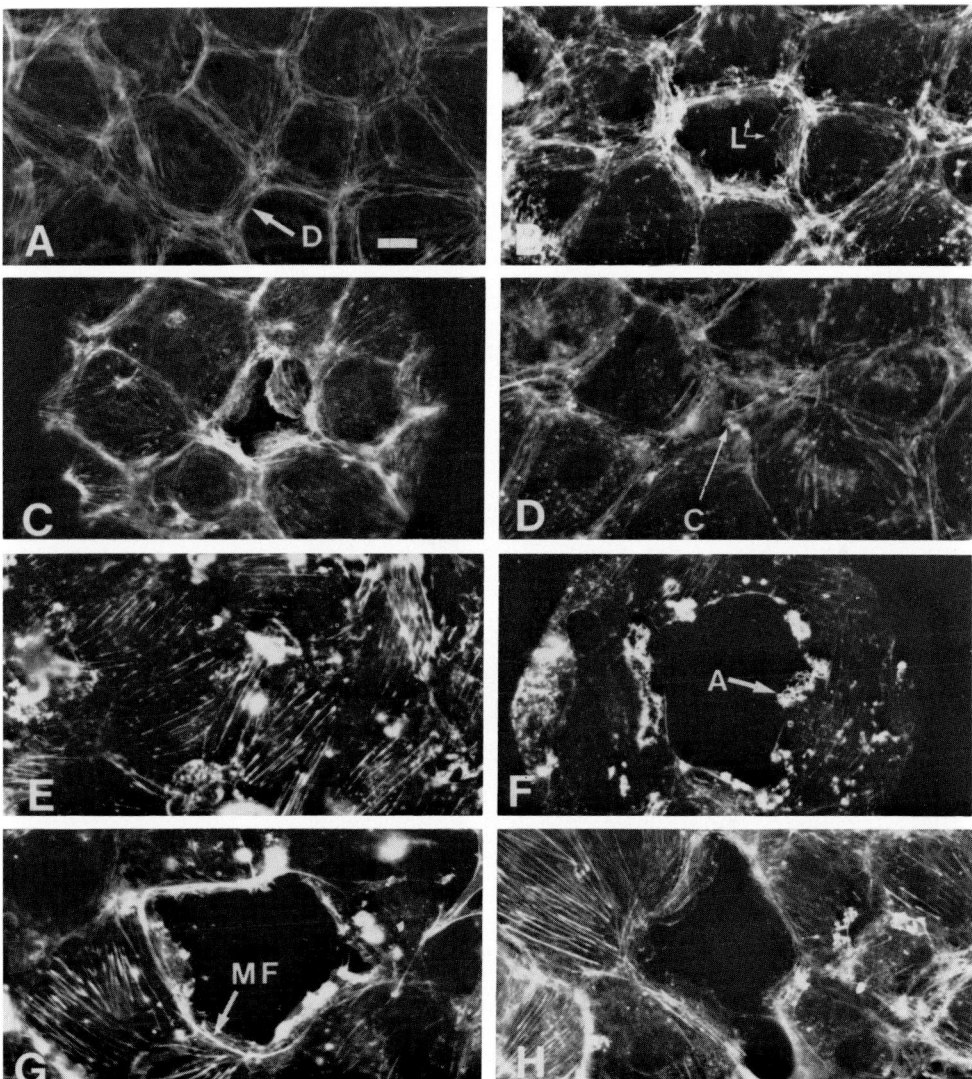

FIGURE 6. Photomicrograph showing F-actin localization during the reendothelialization of single cell wounds from different representative control (A to D), cytochalasin-B treated (E, F), and cytochalasin-B treated and subsequently washed out (G, H) experiments. (A) Intact monolayer with DPBs ("D"). (B) 5 min after the removal of a single cell showing wound retraction and the initiation of lamellipodia extrusion ("L"). (C) 20 min after injury showing several well-developed lamellipodia and splaying of the DPB within them. (D) Closed wound ("C"). (E) Intact monolayer treated with cytochalasin-B for 2 hr and used for wounding experiments; note loss of the DPB. (F) Cytochalasin-B treated culture, 2 hr postinjury and immediately prior to washout, showing failure of reendothelialization; note the lack of DPBs and the presence of F-actin aggregates ("A") at the cell periphery abbuting upon the wound. (G) 5 min after cytochalasin-B washout; note presence of microfilament bundles ("MF") in the reformation of the DPB occurring prior to lamellipodia extrusion. (H) 20 min after washout showing well-developed lamellipodia. Bar = 10 μm. (From Wong, M. K. K. and Gotlieb, A. I., *Lab Invest.*, 51, 75, 1984. With permission.)

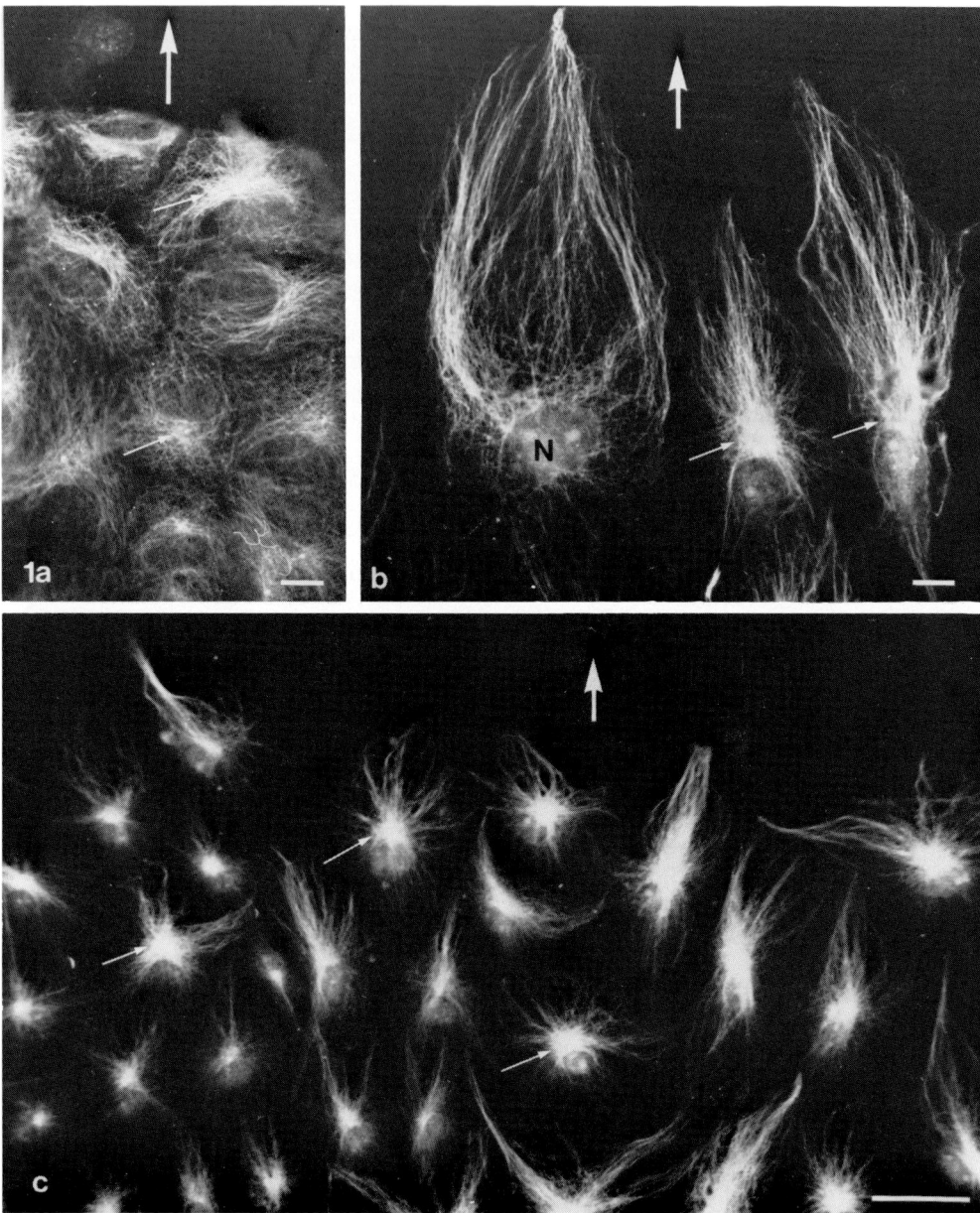

FIGURE 7.   Immunofluorescent photomicrograph of endothelial cells at the wound edge stained with antitubulin serum immediately after wounding before migration begins (a) and 44 hr after wounding (b and c). Note that the centrosome (small arrows) which initially are randomly distributed relative to the wound edge and the position of the nuclei (a) become oriented so that they face the wounded area towards which the endothelial sheet is migrating (b and c). Large arrow is perpendicular to the wound edge and indicates direction of movement of endothelial cell sheet. Bar (a, b) = 10 μm (c) = 50 μm. (From Gotlieb, A. I., McBurnie-May, I., Subrahmanyan, L., and Kalnins, V. I., *J. Cell Biol.*, 91, 589, 1981. With permission.)

Table 1
CENTROSOME DISTRIBUTION IN THE DIFFERENT ZONES OF CELLS AT WOUND EDGE

			Centrosome location[b]											
			22 hr after wounding						44 hr after wounding					
			Toward		Middle		Away		Toward		Middle		Away	
Zone of cells[a]	Cell migration	DPB	X[c]	SE	X	SE	X	SE	X	SE	X	SE	X	SE
Leading edge	Present	Absent	95	1.2	3	0.6	2	1.0	96	1.2	2	0.3	2	0.9
Elongated	Present	Absent	85	1.9	10	2.0	5	1.2	86	2.2	10	1.2	4	1.0
Transition	Absent	Present	64	3.7	22	2.5	14	1.2	76	2.4	16	0.3	8	2.5
Monolayer	Absent	Present	34	1.8	39	0.6	27	2.3	35	2.3	38	1.7	27	0.7

[a] The different zones of cells extending from the wound edge into the monolayer (see Reference 36).
[b] Percent of cells with centrosomes towards the wound edge (i.e., between the nucleus and the wound edge); middle, along the nucleus; and away from the wound edge (i.e., between the nucleus and the monolayer behind the cell).
[c] Mean of three experiments.

redistribution occurred not only in cell culture but also in organ culture,[109] as well as in vivo following wounding.[110] The reorientation occurred most rapidly in the cell culture model, suggesting that the subendothelial matrix and hemodynamic factors may act to modulate the rate of centrosomal distribution. This suggestion has yet to be tested directly. Mascardo and Sherline,[111] however have shown that centrosomal redistribution was enhanced by several factors including serum, multiplication-stimulating factor, and insulin. Although platelet-derived growth factor had no effect on its own, it had a positive synergistic effect with subeffective concentrations of serum, insulin, and multiplication-stimulating factor. These studies are important since they show that cytoskeletal events occurring during endothelial regeneration have the potential of being regulated by external influences.

Centrosomal redistribution has been shown to occur in other migrating cell systems under a variety of conditions.[111-117] In endothelial cells it appears that orientation is important in the initiation and regulation of migration and may thus act as a preprogrammed internal control. Albrecht-Buehler[118] has suggested that the centrosomes may be acting as a type of gyroscope within the cell in order to maintain directionality. Kupfer et al.[114] have suggested that since the Golgi apparatus redistributes along with the centrosome, the function of centrosomal redistribution is to provide new membrane for extruding lamellipodia during cell migration. In cytotoxic lymphocytes the centrosome relocates to the area of the cell facing the target cell, suggesting that the centrosome may be able to change the function of the cell membrane that it faces.[115]

What triggers the centrosome to redistribute? At the wound edge it appears that the loss of contact inhibition is an important stimulus. How this is sensed by the cell at a cellular and molecular level is not known. Is the loss of the DPB triggering centrosomal redistribution? We have observed centrosomal redistribution occurring before the breakdown of the DPB, suggesting that the redistribution is an early cytoskeletal event. In these experiments, examination of the colocalization of F-actin microfilaments and microtubules and centrosomes was carried out in the zone of cells extending from the wound edge into the monolayer in a large linear wound as previously described.[38] The interesting result was that centrosome redistribution had occurred in cells which had undergone minimal shape change but were not migrating and were away from the wound edge (Table 1). These cells still possessed intact DPBs. The nature of the signal establishing centrosomal polarity in the wound is unknown. Mechanisms which can be studied include the idea that a physical stimulus may

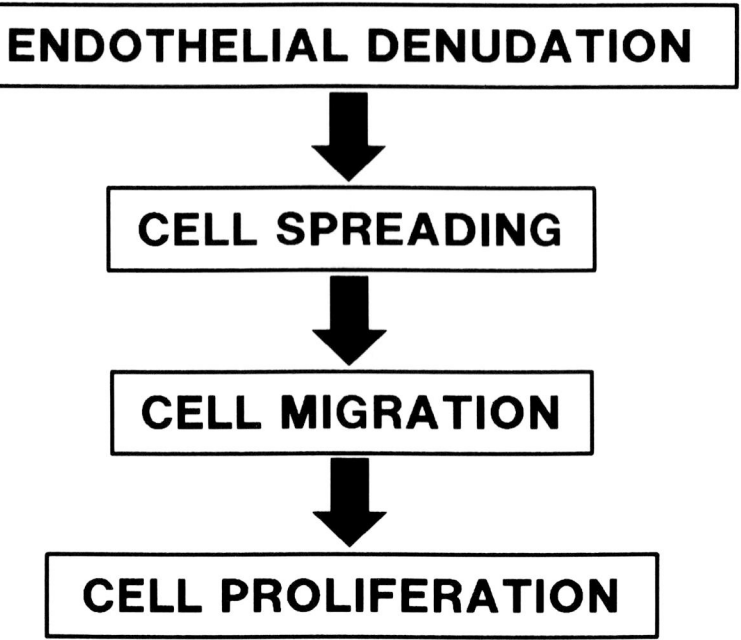

FIGURE 8. Concept of reendothelialization. The repair of the injured endothelium follows a defined sequence of events characterized by cell spreading, cell migration, and cell proliferation. Each of these events involves the cytoskeleton in a specific way (see text).

be conveyed to the cell as the monolayer becomes less tightly packed. Another possibility is that a signal passed through intercellular junctions may be a triggering stimulus. It is well known that small-molecular-weight substances can be passed to adjacent cells.[119]

The mechanism by which the centrosome redistributes toward the front of the cell is also unknown. Centrosomal redistribution requires intact microtubules and appears to be somewhat delayed if the microfilaments are broken down.[108] Thus, one possibility is that there are direct or indirect connections between the microtubules and the DPB with the band acting to anchor the microtubules as the centrosomes redistribute. Euteneur and Schliwa's[120] observations on the motility of centrosomes in polymorphonuclear leukocytes support the idea that actin networks are involved in determining the position of the centrosome through interactions with microtubules. Using time-lapse cinemicrophotography, we have noted that in some cases the centrosome redistributed independent of any detectable nuclear movement. In other instances, however, it appeared that centrosomal redistribution was associated with rotation of the nucleus, so that the centrosome and the nucleus appeared to rotate together in the same direction.[121] We do not have any morphological evidence to suggest that there is prominent depolymerization of microtubules during the redistribution. We have not yet studied the role that might be played by microtubule-associated proteins in establishing centrosomal polarity.

C. Endothelial Repair, a Multistep Process

Based on our studies of the cytoskeleton in the repair of large wound and the single cell wound models, we postulate that the process by which an endothelial cell initiates reendothelialization is a two-step process followed by cell proliferation when necessary (Figure 8). In the first step, the endothelial cell attempts to repair an area of injury by rapidly extruding lamellipodia and spreading. It requires intact microfilament systems to carry this out, including the DPB. If repair has been completed, nothing further occurs. However, if repair is incomplete, then a second set of events occurs characterized by the elongation of the endothelial cell and the redistribution of the centrosome toward the front of the cell in preparation for cell locomotion. DPB breakdown ensues and directed migration occurs. In this manner small wounds are reendothelialized. The role of cell proliferation is also very important; however, it occurs following the onset of migration.[122] Very small wounds are repaired independent of cell proliferation,[123-125] while large wounds require cell proliferation as a means of providing cells to reestablish the monolayer.

V. SUMMARY

The work in the endothelial cytoskeleton is an example of the study of structure-function relationships. Much of the data, however, describes associations, and cause and effect must still be worked out. What is further required is an understanding of the molecular mechanisms which regulate and integrate the dynamic and the stable organization of these endothelial fibrous protein systems. Thus the concepts put forth in this chapter will require revisions as knowledge in the field advances. However, studies designed to further test our current concepts directly will provide important information on the function of the the the endothelial cytoskeleton.

ACKNOWLEDGMENT

We are grateful to Ms. W. Spector for technical assistance, to Ms. B. Libman for the photographs on kidney epithelial cells, and to Ms. Sue Sarju for secretarial assistance.

This work was supported by Medical Research Council grant MT-6485 and Heart and Stroke Foundation of Ontario grant T435. Michael K. K. Wong was a Trainee and is a medical scientist of the Canadian Heart Foundation.

REFERENCES

1. **Kalnins, V. I., Subrahmanyan, L., and Gotlieb, A. I.,** The reorganization of cytoskeletal fibre systems in spreading porcine endothelial cells in culture, *Eur. J. Cell Biol.,* 24, 36, 1981.
2. **Godman, G., Woda, B., Kolberg, R., and Berl, S.,** Redistribution of contractile and cytoskeletal components induced by cytochalasin. I. In Hmf cells, a nontransformed fibroblastoid line. II. In HeLa and HE p-2 cells, *Eur. J. Cell Biol.,* 22, 733, 1980.
3. **Holtzer, H., Croop, J., Dienstman, S., Ishikawa, H., and Somlyo, A. P.,** Effects of cytochalasin B and colcemid on myogenic cultures, *Proc. Natl. Acad. Sci. U.S.A.,* 72, 513, 1975.
4. **Singer, S. J., Ball, E. H., Geiger, B., and Chen, W. T.,** Immunolabeling studies of cytoskeletal associations in cultured cells, *Symp. Quant. Biol.,* 46, 303, 1981.
5. **Pollard, T. D., Selden, S. C., and Manupin, P.,** Interaction of actin filaments with microtubules, *J. Cell Biol.,* 99, 33s, 1984.
6. **Euteneuer, U. and Schliwa, M.,** Evidence for an involvement of actin in the positioning and motility of centrosomes, *J. Cell Biol.,* 101, 96, 1985.
7. **Schliwa, M., Pryzwansky, K.B., and van Blerkom, J.,** Implication of cytoskeletal interactions for cellular architecture and behavior, *Phil. Trans. R. Soc. London,* B299, 199, 1982.
8. **Pollard, T. D.,** Actin-binding protein evolution, *Nature (London),* 312, 403, 1984.
9. **Stossel, T. P., Chaponnier, C., Ezzell, R. M., Hartwig, J. H., Janmey, P. A., Kwiatkowski, D. J., Lind, S. E., Smith, D. B., Southwick, F. S., Yin, H. L., and Zaner, K. S.,** Nonmuscle actin-binding proteins, *Annu. Rev. Cell Biol.,* 1, 353, 1985.
10. **Aamodt, E. J. and Williams, R. C., Jr.,** Microtubule associated proteins connect microtubules and neurofilaments in vitro, *Biochemistry,* 23, 6023, 1984.
11. **Griffith, L. M. and Pollard, T. D.,** The interaction of actin filaments with microtubules and microtubule-associated proteins, *J. Biol. Chem.,* 257, 9143, 1982.

12. **Becker, C. G. and Murphy, G. E.,** Demonstration of contractile protein in endothelium and cells of the heart valve endocardium, intima, arteriosclerotic plaques and Aschoff bodies of rheumatic heart disease, *Am. J. Pathol.,* 55, 1, 1969.

13. **Pollard, T. D. and Weihing, R. R.,** Actin and myosin and cell movement, *CRC Crit. Rev. Biochem.,* 2, 1, 1974.

14. **Korn, E. D.,** Actin polymerization and its regulation by proteins from non-muscle cells, *Physiol. Rev.,* 62, 672, 1983.

15. **Howard, T. H. and Oresajo, C. O.,** The kinetics of chemotatic peptide-induced change in F-actin content, F-actin distribution, and the shape of neutrophils, *J. Cell Biol.,* 101, 1078, 1985.

16. **Gabbiani, G., Gabbiani, F., Heimark, R. L., and Schwartz, S. M.,** Organization of actin cytoskeleton during early endothelial regeneration in vitro, *J. Cell Sci.,* 66, 39, 1984.

17. **Willingham, M. C., Yamada, S. S., Davies, P. J. A., Rutherford, A. V., Gallo, M. G., and Pastan, I.,** Intracellular localization of actin in cultured fibroblasts by electron microscopic immunocytochemistry, *J. Histochem. Cytochem.,* 29, 17, 1981.

18. **Porter, K. R., Claude, A., and Fullam, E. F.,** A study of tissue culture cells by electron microscopy, *J. Exp. Med.,* 81, 233, 1945.

19. **Ishikawa, H., Bischoff, R., and Holtzer, H.,** Formation of arrowhead complexes with heavy meromyosin in a variety of cell types, *J. Cell Biol.,* 43, 312, 1969.

20. **Lazarides, E. and Weber, K.,** Actin antibody: the specific visualization of actin filaments in nonmuscle cells, *Proc. Natl. Acad. Sci. U.S.A.,* 71, 2268, 1974.

21. **Small, J. V., Rinnethaler, G., and Hinssen, H.,** Organization of actin meshworks in cultured cells: the leading edge, *Symp. Quant. Biol.,* 46, 599, 1981.

22. **Stossel, T. P.,** Contribution of actin to the structure of the cytoplasmic matrix, *J. Cell Biol.,* 99, 15s, 1984.

23. **Buckley, I. K. and Porter, K.R.,** Cytoplasmic fibrils in living cultured cells: a light and electron microscope study, *Protoplasma,* 64, 349, 1967.

24. **Byers, H. R. and Fujiwara, K.,** Stress fibers in cells in situ: immunofluorescent visualization with anti-actin, anti-myosin, and anti-alpha-actinin, *J. Cell Biol.,* 93, 804, 1982.

25. **Gordon, W. E.,** Immunofluorescent and ultrastructural studies of ''sarcomeric'' units in stress fibers of cultured non-muscle cells, *Exp. Cell. Res.,* 117, 253, 1978.

26. **Sanger, J. W., Sanger, J. M., and Jockusch, B. M.,** Differences in the stress fibers between fibroblasts and epithelial cells, *J. Cell Biol.,* 96, 961, 1983.

27. **Lazarides, E.,** Actin, α-actinin, and tropomyosin interaction in the structural organization of actin filaments in non-muscle cells, *J. Cell Biol.,* 68, 202, 1976.

28. **Weber, K. and Groeschel-Stewart, U.,** Antibody to myosin: the specific visualization of myosin containing filaments in non-muscle cells, *Proc. Natl. Acad. Sci. U.S.A.,* 71, 4561, 1974.

29. **Kreis, T. E. and Birchmeier, W.,** Stress fiber sarcomeres of fibroblasts are contractile, *Cell,* 22, 555, 1980.

30. **Burridge, K.,** Are stress fibers contractile?, *Nature* London, 294, 691, 1981.

31. **Harris, A. K., Stopak, D., and Wild, P.,** Fibroblast traction as a mechanism for collagen morphogenesis, *Nature (London),* 290, 249, 1981.

32. **Abercrombie, M. J., Heaysman, E. M., and Pegrum, S. M.,** The locomotion of fibroblasts in culture. IV. Electron microscopy of the lamella, *Exp. Cell Res.,* 67, 359, 1971.

33. **Singer, I. I.,** Association of fibronectin and vinculin with focal contacts and stress fibers in stationary hamster fibroblasts, *J. Cell Biol.,* 92, 398, 1982.

34. **Chen, W.-T. and Singer, S. J.,** Immunoelectron microscopic studies of the sites of cell substratum and cell-cell contacts in cultured fibroblasts, *J. Cell Biol.,* 95, 205, 1982.

35. **Bradley, R. A., Coachman, J. R., and Rees, D. A.,** Comparison of the cell cytoskeleton in migratory and stationary chick fibroblast, *J. Muscle Res. Cell Motil.,* 1, 5, 1980.

36. **Herman, I. M., Crisona, N. J., and Pollard, T. D.,** Relation between cell activity and the distribution of cytoplasmic actin and myosin, *J. Cell Biol.,* 90, 84, 1981.

37. **Gotlieb, A. I. and Spector, W.,** Migration into an in vitro experimental wound: a comparison of porcine aortic endothelial and smooth muscle cells and the effect of culture irradiation, *Am. J. Pathol.,* 103, 271, 1981.

38. **Gotlieb, A. I., Spector, W., Wong, M. K. K., and Lacey, C.,** In vitro reendothelialization: microfilament bundle reorganization in migrating porcine endothelial cells, *Arteriosclerosis,* 4, 91, 1984.

39. **Albrecht-Buehler, G.,** Phagokinetic tracks of 3T3 cells: parallels between the orientation of track segments and of cellular structures which contain actin or tubulin, *Cell,* 12, 333, 1977.

40. **Gordon, S. R., Essner, E., and Rothstein, H.,** In situ demonstration of actin in normal and injured ocular tissues using 7-nitrobenz-2-oxa-1,3-diazole phallacidin, *Cell Motility,* 4, 343, 1982.

41. **Yohro, T. and Burnstock, G.,** Filament bundles and contractility of endothelial cells in coronary arteries, *Cell Tissue Res.,* 138, 85, 1973.

42. **Gabbiani, G., Elmer, G., Geulpa, C., Vallotton, M. B., Badonnel, M. C., and Huttner, I.,** Morphologic and functional changes of the aortic intima during experimental hypertension, *Am. J. Pathol.,* 96, 339, 1979.

43. **Gabbiani, G., Badonnel, M. C., and Rona, G.,** Cytoplasmic contractile apparatus in aortic endothelial cells of hypertensive rats, *Lab. Invest.* 32, 227, 1975.

44. **Drenckhahn, S.,** Cell motility and cytoplasmic filaments in vascular endothelium, *Prog. Appl. Microcirc.,* 1, 53, 1983.

45. **Wong, A. J., Pollard, T. D., and Herman, I. M.,** Actin filament stress fibers in vascular endothelial cells in vivo, *Science,* 219, 867, 1983.

46. **White, G. E., Gimbrone, M. A., and Fujiwara, K.,** Factors influencing the expression of stress fibers in vascular endothelial cells in situ, *J. Cell Biol.,* 97, 416, 1983.

47. **Rogers, K. A. and Kalnins, V. I.,** A method for examining the endothelial cytoskeleton in situ using immunofluorescence, *J. Histochem. Cytochem.,* 31, 1317, 1983.

48. **Gabbiani, G., Gabbiani, F., Lombardi, D., and Schwartz, S. M.,** Organization of actin cytoskeleton in normal and regenerating arterial endothelial cells, *Proc. Natl. Acad. Sci. U.S.A.,* 80, 2361, 1983.

49. **Strauss, B. I., Langille, B. L., and Gotlieb, A. I.,** In situ localization of F-actin microfilaments in the vasculature of the porcine retina, *Exp. Eye Res.,* in press.

50. **Barak, L. S., Yocum, R. R., Nothnagel, E. A., and Webb, W. W.,** Fluorescence staining of the actin cytoskeleton in living cells with 7-nitrobenz-2-oxa-1,3-diazole-phallicidin, *Proc. Natl. Acad. Sci. U.S.A.,* 77, 980, 1980.

51. **Faulstich, H., Trischmann, H., and Mayer, D.,** Preparation of tetramethylrhodaminyl-phalloidin and uptake of the toxin into short term cultured hepatocytes by endocytosis, *Exp. Cell Res.,* 144, 73, 1983.

52. **Wong, M. K. K. and Gotlieb, A. I.,** Endothelial cell monolayer integrity. I. Characterization of the dense peripheral band of microfilaments, *Arteriosclerosis,* 6, 212, 1986.

53. **Porter, K. R.,** Cytoplasmic microtubules and their functions, in *Principles of Biomolecular Organizaton,* Wolstenholme, G. E. W. and O'Connor, M., Ed., Little, Brown, Boston, 1966, 308.

54. **Brinkley, B. R.,** Microtubule organizing centers, *Annu. Rev. Cell Biol.,* 1, 145, 1985.

55. **Raff, E. D.,** The control of microtubule assembly in vivo, *Int. Rev. Cytol.,* 59, 1, 1979.

56. **McIntosh, J. R.,** The centrosome as an organizer of the cytoskeleton, in *Spatial Organization of Eukaryotic Cells,* Vol. 2, McIntosh, J. R., Ed., Alan R. Liss, New York, 1983, 115.

57. **Roger, K. A., McKee, N. H., and Kalnins, V. I.,** The preferential orientation of centrioles towards the heart in endothelial cells of major blood vessels is reestablished following reversal of a segment, *Proc. Natl. Acad. Sci. U.S.A.,* 82, 3272, 1985.

58. **Franke, W. W., Schmid, E., Osborn, M., and Weber, K.,** Intermediate-sized filaments of human endothelial cells, *J. Cell Biol.,* 81, 570, 1979.

59. **Bose, S. H. and Meltzer, D. I.,** Visualization of the 10-nm filament vimentin rings in vascular endothelial cells in situ: close resemblance to vimentin cytoskeletons found in monolayers in vitro, *Exp. Cell Res.,* 135, 299, 1981.

60. **Lin, J. J. C. and Feramisco, J. R.,** Disruption of the in vivo distribution of the intermediate filaments in fibroblasts through the microinjection of a specific monoclonal antibody, *Cell,* 24, 185, 1981.

61. **Gawlitta, W., Osborn, M., and Weber, K.,** Coiling of intermediate filaments induced by microinjection of a vimentin specific antibody does not interfere with locomotion and mitosis, *Eur. J. Cell Biol.,* 26, 83, 1981.

62. **Klymkowsky, M. W., Miller, R. H., and Lane, E. B.,** Morphology, behaviour and interaction of cultured epithelial cells after the antibody-induced disruption of keratin filament organization, *J. Cell Biol.,* 96, 494, 1983.

63. **Kocher, O., Skalli, O., Cerutti, D., Gabbiani, F., and Gabbiani, G.,** Cytoskeletal features of rat aortic cells during development, *Circ. Res.,* 56, 829, 1985.

64. **Rogers, K. A. and Kalnins, V. I.,** Comparison of the cytoskeleton in aortic endothelial cells in situ and in vitro, *Lab. Invest.,* 49, 650, 1983.

65. **Wong, M. K. K. and Gotlieb, A. I.,** Control of reendothelialization: the importance of endothelial microfilaments, microtubules and centrosomes in endothelial locomotion, *Surv. Synth. Pathol. Res.,* 4, 341, 1985.

66. **White, G. E. and Fujiwara, K.,** Expression and intracellular distribution of stress fibers in aortic endothelium, *J. Cell Biol.,* 103, 63, 1986.

67. **Langille, B. L. and Adamson, S. L.,** Relationship between blood flow direction and endothelial cell orientation at arterial branch sites in rabbits and mice, *Circ. Res.,* 48, 481, 1981.

68. **Eskin, S. G., Ives, C. L., McIntire, L. V., and Navarro, L.,** Response of cultured endothelial cells to steady flow, *Microvasc. Res.,* 28, 87, 1984.

69. **Langille, B. L., Reidy, M. A., and Klien, R. L.,** Injury and repair of endothelium at sites of flow disturbances near abdominal aortic coarctations in rabbits, *Arteriosclerosis,* 6, 146, 1986.

70. **Franke, R. P., Grafe, M., Schnittle, H., Seiffge, D., Mittermayer, C., and Drenckhahn, D.,** Induction of human vascular endothelial stress fibers by fluid shear stress, *Nature (London),* 307, 648, 1984.

71. **Dewey, C. F., Bussolari, S. R., Gimbrone, M. A., and Davies, P. F.,** The dynamic response of vascular endothelial cells to fluid shear stress, *J. Biomech. Eng.,* 103, 177, 1981.

72. **Madri, J. A. and Stenn, K. S.,** Aortic endothelial cell migration. I. Matrix requirements and composition, *Am. J. Pathol.,* 106, 180, 1982.

73. **Pratt, B. M., Harris, A. S., Morrow, J. S., and Madri, J. A.,** Mechanisms of cytoskeletal regulation: modulation of aortic endothelial cell spectrin by the extracellular matrix, *Am. J. Pathol.,* 117, 349, 1984.

74. **Young, W. C. and Herman, I. M.,** Extracellular matrix modulation of endothelial cell shape and motility following injury in vitro, *J. Cell Sci.,* 73, 19, 1985.

75. **Shasby, M.D., Shasby, S. S., Sullivan, J. M., and Peach, M. J.,** Role of endothelial cell cytoskeleton in control of endothelial permeability, *Circ. Res.,* 51, 657, 1982.

76. **Shasby, M. D., Lind, S. E., Shasby, S. S., Goldsmith, J. C., and Hunninghake, G. W.,** Reversible oxidant-induced increases in albumin transfer across cultured endothelium: alterations in cell shape and calcium homeostasis, *Blood,* 65, 605, 1985.

77. **Shasby, D. M. and Shasby, S. S.,** Effects of calcium on transendothelial albumin transfer and electrical resistance, *J. Appl. Physiol.,* 60, 71, 1986.

78. **Wysolmerski, R. and Lagunoff, D.,** The effect of ethchlorvynol on cultured endothelial cells. A model for the study of the mechanism of increased vascular permeability, *Am. J. Pathol.,* 119, 505, 1985.

79. **Wysolmerski, R. and Lagunoff, D.,** Pulmonary edema in adult respiratory distress syndrome, *Surv. Synth. Pathol. Res.,* 4, 257, 1985.

80. **Galdal, K. S., Evensen, S. A., and Brosstad, F.,** Effects of thrombin on the integrity of monolayers of cultured human endothelial cells, *Thromb. Res.,* 27, 575, 1982.

81. **Laposata, M., Dovnarsky, D. K., and Shin, H. S.,** Thrombin-induced gap formation in confluent endothelial cell monolayers in vitro, *Blood,* 62, 549, 1983.

82. **Fox, P. L., and DiCorleto, P. E.,** Regulation of a production of a platelet derived growth factor-like protein by cultured bovine aortic endothelial cells, *J. Cell Physiol.,* 121, 298, 1984.

83. **Wong, M. K. K. and Gotlieb, A. I.,** The mechanism of thrombin-induced disruption of the confluent endothelial monolayer, *Circulation,* 73, 35, 1985.

84. **Stolpen, A. H., Guinan, E. C., Fiers, W., and Pober, J. S.,** Recombinant tumor necrosis factor and immune interferon act singly and in combination to reorganize human vascular endothelial cell monolayers, *Am. J. Pathol.,* 123, 16, 1986.

85. **Majno, G. and Leventhal, M.,** Pathogenesis of histamine-type vascular leakage, *Lancet,* 2, 99, 1967.

86. **Majno, G., Shea, S. M., and Leventhal, M.,** Endothelial contraction induced by histamine-type mediators: an electron microscopic study, *J. Cell Biol.,* 42, 647, 1969.

87. **Welles, S. L., Shepro, D., and Hechtman, H. B.,** Vasoactive amines modulate actin cables (stress fibers) and surface area in cultured bovine endothelium, *J. Cell Physiol.,* 123, 337, 1985.

88. **Willingham, M. C., Yamada, S. S., Bechtel, P. J., Rutherford, A. V., and Pastan, I.,** Ultrastructural immunochemical localization of myosin in cultured fibroblastic cells, 29, 1289, 1981.

89. **Zigmond, S. H., Otto, J. J., and Bryan, J.,** Organization of myosin in a submembraneous sheath in well spread human fibroblasts, *Exp. Cell Res.,* 119, 205, 1979.

90. **Mooseker, M. S., Bonder, E. M., Conzelman, K. A., Fishkind, D. A., Howe, C. L., and Keller, T. C. S.,** Brush border cytoskeleton and integration of cellular function, *J. Cell Biol.,* 99, 104s, 1984.

91. **Opas, M., Turksen, K., and Kalnins, V. I.,** Adhesiveness and distribution of vinculin and spectrin in retinal pigmented epithelial cells during growth and differentiation in vitro, *Dev. Biol.,* 107, 269, 1985.

92. **Opas, M. and Kalnins, V. I.,** Spatial distribution of cortical proteins in cells of epithelial sheets, *Cell Tissue Res.,* 239, 451, 1985.

93. **Martinez-Palomo, A., Meza, I., Beaty, G., and Cereijido, M.,** Experimental modulation of occluding junctions in a cultured transporting epithelium, *J. Cell Biol.,* 87, 736, 1980.

94. **Meza, I., Ibarra, G., Sabanero, M., Martinez-Polono, A., and Cereijido, M.,** Occluding junctions and cytoskeletal components in a cultured transporting epithelium, *J. Cell Biol.,* 87, 746, 1980.

95. **Geiger, B., Avnur, Z., Kreis, T. E., and Schlessinger, J.,** The dynamics of cytoskeletal organization in areas of cell contact, in *Cell and Muscle Motility,* Vol. 5, Shay, J., Ed., Plenum Press, New York, 1984, 195.

96. **Geiger, B., Schmid, E., and Franke, W. W.,** Spatial distribution of proteins specific for desmosomes and adhaerens junctions in epithelial cells demonstrated by double immunofluorescence microscopy, *Differentiation,* 23, 189, 1983.

97. **Drenckham, K. and Henning, F.,** Identification of actin, α-actinin, and vinculin-containing plaques at the lateral membrane of epithelial cells, *J. Cell Biol.,* 102, 1843, 1986.

98. **Herman, B. and Pledger, W. J.,** Platelet-derived growth factor induced alterations in vinculin and actin distribution in BALB/c-3T3 cells, *J. Cell Biol.,* 100, 1031, 1985.

99. **Herman, B., Harrington, M. A., Olashaw, N. E., and Pledger, W. J.,** Identification of the cellular mechanisms responsible for platelet-derived growth factor induced alterations in cytoplasmic vinculin distribution, *J. Cell Physiol.,* 126, 115, 1986.

100. **Honda, H.,** Geometrical models for cells in tissue, *Int. Rev. Cytol.,* 81, 191, 1983.

101. **Glacomelli, F., Weiner, J., and Spiro, D.,** Cross-striated arrays of filaments in endothelium, *J. Cell Biol.,* 45, 188, 1970.

102. **Byers, H. R., White, G. E., and Fujiwara, K.,** Organization and function of stress fibers in cells in vitro and in situ, in *Cell and Muscle Motility,* Vol. 5, Shay, J., Ed., Plenum Press, New York, 1984, 83.

103. **Groves, H. M., Kinlough-Rathbone, R. L., Richardson, M., Jorgensen, L., Moore, S., and Mustard, J. F.,** Thrombin generation and fibrin formation following injury to rabbit neointima: studies of vessel wall reactivity and platelet survival, *Lab. Invest.,* 64, 605, 1982.

104. **Clowes, A. W., Clowes, M. M., and Reidy, M. A.,** Kinetics of cellular proliferation after arterial injury. III. Endothelial and smooth muscle cell growth in chronically denuded vessels, *Lab. Invest.,* 54, 295, 1986.

105. **Wong, M. K. K. and Gotlieb, A. I.,** In vitro reendothelialization of a single cell wound: role of microfilament bundles in rapid lamellipodia mediated wound closure, *Lab. Invest.,* 51, 75, 1984.

106. **Sholley, M. M., Gimbrone, M. A., and Cotran, R. S.,** Cellular migration and replication in endothelial regeneration: a study using irradiated endothelial cultures, *Lab. Invest.,* 36, 18, 1977.

107. **Gotlieb, A. I., McBurnie-May, I., Subrahmanyan, L., and Kalnins, V. I.,** Distribution of microtubule organizing centers in migrating sheets of endothelial cells, *J. Cell Biol.,* 91, 589, 1981.

108. **Gotlieb, A. I., Subrahmanyan, L., and Kalnins, V. I.,** Microtubule organizing centers and cell migration. Effect of inhibition of migration and microtubule disruption in endothelial culture, *J. Cell Biol.,* 96, 1266, 1983.

109. **Rogers, K. M., Boden, P., Kalnins, V. I., and Gotlieb, A. I.,** The distribution of centrosomes in endothelial cells of non-wounded and wounded aortic organ cultures, *Cell Tissue Res.,* 243, 223, 1986.

110. **Rogers, K. M. and Kalnins, V. I.,** personal communication.

111. **Mascardo, R. N. and Sherline, P.,** Insulin and multiplication-stimulating activity induce a very rapid response to wounding in endothelial cell monolayers, *Diabetes,* 33, 1099, 1984.

112. **Malech, H. L., Root, R. K., and Gallin, J. I.,** Structural analysis of human neutrophil migration, *J. Cell Biol.,* 75, 666, 1977.

113. **Albrecht-Buehler, G. and Bushnell, A.,** The orientation of centrioles in migrating 3T3 cells, *Exp. Cell Res.,* 120, 111, 1979.

114. **Kupfer, A., Louvard, D., and Singer, S. J.,** Polarization of the Golgi apparatus and the microtubule-organizing center in cultured fibroblast at the edge of an experimental wound, *Proc. Natl. Acad. Sci. U.S.A.,* 79, 2603, 1982.

115. **Kupfer, A., Dennert, G., and Singer, S. J.,** Polarization of the Golgi apparatus and the microtubule-organizing center within cloned natural killer cells bound to their targets, *Proc. Natl. Acad. Sci. U.S.A.,* 80, 7224, 1983.

116. **Koonce, M. P., Cloney, R. A., and Berns, M. W.,** Laser irradiation of centrosomes in newt eosinophils: evidence of centriole role in motility, *J. Cell Biol.,* 98, 1990, 1984.

117. **Nemere, I., Kupfer, A., and Singer, S. J.,** Reorientation of the Golgi apparatus and the microtubule-organizing center inside macrophages subjected to a chemotatic gradient, *Cell Motility,* 5, 17, 1985.

118. **Albrecht-Buehler, G.,** Does the geometric design of centrioles imply their function?, *Cell Motility,* 1, 237, 1981.

119. **Larson, D. M. and Sheridan, J. D.,** Intercellular junctions and transfer of small molecules in primary vascular endothelial cultures, *J. Cell Biol.,* 92, 183, 1982.

120. **Euteneuer, U. and Schliwa, M.,** Evidence for an involvement of actin in the positioning and motility of centrosomes, *J. Cell Biol.,* 101, 96, 1985.

121. **Akkor, D. and Gotlieb, A. I.,** The response of single endothelial cells to focal detachment from the substratum, *Fed. Proc.,* 99, 184a, 1984.

122. **Schwartz, S. M., Gajdusek, C. M., and Selden, S. C., III,** Vascular wall growth control: the role of the endothelium, *Arteriosclerosis,* 1, 107, 1981.

123. **Reidy, M. A. and Schwartz, S. M.,** Endothelial regeneration. III. Time course of intimal changes after small defined injury to rat aortic endothelium, *Lab. Invest.,* 43, 233, 1982.

124. **Ramsay, M. M., Walker, L. N., and Bowyer, D. E.,** Narrow superficial injury to rabbit aortic endothelium, *Atherosclerosis,* 43, 233, 1982.

125. **Prescott, M. F. and Muller, K. R.,** Endothelial regeneration in hypertensive and genetically hypercholesterolemic rats, *Arteriosclerosis,* 3, 206, 1983.

Chapter 19

CELLULAR ORGANIZATION OF BLOOD VESSELS IN DEVELOPMENT AND DISEASE

Ronald L. Heimark and Stephen M. Schwartz

TABLE OF CONTENTS

I. INTRODUCTION

The organizational complexity of the vascular tree contrasts with the relative simplicity of the cellular composition of any single blood vessel. Unlike many organs, primary vessels are made up of only two cell types, the smooth muscle cell and the endothelial cell. In theory the problems of vessel formation are limited to understanding how the two cells relate to themselves and each other during vessel growth. We will consider our current knowledge of the role of such interactions in cellular organization and growth control, including interactions between endothelial cells, between smooth muscle cells, and those involving both cell types.

II. GROWTH CONTROL MECHANISMS

We should begin by defining some terms. While "growth" is a common term, it is often poorly defined. At least three distinct processes are called "growth": increase of cell mass, increase of cell number, and formation of new organized tissue structures. The best example of the first and last process may be the effects of one growth factor, nerve growth factor (NGF). NGF is well recognized as a stimulant for nerve cell hypertrophy and development but has no mitogenic effect.[1,2]

Examples of "pure" increase in cell mass are hard to come by. For example, the increase in mass of cardiac or smooth muscle associated with hypertension is complicated by the observation that part of the increase is accompanied by an increase in cell ploidy.[3] Cells can replicate their genomes without increasing the number of cells. Finally, growth of an organ includes phenomena like angiogenesis where not only new cells but new structures are formed. In this review, we will simplify the problem by restricting the discussion to the control of cell replication with an emphasis on those processes which might control formation of new structures, i.e., "morphogenic changes".

Three general mechanisms have been offered for control of cell replication. Examples of this form of cell "growth", i.e., cell replication, are more extensive. The bulk of the work has centered on growth stimulation, usually by polypeptide mitogens. An alternative point of view linking cell growth to the relationships of cells to one another has grown up from studies of mechanisms of growth inhibition. For example, Holley and colleagues[4] and Yen and Pardee[5] have maintained that cells are kept in a nonreplicating, resting state (G_O) by the available supply of nutrients. Interstitial fluid and lymph are likely to contain concentrations of nutrients quite comparable to those seen in medium. Despite this, cells do not normally replicate in vivo. This fact lends importance to the second class of mechanisms — control by specific growth-stimulating or -inhibiting factors.[6] Finally, a number of investigators have attempted to define a mechanism for inhibition of growth by intercellular interactions.

The idea that growth is limited by intercellular contact is usually called contact inhibition,[7] topoinhibition, or density-dependent inhibition of growth.[8] The major evidence in favor of this idea is from studies showing inhibition of cell replication by an established line of density-inhibited mouse fibroblasts, 3T3 cells.[9] For these cells, saturation density depends on serum concentration.[10] It does not depend on viscosity of the medium, implying that the density effect is not simply due to the limited rates of diffusion of nutrients or waste products in the medium.[11] Medium unable to support cells at a high density can support cells at a lower density, and cells will regenerate a high-density culture following an in vitro wound.[8,12] These studies imply a role for cellular interaction.

An alternate possibility is that growth is controlled by a mutual inhibitory activity of secretory products of one cell acting on other cells. For example, the group of interferons are the best characterized of the factors which "inhibit" growth.[13] They act on certain cell types, however, while maintaining a high proportion of the population in the S-phase of the cell cycle. In the case of terminally differentiating hematopoietic cells, an endogenous release of IFN-β functions as an autocrine growth inhibitor with a specific G_0/G_1 arrest.[14] Another endogenously produced growth regulatory protein is transforming growth factor-β (TGF-β).[15] TGF-β is a potent stimulator of growth of cells in soft agar,[16-17] but causes inhibition of certain lung and mammary tumor cell lines.[18] In addition, EGF/insulin-stimulated DNA synthesis is inhibited at 24 hr and a delayed response is observed at 36 hr.[19] A recent study has shown that TGF-β is similar if not identical with a growth inhibitor isolated from media from BSC-1 cells.[20] These results indicate that TGF-β acts as a bifunctional regulator of cell growth. Endothelial regeneration also shows a delayed response with addition of TGF-β, temporarily blocking the cells in the G_1 phase of the cell cycle.[21-23]

III. GROWTH REQUIREMENTS FOR CELL CULTURE

Most of our concepts of growth control are based on studies of the nutritional requirements necessary to establish cells in culture. These requirements are obviously different from those required for normal growth in vivo. To illustrate the point, if the volume of a single cell is about 100 μm^3, then by simple arithmetic only about 60 doublings are required to produce the 60 to 70 kg of mass in an adult human. The 20 or so doublings required to establish a clone in culture probably greatly exceeds the number of doublings required for most biological responses in an adult animal, except those in which cell death plays a major role. Thus, we need to be concerned that the conditions used to obtain a normally quiescent cell in culture may be quite different from the factors controlling growth in vivo.

In addition to defining cell growth requirements, the ability to propagate cells in culture led investigators to examine cellular products released by cells into the medium. Although some products are simply cellular waste, contributed in part by dying cells in the population, other cell products have been found to have specific functions. These extracellular, soluble components generally may be categorized as either nonspecific nutritional,[24] attachment,[25] progression,[26] and spreading[25] factors, or specific polypeptides, the growth factors. While most established and normal cells produce factors of the first category, only a select variety of cell types have been shown to produce growth factors. The endothelial-cell-derived growth factor,[27,28] macrophage-[29] and monocyte-derived growth factors (MDGF),[30] smooth muscle cell-derived growth factors,[31-33] the heparin-binding growth factors,[34-38] and the somatomedins[39] are pertinent to the discussion of the vessel wall.

IV. VASCULAR EMBRYOLOGY AND ANGIOGENESIS

Before uniting these specific mechanisms, we need to review how blood vessels are formed. Most blood vessels originate in the yolk sac of the embryo at multiple foci termed "blood islands".[40,41] These consist of endothelium lining cell masses filled with primitive hematopoetic cells. Apparently, at later stages, smooth muscle cells are recruited from local mesenchyme as the invading capillaries form the vascular tree in the embryo body.[42] New vessels invade the primordia of various ecto- or endodermally derived organs[43,44] which have been used to purify angiogenic factors.[45,46] Even at these early stages there is an exquisite organization of the vessel into a contained blood space, lined by a single cell in a continuous cell sheet, the endothelium, and invested in a supporting thick sheath formed from mesenchymal cells specialized to the mechanical needs of blood vessels — the smooth muscle cell. The development of blood pressure in the endothelial tube correlates with differentiation of the surrounding cells into smooth muscle cells.[47] This vascular neogenesis needs to be contrasted with angiogenesis in later stages of development or during angiogenic responses to inflammation or neoplasia.[48] In all these cases, the existing vessels form "new" vessels by some process that releases the existing endothelial cell and smooth muscle cell from their layered structure and allows the endothelial tube to branch.[49] These branches, along with increases in length, constitute the "new" formation of vessels described as angiogenic.[50] The process of angiogenesis, therefore, requires extensive morphogenic changes in the vessel wall.

V. ENDOTHELIAL REPLICATION

By the time a mammal reaches adulthood, the continuous sheet of endothelium is both stable and very quiescent. While the cells are as thin as about 300 nm, no defects are seen by scanning electron microscopy, even in areas subject to high shear.[51] This is true despite the general absence of endothelial turnover and despite the presence of focal areas where turnover is quite high. In the adult aorta, as few as 1 cell in 1000 replicates each day, but focally turnover may be as high as 1 to 10 cells per 100 per day.[52]

Against this background considerable attention has been paid to endothelial denudation as an initiating factor in the ontogeny of the atherosclerotic lesion.[53] Apparently lesions can begin, at least in experimental animals, without loss of endothelial continuity.[54,55] It is, however, evident that loss of endothelial continuity occurs early in the progression of atherosclerotic lesions. Presumably, this is an important step in lesion progression both because of the release of growth factors into the wall and because a denuded site is likely to serve as the nidus for thrombosis.

This occurrence of denudation is somewhat surprising since the endothelium heals artificial injuries quite rapidly.[56] Endothelial cell turnover in adult animals is as low as 0.1%/day. When the cell layer is wounded, endothelial cells respond to the wound with close to 100% replication. Wounds as wide as ten endothelial cells are small enough to close in about 6 to 8 hr. Wounds of this size do not heal just by endothelial proliferation. A minimum of 10 hr of movement is required before the endothelial cell will replicate. The fact that this stimulation of growth occurs in plasma argues that the critical control is located in the cell layer.

The best evidence available indicates that similar processes occur as part of a normal cycle of cell turnover. Even in the quiescent endothelium, the normal cell layer contains some cells that are dead.[57] Immunocytochemical studies show that some endothelium cells contain autogenous IgG bound to cytoplasmic filaments. Since these filaments are free in the cytoplasm, the membrane of these cells must be sufficiently disrupted to allow access of plasma proteins. Studies in other systems have shown that this degree of loss of membrane integrity is the "point of no return" leading to cell death in several experimental systems. Using cultured cells, we have been able to relate this appearance to other more traditional indicators of cell death, and have been able to follow the process of necrosis in vitro.[57] The neighboring cells undermine the disintegrating cells resulting in maintainance of the cell layer.[57] This mechanism for maintainance of continuity also seems to operate in vivo. Endotoxin causes a large increase in endothelial cell death. The increase in cell death is compensated for by an increase in endothelial cell replication and no discontinuity is seen.[59] Just as in the in vitro studies, we observed what appeared to be detaching endothelial cells being undermined by their living neighbors.

In summary, it now appears that the endothelium maintains its continuity by a continual process of replacement of lost cells. Of the various mechanisms proposed to control cell growth, it is easiest to believe that the stimulation of replication involved in healing a wound or replacing a dying cell would be controlled by localized cell-cell interaction and provide a means for replacement of cells by the undermining mechanism. It is difficult to imagine that this sort of growth, occurring in an environment as large as the entire circulation, would require release of a soluble growth factor.

More direct evidence supports the concept of contact-dependent inhibition of growth and movement in the endothelium. Particularly interesting is the fact that endothelial cell growth in vitro seems independent of concentration of mitogens. While a number of mitogens have been reported to be useful in stimulating growth of endothelial cells from sparse density, the final cell density in culture seems to be constant.[60,61] It is possible, even at postconfluent densities, to reinitiate growth in a monolayer by wounding it or by other techniques that cause disruption of monolayer continuity. Colchicine causes endothelial cells to retract from one another. The stimulated cells will go far enough into the cycle to synthesize DNA, although mitosis itself cannot occur. The kinetics of DNA synthesis for this sort of injury are very similar to the kinetics following a wound.[62] Other agents which stimulate endothelial replication in confluent monolayers also cause a separation of one cell from another. It is important to note that new vessel formation — that is, angiogenesis — can be stimulated by substances that are chemotactic but not necessarily mitogenic for endothelium.[63,64] Indeed substantial amounts of angiogenesis can occur in the absence of cell division.[63,65] It is conceivable that some ''mitogens'' act by causing endothelial cells to move apart from one another, releasing neighboring cells from contact inhibition.

There is one other phenomenon that might be explained by contact inhibition of endothelial growth and movement. Regeneration of the endothelium in vessels that have been mechanically denuded with a catheter is limited to a few millimeters. Areas not covered by regenerated endothelium are covered by a pseudoendothelium comprised of modified smooth muscle cells. This organization persists from 6 weeks to 6 months after denudation.[56] Autoradiography after labeling with ^{3}H-thymidine shows that the endothelial cells have stopped replicating even though extensive areas lack endothelium. An attractive possibility is that interactions between regenerating endothelial cells and the adjacent modified smooth muscle cells cause the endothelial cells to cease replicating (Figure 1A). While we have no direct evidence for this, it is intriguing to note the high level of organization of the endothelial cell cytoskeleton at the interface between the two cell types.[67] Analysis of the actin distribution in the luminal cells by indirect immunofluorescence is shown in Figure 1. The luminal smooth muscle cells (Figure 1B) show a random orientation of actin filaments. Endothelial cells (Figure 1C) show actin cables aligned with blood flow. At the interface where endothelial cells contact smooth muscle cells there is a row of endothelial cells organized perpendicular to blood flow (Figure 1D). Apparently the interface cells are unable to migrate further and prevent migration and regeneration of the cell layer. This behavior is reminiscent of the organization of the vessels formed during early embryogeneis. The endothelium migrates as tubes emigrating from the blood islands in the yolk sac. Once these channels are established, they are surrounded by migrating smooth muscle cells recruited from the mesenchyme. Some mechanism allows these mesenchymal cells to organize alongside each other but not invade the structure each is forming.

Alternative explanations can be constructed until the concepts of cell-cell interaction progress to a molecular level. This means asking what molecules might control the interactions between cells to inhibit endothelial cell growth and movement in the continuous endothelial monolayer, prevent interposing of the two cell types, and allow the smooth muscle cell to become quiescent in a three-dimensional structure. For the endothelium, two mechanisms might be considered: direct growth control by molecules located in the membrane and indirect control of growth by molecules responsible for determining the spatial localization of cells relative to each other.

Growth control, a relatively unexplored area, may be mediated by interactions with the substratum. One might imagine that a system able to maintain cells in a mechanically stable monolayer might, by fact of stability alone, allow the cells to achieve quiescent growth state. For example, agents capable of causing detachment of cells from the endothelium would be expected to be mitogenic. Human umbilical vein endothelial (HUVE) cells and

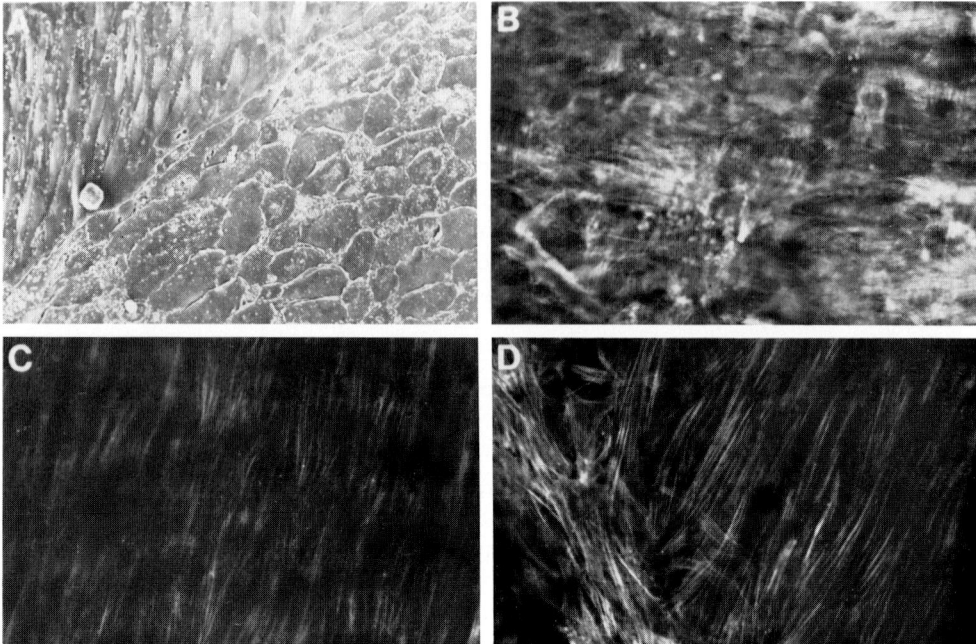

FIGURE 1.   The distribution of actin in cells on the lumen of a rabbit thoracic aorta. Immunofluorescent staining
will be carried out with antiactin antibodies.[67] (A) Scanning electron micrograph 15 days after ballooning of vessel.
The regenerating endothelium is in the top left corner. (B) Smooth muscle cells in the denuded area with thick
stress fibers. (C) Normal endothelial cells with many stress fibers visible and aligned with the flow of blood (top
to bottom). (D) Fifteen days after balloon injury the endothelial cells contain relatively thick stress fibers oriented
with blood flow. However, at the regenerating edge their stress fibers are oriented diagonally or even perpendicularly
to blood flow.

bovine aortic endothelial (BAE) cells synthesize platelet GpIIb/IIIa-like proteins.[68,69] This
is of interest because GpIIb/IIIa on platelets functions in adhesive interactions with fibrinogen
and fibronectin.[70-72] This interested us because of claims that localization of fibronectin in
the endothelial substrate was critical to the maintenance of a normal monolayer,[73] and
suggestions by others that fibrin, binding to some cell-surface receptor, could dramatically
alter endothelial cell shape. We found that the IgG fraction of the rabbit anti-human platelet
GpIIIa and its Fab′ fragments did detach confluent and subconfluent HUVE cells.[74] The
effect was nontoxic. This result certainly suggests that GpIIb/IIIa plays a role in endothelial
morphogenesis. Two other observations, however, make interpretation of the data more
complex. First, we found that anti-GpIIIa IgG does not inhibit cell attachment to extracellular
matrix or purified fibronectin despite the presence of intact GpIIb/IIIa on endothelial cells
detached under various conditions. Thus, unlike the fibronectin receptor,[75,76] it appears that
GpIIIa is not simply an attachment molecule. Second, the time course of detachment is quite
slow, requiring as long as 12 to 24 hr after exposure to the antibody. Since the antibody is
equally effective when washed away after $^1/_2$ hr of incubation or when given to sparsely
plated cells, we conclude that the lag period is not simply secondary to a diffusion artifact,
and the antibodies GpIIb/IIIa must act on cell shape by an indirect mechanism rather than
directly acting at an adhesion site. The possible role of these types of interactions in growth
control remains unexplored. Two published pieces of data, however, suggest that this is a
worthwhile direction. First, we noted that cell replication is inhibited by inhibitors of cell
movement.[12] One might imagine that extraordinarily stable cell-substrate interactions would
inhibit replication, alternatively the absence of an adjacent surface able to support endothelial
cell adhesion might prevent cell replication. Endothelial cells regenerate more estensively

on surfaces coated with type I or III collagen compared to fibronectin.[77] This could explain the failure of endothelial cells to overgrow pseudoendothelial cells. Since the latter, like endothelial cells, lack a thrombogenic surface, one might imagine that nonthrombogenicity correlates with an inability to support overgrowth of one neighbor on the other.[78]

A different approach has been used to look for changes in the cell surface as a function of the extent of cell contact. There is evidence that expression of membrane proteins can be controlled by cell-cell interaction. Vlodavsky et al.[79] showed this using a lactoperoxidase-catalyzed iodination system. A problem with these studies is that their data might represent a difference in the extent of exposure of the upper and lower surfaces of the cells in sparse vs. confluent, continuous cell layers. To eliminate this problem, we applied a technique called ''restrictive iodination''.[80] Using this approach, we repeated the study of Vlodavsky and co-workers.[79] They suggested that a 60-kdalton cell surface protein, CSP-60, was contact dependent — that is, its expression depended on cell contact. When endothelial cells are plated at confluent density, there is a very rapid reappearance of CSP-60 on the cell surface. In contrast, when cells are plated at sparse density, this protein does not reappear for several days. It begins to reappear at a time point in which we see a decreased thymidine index in the culture.[81]

The fact that expression of CSP-60 is modulated by cell density does not tell us anything about its possible role in growth control. It seems unlikely that a protein required to signal each cell of a neighbor's presence would only be expressed on the surface after cell contact is made. The modulation of CSP-60 expression on the cell surface, however, does emphasize the apparent fact that some form of cell-cell interaction is able to dramatically alter endothelial cell organization.

To go further we need direct evidence that a membrane molecule can inhibit the two characteristics of endothelial cell growth, cell movement and cell replication. There is evidence that such molecules are involved in contact inhibition of other cell types in culture. Whittenberger and Glaser[82] studying growth control inhibition in 3T3 cells found that a surface membrane preparation from 3T3 cells was able to mimic the effects of contact inhibition. That is, addition of the cell surface faction was capable of inhibiting replication near 50%. In very similar studies, Natraj and Datta[83] showed that urea could extract a glycoprotein fraction from the membrane that was again capable of inhibiting 3T3 replication.

Since the biology of endothelium suggests that cell-cell interactions may regulate endothelial cell growth, we prepared a cell surface fraction from confluent endothelial cells by treatment with a low concentration of urea.[84] Addition of the cell surface fraction after exhaustive dialysis to subconfluent growing cells inhibits DNA synthesis. Growth was inhibited approximately 60% in a concentration-dependent manner. The inhibition was transient when added to growing cells, and after 48 hr the rate of growth has returned to that of the control. Both cell migration and replication are blocked by the cell surface membrane fraction in a wound edge assay. The activity was labile to proteases, heat treatment, and reduction, suggesting that it was a protein (EGIP, endothelial growth inhibitory protein). This was in contrast to the heparin-like inhibitor of smooth muscle cell growth described by Castellot and co-workers.[85] The EGIP fraction was enriched in enzyme markers for the plasma membrane and endoplasmic reticulum, in addition to the 60-kdalton density-dependent cell surface protein CSP-60. The inhibitor was solubilized by octyl β-glucoside, but it was not dissociated by treatment with 0.1 M carbonate, pH 11.5, which removed extrinsic membrane proteins, suggesting it is an intrinsic membrane protein. Recently, we have obtained the growth inhibitor using a postnuclear membrane fraction[86] from confluent cultured endothelial cells. Addition of enriched plasma membranes from endothelial cells, but not red blood cells, inhibits endothelial cell growth in a concentration-dependent manner. The endothelial growth inhibitory components could be extracted from the membrane fraction in an active form by the use of octyl β-glucoside. Cell surface glycoproteins have been implicated in the inhibition of growth in BHK and CHO cells[87] and in GH$_3$ cells.[88] In preliminary studies, the solubilized EGIP binds to wheat germ agglutinin-agarose. After extensive washing, the activity is eluted with 0.45 M N-acetyglucosamine. The inhibitor is apparently a cell surface glycoprotein.

In summary, the growth behavior of endothelium both in vivo and in vitro is consistent with contact inhibition as a fundamental control mechanism. In this view, various growth phenomena — that is, regeneration of a wound, angiogenesis, and even growth in culture — would depend on disruption of cell-cell contacts. Growth factors might be required to support this response or might themselves operate by "morphogenic" mechanisms involving breakdown of cell-cell interactions.

VI. SMOOTH MUSCLE REPLICATION

The approach to control of endothelial cell growth is very different from the usual view of smooth muscle growth. The current concept of smooth muscle cell growth is an outgrowth of the in vitro observation by Ross and co-workers that platelets release a peptide growth factor (platelet-derived growth factor, PDGF) able to stimulate smooth muscle growth.[89] They proposed that denuding injuries to the endothelium led to platelet thrombosis, release of PDGF, and smooth muscle proliferation. This was seen as the critical step in evolution of the atherosclerotic plaque. This hypothesis was supported by experiments by Stemerman et al.[90] and Baumgartner et al.,[91] demonstrating that mechanical abrasion of the endothelium could stimulate smooth muscle cell proliferation. Walker et al.[92] have also shown that as lesions in the fat-fed animal progress, there are focal areas where the endothelium has lost integrity. Gerrity[93] and Faggiotto et al.[94] have reported similar results. We begin with two critical ideas: (1) failure of endothelial-endothelial interactions might initiate lesions and (2) specific polypeptides can control smooth muscle cell growth.

We have explored these ideas in vivo by studying the cell kinetics of the response of the carotid artery to denudation.[95] DNA replication can be measured either by a single dose of tritiated thymidine if we are measuring those cells in S at one time, or by continuous labeling which measures the integral entry into S over the duration of labeling. Following abrasion with the balloon catheter, smooth muscle cells migrate across the internal elastic lamina and proliferate. The result is the now classical, intimal proliferative response already mentioned. In the carotid artery, however, this response is readily quantified because of the absence of vascular branches and the absence of smooth muscle in the normal intima. If one continuously labels the animal with tritiated thymidine so that every cell that enters the cell cycle becomes labeled, it turns out that almost every cell in the lesion does become labeled. A few cells, however, remain unlabeled. By working backward from the increase in cell number, figuring the average doubling time and counting labeled cells, we could determine that about 50% of the cells that migrate across the internal elastic lamina never label. The bulk of the lesion is formed by about three doublings of cells that do cross.

This suggests that the proliferative response is an amplification by division of a limited portion of the total cell population. It remains to be seen whether the replicative rate of each cell is random or genetically fixed. The second implication of these data is that once vascular smooth muscle cells begin to replicate, they continue to do so for some time, independently of the initial stimulus.[96]

These data are consistent with the possibility that the population of vessel wall smooth muscle cells, like a striated muscle population, includes postreplicative cells. This conclusion needs to be put in the context of other events that may control proliferation. One might think that platelets would continue to interact as long as the surface remains denuded. Instead, while the deendothelialized vessel wall remains denuded for months or even years, the platelet response only lasts about 4 to 24 hr. By 4 hr the level of interaction is greatly reduced and at 24 hr, it is barely measurable above the background.[97] Using continuous labeling with tritiated thymidine, we find that the number of cells which are going to enter the cell cycle is essentially complete by 3 to 7 days. All of the thickening following the balloon catheter is by continued replication of this group of cells that have initially become committed to replicate.

These models beg the central question of the initiating event. Is the transient platelet response sufficient to initiate replication? The answer is no. Using a very narrow catheter, Reidy and Silver[98] were able to restrict denudation to strips as narrow as ten cells wide or sufficiently wide as to require up to 7 days of regenerative response. As with the balloon procedure,[90,91] platelets adhere to the more selectively denuded swath. A few days later, regenerative endothelial cells can be seen. It is, perhaps, not surprising that there is no smooth muscle proliferation when wounds are closed within 2 to 3 days. There is, however, no evidence of smooth muscle proliferation even when as much as 7 days is required to regenerate the endothelium.[98]

At a minimum, this means that endothelial wounds are sufficient to cause smooth muscle proliferation. Furthermore, the fact that wounds as large as the 7-day wound do not stimulate proliferation implies that platelet adherence, while it could be necessary for the commitment of smooth muscle cells to replication, is not sufficient. Other changes caused by the balloon catheter must be important in the initiation of replication.

Another approach to the question of possible controls of smooth muscle replication is to consider the response of vessels to another form of injury. In hypertension, smooth muscle replication accounts for the bulk of the vascular wall thickening and, therefore, presumably for the characteristic increases in "structural" resistance as a common feature of most forms of chronic hypertension.[3,99-101] The stimulus for this form of replication is unknown. In small vessels it could be secondary to denudation since loss of endothelium is a prominent feature of hypertensive changes in the microvasculature.[102,103] Thus the balloon model may be particularly apt as a model for the proliferative changes seen in hypertensive microvessels. In large vessels, however, smooth muscle replication begins without denuding injury,[104] and it is not clear what mechanism is present to commit the smooth muscle cells to replicate.

It is important to remember that the sequence of events in smooth muscle cell proliferation in hypertension is probably quite different from the events seen in atherosclerosis. The characteristic pathology of small arteries with malignant hypertension includes deposition of plasma proteins in the wall, fibrin deposition, and marked proliferative responses of smooth muscle.[105] Although there is argument about these lesions, it is probably accurate to say that smooth muscle proliferation occurs in both the media and the intima. The pathology of mild, chronic hypertension in humans is more ambiguous. Folkow[99] has studied the "passive resistance" of peripheral vessels in hypertension. The outcome of these studies is clear. Resistance to flow through peripheral vessels in hypertension is increased even in the presence of vasodilators. This implies either a very large increase in the length of resistance vessels or a decrease in their lumen size. The latter effect is exponential; that is, resistance to flow is proportional to the reciprocal of the lumen radium squared. This means that very small changes in lumen size can produce very large changes in vascular resistance.

We know from the physiologic data[99] as well as from careful morphometric studies of hypertensive vessels in animals and people with long-term blood pressure elevation that smooth muscle cell mass increases.[51] The increase could be in protein alone (that is, hypertrophy), or the increase could be at the level of DNA (that is, some form of cell replication). These responses are not mutually exclusive; rather both changes occur. The cell mass of hypertensive vessels is increased as is the cellular content of contractile proteins.[3] The DNA mass is also increased, although in a somewhat surprising fashion. Increases in DNA have been shown both by tritiated thymidine autoradiography and measurement of DNA mass.[102] For large vessels, the increase in DNA occurs by endoreplication — that is, these vessels show an increase in ploidy. The increased mass in these vessels can be largely accounted for by the very high mass of the tetraploid and octaploid cells.[3] Apparently, when the endothelium is intact, the response of the vessel to hypertension is to undergo endoreplication. Returning to the relationship of morphogenesis to growth control, it is interesting to note that hypertension produces an increase in DNA while maintaining the normal structure of the vessel wall. The stimuli for this change are apparently intrinsic to the vessel wall since DNA synthesis is stimulated in the absence of denudation or inflammatory changes. In contrast, in both atherosclerosis and the balloon catheter model, some factor is responsible not only for stimulating replication but for disordering the vessel to the point where massive cell accumulation occurs in the intima.[95]

VII. SMOOTH MUSCLE-ENDOTHELIAL INTERACTIONS

Denudation as already discussed is not a sufficient stimulus by itself. Smooth muscle cell replication does not require denudation, and denudation, even with thrombosis, is insufficient to stimulate smooth muscle growth. One possible endogenous mechanism to commit smooth muscle cells to replication is dependence of growth on release from a normal endogenous inhibitor. The evidence for endogenous growth inhibitors is quite old. Reports that extracts of aorta are able to inhibit smooth muscle replication in vivo as well as in vitro, have been published for 20 years.[106] This concept was greatly strengthened by three observations. First, Clowes and colleagues found that heparin was able to inhibit smooth muscle proliferation following denudation with the balloon catheter.[102] Castellot and colleagues found that this effect could be mimicked in vitro by an endothelial-derived factor that was able to inhibit smooth muscle replication in vitro.[85] This factor, they suggested, was a fraction of heparan sulfate released by a platelet endoglycosidase present in serum.[107] Later studies from the same laboratory described an active heparan fraction could also be derived from smooth muscle cells.[108]

The importance of these in vitro experiments, combined with evidence that the inhibitor isolated from whole tissue is also a heparan sulfate,[109] is the implication that the cells of the vessel wall can themselves produce inhibitors of smooth muscle proliferation. In other words, we might imagine that both cells of the normal vessel wall exist in a growth-inhibited state rather than in a passive state awaiting arrival of a mitogen. "Mitogenesis" then could consist either of the removal of neutralization of inhibitory activity or presentation of some traditional growth factor able to overcome the inhibitor.

This hypothesis is supported by a comparison of Castellot's original experiment on a growth inhibitor with our studies showing that endothelial cells synthesize a growth stimulant.[27] The existence of a growth stimulant was later confirmed by the discovery that part of this activity was identical to PDGF[110] and by evidence that the cultured endothelial cell expresses high levels of the gene for PDGF, c-sis.[111,112] How could Castellot[85] have detected an inhibitor of the cells were making a growth stimulant? The answer may lie in the design of the experiments and could be important to developing a hypothesis based on control of smooth muscle growth by inhibition. Castellot's experiments were performed in serum containing PDGF. After incubation in endothelial cells, culture medium containing PDGF lost its ability to stimulate smooth muscle proliferation. In contrast, our demonstration of growth stimulation entailed conditioning of plasma-derived serum, that is, serum lacking PDGF. This suggests the possibility that cultured cells of the vessel wall in the absence of exogenous PDGF secrete an excess of PDGF while they produce an excess of inhibitor in its presence. It is now worthwhile considering the possibility that this sort of balance between synthesis of inhibitors and stimulants is operative in vivo.

Support for this idea comes from the discovery that smooth muscle cells as well as endothelial cells can synthesize PDGF.[31,32] Cultured neonatal rat smooth muscle cells synthesize PDGF and are able to grow in the absence of exogenous mitogens. In contrast, smooth muscle cells cultured from the aorta of adult rats show very little or no production of PDGF. This suggests that PDGF produced by vessel wall cells may play a role in embryogenesis. Consistent with that idea, Maciag and collaborators[37] showed that cultured endothelial cells in the sheet form produce the message for PDGF-b chain, i.e., c-*sis*. When the cells were induced to change from a monolayer to a tube form by removal of endothelial cell growth factor (ECGF), c-*sis* levels disappeared, suggesting that PDGF production may play a role in morphogenesis of the primitive vessel tube.[113] Walker and collaborators have shown in preliminary studies that smooth muscle cells cultured from balloon denuded vessels show similar, though lower, levels of PDGF production.[114] It is intriguing to consider the possibility that cells in the adult wall may retain a residual ability to synthesize PDGF and that this ability may somehow be stimulated by the balloon injury. If this were confined to a subset of cells, it might explain both the commitment kinetics discussed above and the observation that only a portion, about 50% of the cells in the lesion, ever enter the cell cycle.[95] Perhaps more importantly, atherosclerotic plaques are monoclonal. It is intriguing to consider the possibility that this monoclonality could represent the overgrowth of a fetal cell, perhaps one retaining the ability to synthesize the *sis* oncogene. Barrett and collaborators have recently found that atherosclerotic plaques contain higher levels of *sis* message than are found in normal vessels.[115] Interpretation of this exciting result will have to await new techniques that better define which cell in the plaque is making PDGF.

To return to the analogy between hypertension and atherosclerosis, in both the balloon model of atherosclerosis and in atherosclerosis itself, the characteristic response is a combination of migration and proliferation. In contrast, proliferation in hypertension is local — that is, cells synthesize DNA in place. In fact, this occurs without cell division, resulting instead in polyploid cells. We can imagine that the replicated smooth muscle cells retain normal mechanical and electrical connections required for function, and that the extracellular environment is generally unchanged.

The role of cell-cell interactions in the regulation of smooth muscle cell proliferation has yet to be explored. Schwartz et al.[116] and Spagnoli et al.[117] have shown that regenerating endothelial cells lose their gap junctions. This is a common feature of cell regeneration in epithelia and has been suggested as a major controlling factor in the stimulation of growth by release from quiescence.[118] Recent studies have shown, in addition to gap junction communication between endothelial cells, that heterocellular communication between endothelial and smooth muscle cells can also occur.[119,120]

One simple hypothesis which could account for the quite different series of events is that the trauma caused by the balloon, or similar processes occurring during lipid insudation and monocyte accumulation, stimulates a morphogenic change. Consistent with this hypothesis, PDGF has been shown to be chemotactic as well as mitogenic for smooth muscle.[121] Similarly, heparin not only inhibits growth, but it also inhibits motility.[122] Platelets also contain a factor that is able to control the type of growth. TGF-β inhibits growth of attached cells; however, in combination with other growth factors as epidermal growth factor, it will permit growth of nontransformed cells in soft agar. In nonneoplastic terms, however, it is intriguing to consider TGF-β as a morphogenic factor that changes a normal cell's dependence or extracellular matrix for growth control. Normal tissues contain high levels of TGF-β in an inactive form.[123] We have shown that this factor inhibits endothelial regeneration at least in vitro, while Assoian has shown that it stimulates smooth muscle growth in soft agar.[124] This combination of actions could contribute to plaque ontogeny, with platelet PDGF stimulating smooth muscle cells to migrate from the normal media with its high levels of collagen and elastin to the intima, while platelet TGF-β permits the cells to grow in the less structured environment of the intima and, perhaps, stimulates them to synthesize their own PDGF;[114] at the same time endothelial regeneration would be inhibited. It is intriguing to consider the possibility that some sequence of this sort might depend on the extent of disruption of heparin species in the extracellular matrix.

VIII. SUMMARY

We would like to suggest that the current focus on growth factors may neglect the possibility that the critical events in formation of vessel wall lesions are morphogenic changes rather than mitogenic stimulation. For the endothelium the maintenance of a characteristic flattened cell sheet and formation of branched tubes is obligatory. In normal growth responses, for example, angiogenesis, one would imagine that formation of new vessels would depend on factors that release the endothelial cell from normal mechanisms of cell-cell interaction. Consisent with this view is the observation that chemotactic factors can be angiogenic without being mitogenic for cultured endothelium[63,64] and the observation that withdrawal of factors required for maintenance of proliferation of endothelium in monolayer culture encourages endothelial cells to differentiate into tube forms.[113]

We would like to consider three possible sets of cell surface molecules that might be critical to control the morphogenic changes underlying these changes in cell behavior. The first is the set of molecules already identified as critical to cell substrate adhesion, that is, the receptors for laminin, fibronectins, and collagens.[75,125] The relationship of these receptors to control of cell function remains to be studied. Second, it is easy to imagine that a set of molecules exists that indirectly control cell shape. Our data would argue that the GpIIb/IIIa complex may be in this class.[74] The way GpIIb/IIIa controls endothelial shape needs to be studied. Finally, the plasma membrane also contains a distinct set of molecules responsible for control of cell growth.

Thinking about smooth muscle growth in "morphogenic" terms is less obvious. The normal quiescent state could represent the results of a high level of endogenous growth inhibition. Atherosclerotic proliferation would then involve two changes: a morphogenic change required to remove the cells from their normal environment permitting migration as well as proliferation, and a proliferative state that might represent either the response to exogenous growth factors or synthesis of endogenous factors by cells released from normal control. Actions of growth factors other than their ability to stimulate replication may be critical to lesion formation. In contrast, the proliferation seen in hypertension would be a pure proliferative change, occurring without loss of cell-cell interactions or change in vessel wall morphogenesis. The intriguing question for hypertension is what the nature of the initial stimulus is if replication occurs without either disruption of extracellular matrix or release of exogenous mitogens.

REFERENCES

1. **Thoenen, H. and Barde, Y. A.,** Physiology of nerve growth factor, *Physiol. Rev.,* 60, 1284, 1980.
2. **Berger, E. A. and Shooter, E. M.,** Nerve growth factor: studies on the localization, regulation and mechanism of its biosynthesis, in *Molecular Control of Proliferation and Differentiation,* Papaconstantinou, J. and Rutter, W. J., Eds., Academic Press, New York, 1978, 83.
3. **Owens, G. K. and Schwartz, S. M.,** Vascular smooth muscle cell hypertrophy and hyperploidy in the Goldblatt hypertensive rat, *Circ. Res.,* 56, 525, 1983.
4. **Holley, R. W., Armour, R., and Baldwin, J. H.,** Density-dependent regulation of growth by low molecular weight nutrients, *Proc. Natl. Acad. Sci. U.S.A.,* 75, 339, 1978.
5. **Yen, A. and Pardee, A.,** Arrested states produced by isoleucine deprivation and their relationship to the low serum produced arrested state in Swiss 3T3 cells, *Exp. Cell Res.,* 114, 389, 1978.
6. **Rytömaa, T.,** The chalone concept, *Int. Rev. Exp. Pathol.,* 16, 156, 1976.

7. **Lieberman, M. A. and Glaser, L.,** Density-dependent regulation of cell growth: an example of a cell-cell recognition phenomenon, *J. Membr. Biol.,* 63, 1, 1981.

8. **Dulbecco, R.,** Topoinhibition and serum requirement of transformed and ultratransformed cells, *Nature (London),* 227, 802, 1970.

9. **Todaro, G. J., Lazar, G. K., and Green, H.,** The initiation of cell division in a contact-inhibited mammalian cell line, *J. Cell. Comp. Physiol.,* 56, 325, 1965.

10. **Vogel, A., Ross, R., and Raines, E.,** Role of serum components in density-inhibition of growth of cells in culture. Platelet-derived growth factor is the major serum determinant of saturation density, *J. Cell Biol.,* 85, 377, 1980.

11. **Maroudas, N. G.,** Diffusion-limitation of cell growth, *Nature (London),* 274, 722, 1978.

12. **Selden, S. C., III and Schwartz, S. M.,** Cytochalasin B inhibition of endothelial proliferation at wound edges in vitro, *J. Cell Biol.,* 81, 348, 1979.

13. **du Heynes, A. D., Eldor, A., Vlodavsky, I., Fridman, R., and Panet, A.,** The antiproliferative effect of interferon and the mitogenic activity of growth factors are independent cell cycle events. Studies with vascular smooth muscle cells and endothelial cells, *Exp. Cell Res.,* 161, 297, 1985.

14. **Resnitzkey, D., Yarden, A., Zipori, D., and Kimchi, A.,** Autocrine β-related interferon controls c-myc suppression and growth arrest during hematopoietic cell differentiation, *Cell,* 46, 31, 1986.

15. **DeLarco, J. and Todaro, G.,** Growth factors from murine sarcoma-virus transformed cells, *Proc. Natl. Acad. Sci. U.S.A.,* 75, 4001, 1978.

16. **Assoian, R. K., Grotendorst, G. R., Miller, D. M., and Sporn, M. B.,** Cellular transformation by coordinated action of three peptide growth factors from human platelets, *Nature (London),* 309, 804, 1984.

17. **Anzano, M., Roberts, A., Smith, J., Sporn, M., and DeLarco, J.,** Sarcoma growth factor from conditioned medium of virally transformed cells is composed of both type α and β transforming growth factors, *Proc. Natl. Acad. Sci. U.S.A.,* 80, 6264, 1983.

18. **Roberts, A. B., Anzano, M. A., Wakefield, L. M., Riche, N. S., Stern, D. F., and Sporn, M. B.,** Type β transforming growth: a bifunctional regulators of cellular growth, *Proc. Natl. Acad. Sci. U.S.A.,* 82, 119, 1985.

19. **Shipley, G. D., Tucker, R. F., and Moses, H. L.,** Type β transforming growth factor/growth inhibitor stimulates entry of monolayer cultures of AKR-2B cells into S phase after a prolonged prereplicative interval, *Proc. Natl. Acad. Sci. U.S.A.,* 82, 4147, 1985.

20. **Tucker, R. F., Shipley, G. D., Moses, H. L., and Holley, R. W.,** Growth inhibitor from BSC-1 cells closely related to platelet type β transforming growth factor, *Science,* 226, 705, 1984.

21. **Heimark, R. L., Twardzik, D., and Schwartz, S. M.,** Inhibition of endothelial regeneration by type beta transforming growth factor from platelets, *Science,* 233, 1078, 1986.

22. **Baird, A. and Durkin, T.,** Inhibition of endothelial cell proliferation by type beta-transforming growth factor: interactions with acidic and basic fibroblasts growth factors, *Biochem. Biophys. Res. Commun.,* 138, 476, 1986.

23. **Frater-Schröder, M., Müller, G., Birchmeier, W., and Böhlen, P.,** Transforming growth factor-beta inhibits endothelial cell proliferation, *Biochem. Biophys. Res. Commun.,* 137, 295, 1986.

24. **Ollerman, R. A. and Miller, E. S.,** The influence of conditioned media and nonessential amino acid supplementation on the growth of cells in vitro, *J. Cell. Physiol.,* 74, 209, 1969.

25. **Grinnell, F.,** Cellular adhesiveness and extracellular substrata, *Int. Rev. Cytol.,* 53, 65, 1978.

26. **Pledger, W. J., Stiles, C. D., Antoniades, H. N., and Scher, C. D.,** An ordered sequence of events is required before Balb/C-3T3 cells become committed to DNA synthesis, *Proc. Natl. Acad. Sci. U.S.A.,* 75, 2839, 1978.

27. **Gajdusek, C., DiCorleto, P., Ross, R., and Schwartz, S. M.,** An endothelial cell-derived growth factor, *J. Cell Biol.,* 85, 467, 1980.

28. **DiCorleto, P. E.,** Cultured endothelial cells produce multiple growth factors for connective tissue cells, *Exp. Cell Res.,* 153, 167, 1984.

29. **Martin, B. M., Gimbrone, M. A., Jr., Unanue, E. R., and Cotran, R. S.,** Stimulation of nonlymphoid mesenchymal cell proliferation by a macrophage-derived growth factor, *J. Immunol.,* 126, 1510, 1981.

30. **Glenn, K. C. and Ross, R.,** Human monocyte-derived growth factor(s) for mesenchymal cells. Activation of secretion by endotoxin and concanavalin A (Con-A), *Cell,* 25, 603, 1981.

31. **Seifert, R. A., Schwartz, S. M., and Bowen-Pope, D. F.,** Developmental regulation of platelet-derived growth factor-like molecules, *Nature (London),* 311, 669, 1984.

32. **Nilsson, J., Sjolund, M., Palmberg, C., Thyberg, T., and Heldin, C. H.,** Arterial smooth muscle cells in primary culture produce a platelet-derived growth factor-like protein, *Proc. Natl. Acad. Sci. U.S.A.,* 82, 4418, 1985.

33. **Clemmons, D. R.,** Variables controlling the secretion of a somatomedin-like peptide by cultured porcine smooth muscle cells, *Circ. Res.,* 56, 418, 1985.

34. **Gospodarowicz, D., Cheng, J., Lui, G. M., Baird, A., and Bohlen, A.,** Isolation of brain fibroblast growth factor by heparin-Sepharose affinity chromatography: identity with pituitary fibroblast growth factor, *Proc. Natl. Acad. Sci. U.S.A.,* 81, 6963, 1984.

35. **Klagsbrun, M. and Shing, Y.,** Heparin affinity of anionic and cationic capillary endothelial cell growth factors: analysis of hypothalamus-derived growth factors and fibroblast growth factors, *Proc. Natl. Acad. Sci. U.S.A.,* 82, 805, 1985.

36. **Lobb, R. R. and Fett, J. W.,** Purification of two distinct growth factors from bovine neural tissue by heparin affinity chromatography, *Biochemistry,* 23, 6295, 1984.

37. **Maciag, T., Mehlman, T., Friesel, R., and Schreiber, A. B.,** Heparin binds endothelial cell growth factor, the principal endothelial cell mitogen in bovine brain, *Science,* 225, 932, 1984.

38. **D'Amore, P. A., Glaser, B., Brunson, S. K., and Fenselau, A. H.,** Angiogenic activity from bovine retina: partial purification and characterization, *Proc. Natl. Acad. Sci. U.S.A.,* 78, 3068, 1981.

39. **Tiell, M. L., Stemerman, M. B., and Spaet, T. H.,** The influence of the pituitary on arterial intimal proliferation in the rat, *Circ. Res.,* 42, 644, 1978.

40. **Wagner, R. C.,** Endothelial cell embryology and growth, *Adv. Microcirc.,* 9, 45, 1980.

41. **Haar, J. L., and Ackerman, G. A.,** A phase and electron microscopic study of vasculogenesis and erythropoiesis in the yolk sac of the mouse, *Anat. Rec.,* 170, 199, 1971.

42. **LeLievre, C. S. and LeDouarin, N. M.,** Mesenchymal derivatives of neural crest: analysis of chimaeric quail and chick embryos, *J. Embryol. Exp. Morphol.,* 34, 125, 1975.

43. **Ekblom, P., Sariola, H., Karkinen, M., and Saxen, L.,** The origin of the glomerular endothelium, *Cell Differ.,* 11, 35, 1982.

44. **Stewart, P. A. and Willey, M. J.,** Developing nervous tissue induces formation of blood-brain barrier characteristics in invading endothelial cells: a study using quail-chick transplantation, *Dev. Biol.,* 84, 183, 1981.

45. **Risau, W.,** Developing brain produces an angiogenesis factor, *Proc. Natl. Acad. Sci. U.S.A.,* 83, 3855, 1986.

46. **Risau, W. and Ekblom, P.,** Production of a heparin-binding angiogenesis factor by embryonic kidney, *J. Cell Biol.,* 103, 1101, 1986.

47. **Girard, H.,** Arterial pressure in the chick embryo, *Am. J. Physiol.,* 224, 454, 1973.

48. **Folkman, J. and Cotran, R.,** Relation of vascular proliferation to tumor growth, *Int. Rev. Exp. Pathol.,* 16, 207, 1976.

49. **Ausprunk, D. H. and Folkman, J.,** Migration and proliferation of endothelial cells in preformed and newly formed blood vessels during tumor angiogenesis, *Microvasc. Res.,* 14, 53, 1977.

50. **Harris-Hooker, S. A., Gajdusek, C. M., Wight, T. N., and Schwartz, S. M.,** Neovascular responses induced by cultured aortic endothelial cells, *J. Cell. Physiol.,* 114, 302, 1983.

51. **Weiner, J., Loud, A. V., Giacomelli, F., and Anversa, P.,** Morphometric analysis of hypertension-induced hypertrophy of rat thoracic aorta, *Am. J. Pathol.,* 88, 619, 1977.

52. **Schwartz, S. M. and Benditt, E. P.,** Aortic endothelial cell replication. I. Effects of age and hypertension in the rat, *Circ. Res.,* 41, 248, 1977.

53. **Ross, R.,** The pathogenesis of atherosclerosis. An update, *N. Engl. J. Med.,* 314, 488, 1986.

54. **Majesky, M. W., Reidy, M. A., Benditt, E. P., and Juchau, M. R.,** Focal smooth muscle hyperplasia in the aortic intima produced by an initiation-promotion procedure, *Proc. Natl. Acad. Sci. U.S.A.,* 82, 3450, 1985.

55. **Scott, R. F., Kim, D. N., and Schmee, J.,** Endothelial and lesion cell growth patterns of early smooth-muscle cell atherosclerotic lesions in swine, *Arch. Pathol. Lab. Med.,* 109, 450, 1985.

56. **Reidy, M. A. and Schwartz, S. M.,** Endothelial regeneration. III. Time course of intimal changes after small defined injury to rat aortic endothelium, *Lab. Invest.,* 44, 301, 1981.

57. **Hansson, G. K. and Schwartz, S. M.,** Evidence for cell death in the vascular endothlium *in vivo* and *in vitro,* *Am. J. Pathol.,* 112, 278, 1983.

58. **Hansson, G. K. and Schwartz, S. M.,** Endothelial dysfunction without cell loss, in *Biochemical-Interactions at the Endothelium,* Cryer, A., Ed., Elsevier, London, 1983, 343.

59. **Hansson, G., Chao, S., Schwartz, S. M., and Reidy, M. A.,** Aortic endothelial cell death and replication in normal and lipopolysaccharide-treated rats, *Am. J. Pathol.,* 121, 123, 1985.

60. **Haudenschild, C. C., Zahniser, D., Folkman, J., and Klagsbrun, M.,** Human vascular endothelial cells in culture. Lack of response to serum growth factors, *Exp. Cell Res.,* 98, 175, 1976.

61. **Schwartz, S. M., Selden, S. C., III, and Bowman, P.,** Growth control in aortic endothelium at wound edges, Cold Spring Harbor 3rd Conference on Cell Proliferation, in *Hormones and Cell Culture,* Vol. 6, Ross, R. and Sato, G., Eds., Cold Spring Harbor, New York, 1979, 593.

62. **Selden, S. C., III, Rabinovitch, P. S., and Schwartz, S. M.,** Effects of cytoskeletal disrupting agents on replication of bovine endothelium, *J. Cell. Physiol.,* 108, 195, 1981.

63. **Banda, M. J., Knighton, D. R., Hunt, T. K., and Werb, Z.,** Isolation of a nonmitogenic factor from wound fluid, *Proc. Natl. Acad. Sci. U.S.A.,* 79, 7773, 1982.
64. **Fett, J. W., Strydan, D. J., Lobb, R. R., Alderman, E. M., Bethune, J. L., Riordan, J. F., and Vallee, B. L.,** Isolation and characterization of angiogenin, an anionic protein from human carcinoma cells, *Biochemisry,* 24, 5480, 1985.
65. **Sholley, M. M., Ferguson, G. P., Seibel, H. R., Montour, J. L., and Wilson, J. D.,** Mechanism of neovascularization. Vascular sprouting can occur without proliferation of endothelial cells, *Lab. Invest.,* 51, 624, 1984.
66. **Reidy, M. A.,** A reassessment of endothelial injury and arterial lesion formation, *Lab. Invest.,* 53, 513, 1985.
67. **Gabbiani, G., Gabbiani, F., Lombardi, D., and Schwartz, S. M.,** Organization of actin cytoskeleton in normal and regenerating arterial endothelial cells, *Proc. Natl. Acad. Sci. U.S.A.,* 80, 2361, 1983.
68. **Fitzgerald, L. A., Charo, I. F., and Phillips, D. R.,** Human and bovine endothelial cells synthesize membrane proteins similar to human platelet glycoproteins IIb and IIIa, *J. Biol. Chem.,* 260, 10893, 1985.
69. **Thiagarajan, P., Shapiro, S. S., Levine, E., DeMarco, L., and Yalcin, A.,** Monoclonal antibody to human platelet glycoprotein IIIa detects a related protein in cultured human endothelial cells, *J. Clin. Invest.,* 75, 896, 1985.
70. **Nachman, R. L. and Leung, L. L.,** Complex formation of platelet membrane GpIIb and IIa with fibrinogen, *J. Clin. Invest.,* 69, 263, 1982.
71. **Plow, E. F. and Ginsberg, M. H.,** Specific and saturable binding of plasma fibronectin to thrombin stimulated platelets, *J. Biol. Chem.,* 256, 9477, 1981.
72. **Ginsberg, M. H.,** The effect of Arg-Gly-Asp-containing peptides on fibrinogen and von Willebrand factor binding to platelets, *Proc. Natl. Acad. Sci. U.S.A.,* 82, 8057, 1985.
73. **Birdwell, C. R., Gospodarowicz, D., and Nicolson, G. L.,** Identification, localization and role of fibronectin in cultured bovine endothelial cells, *Proc. Natl. Acad. Sci. U.S.A.,* 75, 3273, 1978.
74. **Chen, C. S., Thiagarajan, P., Schwartz, S. M., Harlan, J. M., and Heimark, R. L.,** The platelet glycoprotein IIIa-like protein on human endothelial cells promotes adhesion not attachment, *J. Cell Biol.,* in press.
75. **Pytela, R., Pierschbacher, M. D., and Rouslahti, E.,** Identification and isolation of a 140 kd cell surface glycoprotein with properties expected of a fibronectin receptor, *Cell,* 40, 191, 1985.
76. **Damsky, C. H., Knudsen, K. A., Bradley, D., Buck, C. A., and Horowitz, A. F.,** Distribution of the cell substratum attachment (CSAT) antigen on myogenic and fibroblastic cells in culture, *J. Cell Biol.,* 100, 1528, 1985.
77. **Pratt, B. M., Harris, A. S., Morrow, J. S., and Madri, J. A.,** Mechanisms of cytoskeletal regulation. Modulation of aortic endothelial cell spectrin by the extracellular matrix, *Am. J. Pathol.,* 117, 349, 1984.
78. **Margolis, L. B., Vasilieva, E. J., Vasilieva, J. M., and Gelfand, J. M.,** Upper surfaces of epithelial sheets and of fluid liquid films are nonadhesive for platelets, *Proc. Natl. Acad. Sci. U.S.A.,* 76, 2303, 1979.
79. **Vlodavsky, I., Johnson, L. K., and Gospodarowicz, D.,** Appearance in confluent vascular endothelial cell monolayers of a specific cell-surface protein (CSP-60) not detected in actively growing endothelial cells or in cell types growing in multiple layers, *Proc. Natl. Acad. Sci. U.S.A.,* 76, 2306, 1979.
80. **Heimark, R. L. and Schwartz, S. M.,** Characterization of the luminal surface domain of aortic endothelial cells by restrictive iodination, *J. Cell. Physiol.,* in press.
81. **Kenagy, R., Bierman, E. L., and Schwartz, S.,** Regulation of flow-density lipoprotein metabolism by ccll density and proliferative state, *J. Cell. Physiol.,* 116, 404, 1983.
82. **Whittenberger, B. and Glaser, L.,** Inhibition of DNA synthesis in culture of 3T3 cells by isolated surface membranes, *Proc. Natl. Acad. Sci. U.S.A.,* 75, 2251, 1977.
83. **Natraj, C. V. and Datta, P.,** Control of DNA synthesis in growing Balb/c 3T3 mouse cells by a fibroblast growth regulatory factor, *Proc. Natl. Acad. Sci. U.S.A.,* 75, 6115, 1978.
84. **Heimark, R. L. and Schwartz, S. M.,** The role of membrane:membrane interactions in the regulation of endothelial cell growth, *J. Cell Biol.,* 100, 1934, 1985.
85. **Castellot, J. J., Addonizio, M. L., Rosenberg, R., and Karnovsky, M. J.,** Cultured endothelial cells produce a heparin-like inhibitor of smooth muscle cell growth, *J. Cell Biol.,* 90, 372, 1981.
86. **Margargal, W. W., Dickinson, E. S., and Slakey, L. L.,** Distribution of membrane marker enzymes in cultured arterial endothelial and smooth muscle cells, *J. Biol. Chem.,* 253, 8311, 1978.
87. **Charp, P. A., Kinders, R. J., and Johnson, T. C.,** G_2 cell cycle arrest induced by glycopeptides isolated from bovine cerebral cortex, *J. Cell Biol.,* 97, 311, 1983.
88. **Wieser, R. and Brunner, G.,** Imitation of contact inhibition by substrate-bound plasma membrane glycoproteins and lectins in serum-free hormone-supplemented cultures of GH_3 cells, *Exp. Cell Res.,* 147, 23, 1983.
89. **Ross, R., Glomset, J., Kariya, B., and Harker, L.,** A platelet-dependent serum factor that stimulates the proliferation of arterial smooth muscle cells *in vitro, Proc. Natl. Acad. Sci. U.S.A.,* 71, 1207, 1974.

90. **Stemerman, M. B., Spaet, T. H., Pitlick, F., Cintron, J., Lejnicks, I., and Tiell, M. L.,** Intimal healing: the pattern of re-endothelialization and intimal thickening, *Am. J. Pathol.,* 87, 125, 1977.

91. **Baumgartner, H. R., Muggli, R., and Tschoff, T. B.,** Platelet adhesion, release and aggregation in flowing blood: effects of surface properties and platelet function, *Thromb. Haemostasis,* 35, 124, 1976.

92. **Walker, L. N., Reidy, M. A., and Bowyer, D. E.,** Morphology and cell kinetics of fatty streak lesion formation in the hypercholesteolemic rabbit, *Am. J. Pathol.,* 125, 450, 1986.

93. **Gerrity, R. G.,** The role of the monocyte in atherogenesis. I. Transition of blood-borne monocytes into foam cells in fatty lesions, *Am. J. Pathol.,* 103, 181, 1981.

94. **Faggiotto, A., Ross, R., and Harker, L.,** Studies of hypercholesterolemia in the nonhuman primate. I. Changes that lead to fatty streak formation, *Arteriosclerosis,* 4, 323, 1984.

95. **Clowes, A. W. and Schwartz, S. M.,** Significance of quiescent smooth muscle cell proliferation in the injured rat carotid artery, *Circ. Res.,* 56, 139, 1985.

96. **Schwartz, S. M., Reidy, M. A., and Clowes, A.,** Kinetics of atherosclerosis, a stem cell model, *Ann. N.Y. Acad. Sci.,* 454, 292, 1985.

97. **Groves, H. M., Kinlough-Rathbone, R. L., Richardson, M., Moore, S., and Mustard, J. F.,** Platelet interaction with damaged rabbit aorta, *Lab. Invest.,* 40, 194, 1983.

98. **Reidy, M. A. and Silver, M.,** Endothelial regeneration. VII. Lack of intimal proliferation after defined injury to rat aorta, *Am. J. Pathol.,* 118, 173, 1985.

99. **Folkow, B.,** Physiologic aspects of primary hypertension, *Physiol. Res.,* 62, 347, 1984.

100. **Mulvaney, M. J., Hansen, P. K., and Aalkjaer, C.,** Direct evidence that the greater contractility of resistance vessels in spontaneously hypertensive rats is associated with a narrowed lumen, a thickened media and an increased number of smooth muscle cell layers, *Circ. Res.,* 43, 854, 1978.

101. **Olivetti, G., Anversa, P., Melissari, M. M., and Loud, A. V.,** Morphometry of medial hypertrophy in the rat thoracic aorta, *Lab. Invest.,* 42, 559, 1980.

102. **Clowes, A. W. and Clowes, M. M.,** Heparin inhibits injury induced arterial smooth muscle migration and proliferation, *Fed. Proc.,* 44, 1138, 1983.

103. **Reidy, M. A. and Schwartz, S. M.,** A technique to investigate surface morphology and endothelial cell replication of small arteries. A study in acute angiotensin-induced hypertensive rats, *Microvasc. Res.,* 24, 158, 1982.

104. **Owens, G. K. and Reidy, M. A.,** Hyperplastic growth response of vascular smooth muscle cells following induction of acute hypertension in rats by aortic coarctation, *Circ. Res.,* 57, 695, 1986.

105. **Giese, J.,** Renin, angiotensin and hypertensive vascular damage: a review, *Am. J. Med.,* 55, 315, 1973.

106. **Thomas, W. A. and Kim, D. N.,** Atherosclerosis as a hyperplastic and/or neoplastic process, *Lab. Invest.,* 48, 245, 1983.

107. **Castellot, J. J., Jr., Favreau, L. V., Karnovsky, M. J., and Rosenberg, R. D.,** Inhibition of vascular smooth muscle cell growth by endothelial cell-derived heparin. Possible role of a platelet endoglycosidase, *J. Biol. Chem.,* 257, 11256, 1982.

108. **Fritze, L. M. S., Reilly, L. F., and Rosenberg, R. D.,** An antiproliferative heparan sulfate species produced by postconfluent smooth muscle cells, *J. Cell Biol.,* 100, 1041, 1985.

109. **Campbell, J. H. and Campbell, G. R.,** cellular interactions in the artery wall, in *The Peripheral Circulation,* Hunyor, S., Ludbrook, J., Shaw, J., and McGrath, M., Eds., Elsevier, New York, 1984, 33.

110. **Gajdusek, C. M. and Schwartz, S. M.,** Comparison of intercellular and extracellular mitogen activity, *J. Cell. Physiol.,* 121, 316, 1984.

111. **DiCorletto, P. E. and Bowen-Pope, D. F.,** Cultured endothelial cells produce a platelet-derived growth factor-like protein, *Proc. Natl. Acad. Sci. U.S.A.,* 80, 1919, 1983.

112. **Barrett, T. B., Gajdusek, C. M., Schwartz, S. M., McDougall, J. K., and Benditt, E. P.,** Expression of the sis gene by endothelial cells in culture and in vivo, *Proc. Natl. Acad. Sci. U.S.A.,* 81, 6772, 1984.

113. **Jaye, M., McConathy, E., Drohan, W., Tong, B., Duel, T., and Maciag, T.,** Modulation of the sis gene transcript during endothelial cell differentiation in vitro, *Science,* 228, 882, 1985.

114. **Walker, L. N., Bowen-Pope, D. F., Ross, R., and Reidy, M. A.,** Production of platelet-derived growth factor-like molecules by cultured arterial smooth muscle cells accompanies proliferation after arterial injury, *Proc. Natl. Acad. Sci. U.S.A.,* 83, 7311, 1986.

115. **Barrett, T. B., and Benditt, E. P.,** Sis (PDGF-B) gene transcripts are elevated in human atherosclerotic lesions compared to normal artery, *Proc. Natl. Acad. Sci. U.S.A.,* 84, 1099, 1987.

116. **Schwartz, S. M., Stemerman, M. B., and Benditt, E. P.,** The aortic intima. II. Repair of the aortic lining after mechanical denudation, *Am. J. Pathol.,* 81, 1, 1975.

117. **Spagnoli, L. G., Pietra, G. G., Villaschi, S., and Johns, L. W.,** Morphometric analysis of gap junctions in regenerating arterial endothelium, *Lab. Invest.,* 46, 139, 1982.

118. **Lowenstein, W. R.,** Junctional intercellular communication and the control of growth, *Biochim. Biophys. Acta,* 560, 1, 1979.

119. **Larson, D. M. and Sheridan, J. D.,** Intercellular junctions and transfer of small molecules in primary vascular endothelial cultures, *J. Cell Biol.,* 92, 183, 1982.

120. **Davies, P. F., Ganz, D., and Diehl, P. S.,** Reversible microcarrier-mediated junctional communication between endotheial and smooth muscle cell monolayers: an in vitro model for vascular cell interactions, *Lab. Invest.,* 85, 710, 1985.

121. **Grotendorst, G. R., Seppa, H. E. J., Keinman, H. K., and Martin, G. R.,** Attachment of smooth muscle cells to collagen and their migration toward platelet-derived growth factor, *Proc. Natl. Acad. Sci. U.S.A.,* 78, 3669, 1981.

122. **Majack, R. A. and Clowes, A. W.,** Inhibition of vascular smooth muscle cell migration by heparin-like glycosaminoglycans, *J. Cell. Physiol.,* 118, 253, 1984.

123. **Dernck, R., Jarrett, J. A., Chen, E. Y., Eaton, D. H., Bell, J. R., Assoian, R. K., Roberts, A. B., Sporn, M. B., and Goeddel, D. V.,** Human transforming growth factor-beta: complementary DNA sequence and expression in normal and transformed cells, *Nature (London),* 316, 701, 1985.

124. **Assoian, R. K. and Sporn, M. B.,** Type β transforming growth factor in human platelets: release during platelet degranulation and action on vascular smooth muscle cells, *J. Cell Biol.,* 102, 1217, 1986.

125. **Malinoff, H. L. and Wicha, M. S.,** Isolation of a cell surface receptor protein for laminin from murine fibrosarcoma cells, *J. Cell Biol.,* 96, 1475, 1983.

Hemodynamic Forces and Interactions with Blood Cells

Chapter 20

ENDOTHELIAL CELLS, HEMODYNAMIC FORCES, AND THE LOCALIZATION OF ATHEROSCLEROSIS

Peter F. Davies

TABLE OF CONTENTS

I. INTRODUCTION

In 1951, Duff and McMillan[25] wryly observed that "the casual reader might wonder whether some authors conceive of an atherosclerosis so independent of the substrate of the vessel wall, that it may occur in the absence of the blood vessels themselves".

Throughout the past 30 years and particularly the past 15 years, there has been an explosion of knowledge concerning the cells which constitute the arterial wall, their morphology, cell biology, metabolism, and the interrelationships which exist between blood and vessel wall cells. More recently, the pathology of vessel wall diseases has been considered in relation to the interactions between endothelial and smooth muscle cells[16] and endothelial-leukocyte interactions in the context of the inflammatory response.[15,41,48] While some of this vast amount of experimental work has been conducted in vivo, a large amount, perhaps the majority, has been conducted with isolated cell cultures where the experimental environment can be controlled and manipulated. In vivo, however, the presence of blood flow is likely to modify endothelial structure and function in ways which will vary from location to location throughout the circulatory system. The need to understand local hemodynamic forces acting upon endothelium is nowhere more important than in the arterial circulation in regions susceptible to the development of atherosclerosis. This chapter briefly summarizes the principal evidence linking hemodynamics to vessel wall pathology and, since the endothelium is subjected to most of the tangential forces associated with flow, reviews the effects of shear stresses upon this cell layer in vivo and in vitro.

II. NORMAL ENDOTHELIAL CELL FUNCTION

The properties of normal endothelium in vivo are summarized in Figure 1. Arterial endothelium exists as a single continuous layer of flattened cells which in some species directly overlies the internal elastic lamella and thus constitutes the entire intima. In most species, including man, aging is accompanied by migration of medial smooth muscle cells through a fenestrated internal elastic lamella into the subendothelial space to form a diffusely thickened intima.[33,40,73] The endothelium can be considered as a barrier, both structural and metabolic, between the blood and the rest of the vessel wall. A large body of experimental evidence strongly implicates it as a key regulator in the passage of substances between blood and artery in both directions and in the prevention of thrombosis in the vessel lumen (for a review see Reference 44). Anticoagulant properties of the endothelium are maintained by a combination of the nature of the surface components of the cell and its active metabolism (see Figure 1). While the nature of many of the biochemicals synthesized and released from endothelium in maintaining its antithrombotic properties have been characterized, virtually nothing is known concerning the local concentrations of these molecules adjacent to the cell surface when the cell is subjected to blood flow, or the effects of mechanical forces associated with blood flow upon the synthesis and secretion of anti- and procoagulants.

The permeability barrier function of normal endothelium controls the movement of plasma proteins into the subendothelial interstitium. Transendothelial transport of macromolecules by passage through the junctions between endothelial cells[33,53] and by increased pinocytosis and receptor-mediated endocytosis through the cytoplasm of the endothelial cells[39,80,88] is subject to regulation by a number of humoral mediators. The importance of such a permeability barrier has been recognized since the times of Virchow[89] in relation to the "imbibition of lipid" theory of atherogenesis.

A third key property of endothelium is its interactions with other cells, particularly circulating monocytes and polymorphonuclear leukocytes which, under certain pathological (and probably physiological) conditions, can recognize specific receptors on the endothelial cell surface, resulting in leukocyte adhesion and subsequent migration into the subendothelial space.[15,41,48,93]

KEY ENDOTHELIAL CELL PROPERTIES

BLOOD COMPATIBILITY   PERMEABILITY BARRIER   INTERACTIONS WITH

OTHER CELLS

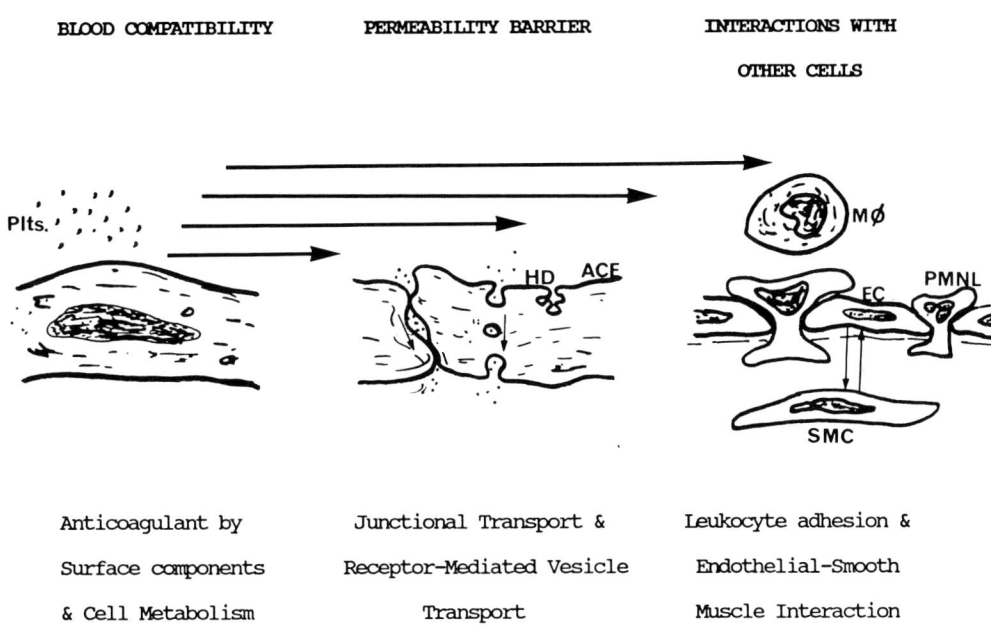

Anticoagulant by   Junctional Transport &   Leukocyte adhesion &

Surface components   Receptor-Mediated Vesicle   Endothelial-Smooth

& Cell Metabolism   Transport   Muscle Interaction

FIGURE 1.   Major functional properties of vascular endothelium.   Endothelium is antithrombotic, behaves as a physical and metabolic permeability barrier, and can interact in a highly regulated fashion with circulating blood cells as well as with smooth muscle cells of the vessel wall. Its anticoagulant properties involve cell surface components such as heparin sulfate and molecules such as plasminogen activators which assist in fibrinolysis of a thrombus by activating plasminogen to plasmin. Anticoagulant metabolic products include prostacyclin, a potent inhibitor of platelet aggregation, and other mediators of thrombus formation. Under certain pathological circumstances, endothelium expresses procoagulant activity via tissue factor, von Willebrand Factor (vWF), and by inhibition of plasminogen conversion to plasmin. Subendothelial collagen is a potent surface for platelet aggregation if the cell is significantly injured. Plasma macromolecules traverse the endothelium via cell junctions and through vesicle pathways across the cell. Locally enhanced concentrations of vasoactive mediators such as angiotensin II, histamine, and serotonin may influence junctional permeability. Enzyme systems at the endothelial surface influence the generation of these mediators, e.g., angiotensin converting enzyme (ACE), and histidine decarboxylase (HD). Monocytes (Mϕ), polymorphonuclear leukocytes (PMNL) and lymphocytes can recognize specific receptors at the endothelial cell surface, adhere, and emigrate into the subendothelial space, particularly as part of an inflammatory response. Directed migration occurs in response to chemotactic gradients across the vessel wall. Many of these events are likely to be influenced by blood flow, particularly the local concentration of active metabolites adjacent to the lumenal cell surface.

The great pathologists of the 19th and early 20th centuries, Virchow, Rokitansky, Aschoff[1,72,89] and others recognized the focal nature of atherosclerotic lesions and their distribution and remarked that this might be determined by the different mechanical forces operating in different locations throughout the arterial tree. Rupture, degradation, and lysis of elastic membranes and laminae induced by abnormal hemodynamic and mechanical stresses was considered to be a prerequisite for intimal thickening, fibrosis, and lipid deposition. Such a view was supported by many observations that pulsatile and rapid arterial flow are major factors which promote intimal thickening, favor the development of lipid-rich lesions, and promote thrombus formation. The original 19th century imbibition concept, that increased blood pressure led to trapping of blood lipids followed by focal intimal thickening, has been modified to take into account the presence of the endothelium, the flow of blood, and, more recently, hemodynamic shear stresses. In certain areas endothelium is constantly subjected to disturbed (rather than unidirectional laminar) blood flow; furthermore, the transfer of blood constituents into the artery wall is likely to vary with the flow patterns. Thus, intramural transfer of substances from the lumen is related to endothelial integrity, which itself is influenced by flow. The mass transfer of macromolecules may be enhanced in certain regions of disturbed flow where residence time adjacent to the vessel wall is greatly increased, e.g., in annular vortices.[55]

III. ENDOTHELIAL INJURY AND ALTERED FUNCTION IN THE LOCALIZATION OF ATHEROSCLEROSIS

The role of endothelial injury or "dysfunction" in the initiation of atherosclerosis was developed from morphological and permeability studies of endothelium (1) in areas susceptible to endothelial damage resulting from naturally occuring hemodynamic stresses, and (2) following experimental disruption of arterial endothelium.

In investigations of the permeability of arterial endothelium, the staining of artery wall by colloidal dyes such as Evans blue or Trypan blue has been accepted as evidence that proteins to which the dye is bound can enter the wall from the lumen. The intensity of staining is approximately proportional to the permeability of the endothelium.[35] This relationship has been used to locate areas of the normal arterial tree which may be susceptible to the development of focal lesions because of locally increased permeability to plasma proteins in areas of sponaneous hemodynamic injury or dysfunction.[83-85] Areas of "natural" local injury or dysfunction in young swine aorta as determined by injection of Evans blue dye correlated with an increased uptake of ^{131}I-albumin,[4] ^{3}H-cholesterol,[84] fibrinogen,[3] and with an altered endothelial cell morphology.[10] Caplan and Schwartz[11] attributed increased ^{3}H-thymidine labeling of endothelial cells in Evans blue-stained regions of pig aorta to increased endothelial regeneration resulting from hemodynamic injury. These results were in agreement with Wright's observations[94] that endothelial cell mitoses, representing nuclear proliferation during the repair of localized injury, were most frequent in the aortic arch and/ or around branches of the aorta of guinea pigs. In the same study, she also noted frequent mitoses upstream of arterial coarctations. Somer et al.[83] reported increased uptake of Evans blue dye in upstream areas following experimental coarctation in swine, and Schwartz and Benditt[75] reported similar studies in rat aorta where they noted the nonrandom distribution of ^{3}H-thymidine labeled endothelial cell nuclei in normal animals. The association of atherosclerotic lesions, increased endothelial cell turnover, and sites associated with atypical hemodynamic forces has long suggested a cause and effect relationship in the pathogenesis of this disease.

The stresses to which the endothelium and arterial wall are exposed can be resolved into two components: (1) normal stress, perpendicular to the wall, and (2) shear stress, parallel to the wall. From studies in dogs, Fry[38] concluded that the permeability of the endothelial surface in regions associated with branch vessels appeared to be nonuniform and was related to the shearing stress of the blood flow. Increased fluxes of protein across the endothelium were associated with high shear stress, suggesting denudation or injury to the endothelial barrier. Caro and colleagues[12] however, noted that fatty atherosclerotic lesions often occurred in regions of low wall shear rate, and subsequent studies have reported elevated transendothelial permeability at low wall shear stress.[68] Friedman's studies in human vascular casts are also consistent with these observations.[37] Thus, both high and low shear stresses have been implicated in the localization of atherosclerotic lesions.

Deliberate endothelial injury compromises the control of vascular permeability. Because this may simply be via the presence or absence of endothelial cells, experiments have been reported which seek to demonstrate the effects of damage and removal of endothelium and to relate such effects to the development of atherosclerosis. Many methods of deliberate vascular injury have been explored including direct mechanical injury,[5,6,69] freezing,[87] noradrenaline infusion,[49] endotoxin-induced damage,[86] hypertension associated with aortic coarctation,[83,84] anaphylactic shock,[95] thrombus induction,[34] rejection of transplanted artery,[8] and many others. Although some of the techniques which have been employed are unlike any conditions which initiate or accompany the development of atherosclerotic lesions in vivo, the short-term effects of gross damage may be similar to the effects of more subtle endothelial damage over a long period. Deliberate arterial injury also leads to arterial lipid accumulation in the absence of hypercholesterolemia,[5] although the type and extent of the endothelialization determines the subsequent pathological response.[7] Certainly in the presence of hypercholesterolemia, endothelial injury rapidly produces necrotic lesions which are morphologically similar to advanced lesions in man.[66] Paradoxically, however, it is clear that hypercholesterolemia-induced atherosclerosis in experimental animals is initiated in the presence of an intact endothelial monolayer.[19,28,45] suggesting that subtle injury or altered cell function will initiate a focal lesion.[19,43,45] Hemodynamically induced endothelial cell turnover may be a manifestation of a subtle, nondenuding endothelial injury predisposing such locations to the development of atherosclerotic lesions. It is, therefore, of importance to obtain information concerning the effects of defined hemodynamic shear stresses upon endothelial structure and function both in vivo and in vitro.

IV. BLOOD FLOW PATTERNS IN MODEL AND NATURAL ARTERIES

Despite advances in the measurement of the nature of blood flow in vivo using anemometry, pulsed Doppler, and laser Doppler systems,[2,13,24,50,77] techniques have yet to be devised which can accurately describe the complexities of flow in very small regions (perhaps with an area of less than a few endothelial cells) susceptible to the initiation of atherogenesis. The fluid mechanics of blood flow extends beyond classical Poiseuille flow considerations which apply only to a rigid tube of circular cross-section subjected to a steady flow rate. In the arterial circulation, blood flow is pulsatile, blood is a non-Newtonian fluid, and the blood vessel is a compliant tube of changing cross-sectional shape and area, and is of complex and tortuous length with many side branches and bifurcations. Thus, vascular geometry and flow velocity determine the detailed hemodynamics of the arterial circulation.

Based on the average geometry of several branching sites in the human arterial circulation, a number of investigators have constructed transparent models and vascular casts which allow flow patterns to be observed and analyzed.[24,36,51,55,56,61,62,64,82] In addition, Karino, Goldsmith, and colleagues at McGill University have worked with post-mortem human arteries made transparent by treatment with oil of wintergreen[52] and have confirmed much of the information obtained for steady flow in glass models. In summary, the major conclusions concerning the nature of pulsatile flow patterns in these models (as illustrated in Figure 2) are

1. Flow separation occurs on the outer wall of the major vessel in early systole (the minor vessel being a branch artery or a bifurcation of smaller diameter).
2. Just after peak systole, the separation point moves a short distance proximally and helical patterns of secondary laminar flow develop within the zone of separation.
3. Just before diastole, the separation zone expands towards the inner wall where it can induce vortices. Some turbulent characteristics may occur transiently at the inner wall.
4. Throughout the cardiac cycle, helices of complex secondary laminar flow are present within the zone of separation. To some extent, they may trap particles, macromolecules, and fluid by recirculation of the vortices which eventually rejoin the main flow.
5. Measurements of wall shear stress indicate a low average shear stress in the region of flow separation adjacent to the outer wall accompanied by flow reversal associated with the annular vortices. The stagnation point that exists at the boundary of the flow

A. Abdominal aorta

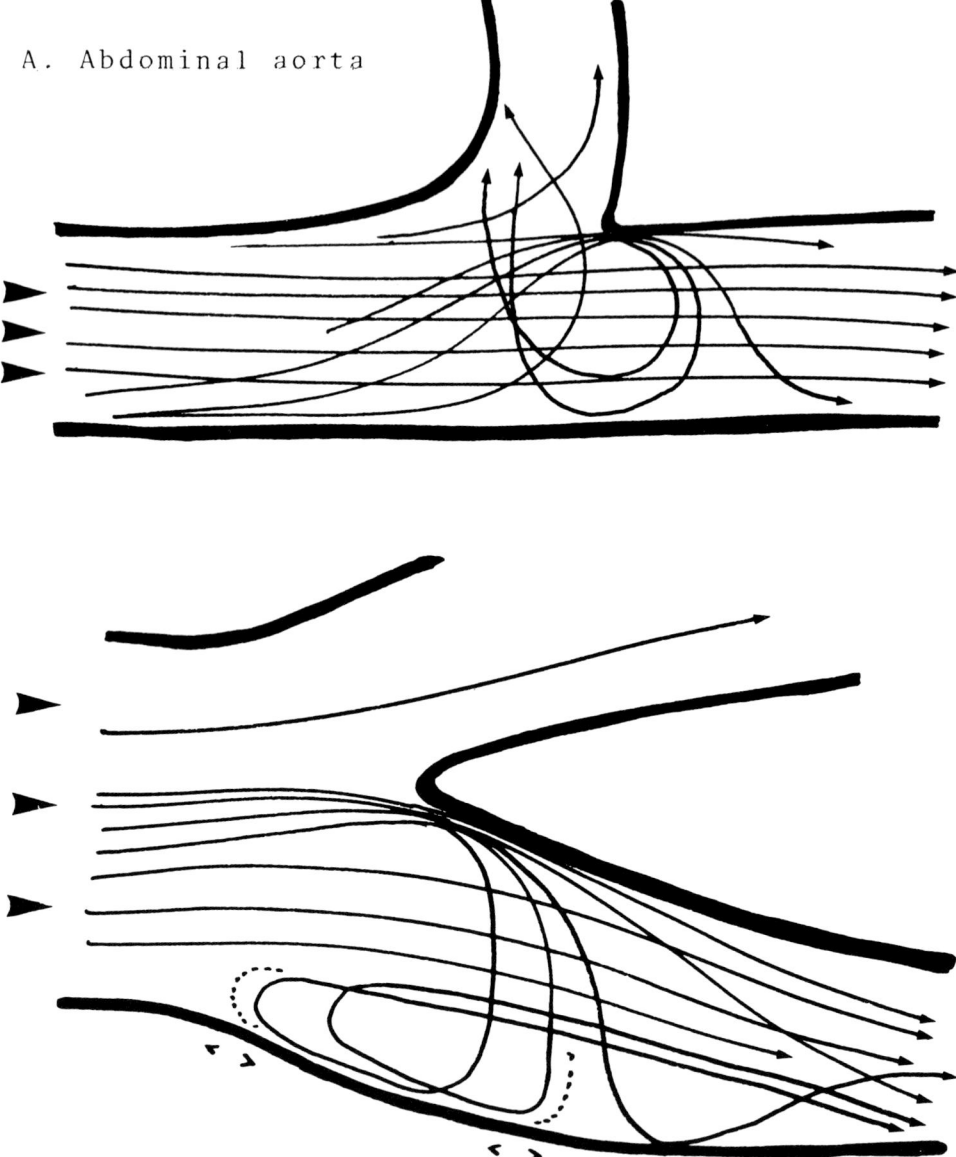

B. Carotid artery bifurcation

FIGURE 2. Flow patterns at arterial branches and bifurcations. (A) At branches of the abdominal aorta, and (B) at the carotid artery bifurcation at moderate Reynolds number (approximately 600). (A) Vortex formation swirling around the mainstream is induced by the flow divider. These secondary laminar flow patterns increase in size with Reynolds number until turbulence occurs. (B) At the carotid artery bifurcation, standing vortices occur at a range of Reynolds numbers, creating a zone of secondary laminar flow in which the residence times of particles is likely to be significantly increased. During the cardiac cycle, the limits of this region of disturbed flow extend proximally and distally (< >) creating narrow zones of the lumenal surface which are subjected to changing shear stress magnitude and direction as the separation and reattachment points oscillate. Atherosclerotic lesions typically develop in this region. (Condensed from glass and transparent artery models of flow reported in References 51, 52, 55, 56, 64 and 96.)

separation must move proximally and distally in concert with the cardiac cycle. The distances involved in such movements may be of the order of several endothelial cell lengths (0.5 to 1.0 mm in the carotid sinus). Thus, not only are adjacent regions of the vessel lumen subjected to high and low shear stresses, but there appears to be a narrow zone adjacent to the separation region which is subjected to relatively high shear (10 to 20 dynes/cm^2), stagnation (zero shear stress), and low shear in reverse flow (-6 to $+6$ dynes/cm^2) within the same cardiac cycle.[55,56,96]

V. DO ATHEROSCLEROTIC LESIONS CORRELATE WITH HIGH OR LOW WALL SHEAR STRESS?

In humans, atherosclerosis is generally localized to regions of predicted low shear stress, although exceptions have been reported in the fetus and neonate[81] and in some experimental animals where lesions are also found associated with zones of predicted high shear stress.[14] This general statement, however, neglects an important consideration which is that shear stresses vary considerably over very small distances in major arteries.[62] The *initiation* of the lesion may occur over a remarkably small distance, perhaps one or two endothelial cells in area. Furthermore, the early developing lesion may itself alter the local hemodynamic pattern acting on several endothelial cells distal to this location. Thus, when macroscopically visible atherosclerosis is correlated with predicted flow patterns in post-mortem specimens, the interpretation is likely to be several years removed from the initiation events whose localization is difficult to determine. It is necessary, therefore, to examine the earliest focal changes of endothelial morphology associated with lesion development in order to pinpoint the beginning of atherosclerosis and relate it to the local hemodynamic patterns. Evidence to date strongly implicates the regions of distributed flow in atherogenesis; it is the precise definition of such flow pattern together with the shear stress effects upon the endothelial cells which currently challenge both fluid dynamicists and vascular biologists.

VI. METHODS FOR APPLYING FLUID SHEAR STRESSES TO ENDOTHELIAL CELLS IN VITRO

In the absence of ultra-high-resolution techniques for the determination in vivo of flow dynamics and force magnitudes acting at the lumenal endothelial cell surface, a number of in vitro systems have been developed to investigate endothelial biology in a well defined flow environment in order to study the influence of defined physical forces upon cell structure and function.

The most commonly used devices are a parallel plate flow channel[31,46,90] and a cone-plate Couette flow apparatus.[9,13,18,20-23,29,32,71] The flow channel device is usually constructed of lucite or other transparent material to facilitate continuous microscopic viewing and recording of cell morphology. The flow geometry is that of parallel plates, being the upper and lower sides of a chamber of rectangular cross-section, producing uniform fully developed two-dimensional flow. Streamlining of the entry and exit sections of the device ensures uniformity. Cells cultured at the plate surface are subjected to unidirectional shear stress in steady or pulsatile laminar flow (reported range: 0 to 85 dynes/cm^2). The channel height and flow rate, measured directly or calculated from the pressure gradient between entrance and exit, determine the shear stress imposed upon the endothelial monolayer. In the Couette flow apparatus, which utilizes the geometry of a cone-plate viscometer, shear stress is produced in the fluid contained between a stationary plate and a rotating cone. The flow is defined by a dimensionless parameter (analogous to Reynold's number) which measures the ratio of centrifugal to viscous forces acting on the moving fluid;[29] both laminar and turbulent flows can be readily induced in this system. Monolayers of endothelial cells grown on coverslips can be placed at various radial distances from the center of the cone where they experience the same shear stress magnitudes in laminar flow but quite different average stresses when flow conditions include turbulence or a combination of laminar and turbulence at varying radial distances. The shear stress forces in this system are determined by the rotational velocity of the cone, the viscosity of the medium, and the angle between the cone and the stationary plate. The cone plate system has been constructed in both stainless steel[9] and transparent Plexiglass.[32]

Other in vitro systems include a closed loop of circular cross-section constructed of various materials,[26,27] the introduction of microcarrier cultures of endothelial cells suspended between a rotating cone and a stationary plate,[54] and the use of pieces of artery with the endothelial surface mounted a short distance from a rotating disk to induce controlled high shear stresses.[57]

VII. STRUCTURAL AND FUNCTIONAL RESPONSES OF CULTURED VASCULAR ENDOTHELIAL CELLS TO FLUID SHEAR STRESS

A. Laminar Shear Stress

Numerous studies in vivo have linked the shape of arterial endothelial cells with regionally predicted flow patterns.[10,17,30,42,58-60,66,67,70,79] Hemodynamic shear stress as a major determinant of endothelial cell shape in vitro was first clearly shown in the cone plate device by Dewey and colleagues.[22,71] Upon exposure to a uniform fluid shear stress of >5 dynes/cm^2 for 18 to 24 hr, endothelial cells within a confluent monolayer changed from a polygonal to an ellipsoidal shape and aligned uniformly with the direction of flow (Figure 3). Cell reorientation is time and shear dependent (e.g., Nerem and colleagues[97] have reported realignment within a few hours at 85 dynes/cm^2), and quantitative measurements of eccentricity[71] and shape index[66] have been reported. Remuzzi et al.[71] used image analysis to study the reversal of alignment (relaxation) following exposure to laminar shear stress. They quantitated relaxation of alignment as well as loss of elongation of individual cells (eccentricity) and noted that morphological relaxation only began 3 hr after removal of shear stress, that it was not spatially or temporally uniform (as judged qualitatively by video recording), and that by 72 hr the morphology of the monolayer was comparable to unstressed control cells.

Change of cell shape associated with laminar shear stresses is accompanied by distinct changes in the organization of the cytoskeleton. In vitro, stress fibers consisting of bundles of microfilaments underwent reorientation and coaxial alignment with the major axis of the cell as it in turn aligned with the direction of the flow.[91] The cytoskeleton was visualized using fluorescent-labeled antibodies to actin and myosin. These results strongly suggest that the organization of actinomyosin in vascular endothelial cells *in situ* when studied by immunofluorescence microscopy reflects cell alignment induced by the local hemodynamic patterns of flow.[92]

Using the cone plate device, Davies and colleagues[18] investigated the relationship between laminar shear stress and endothelial pinocytosis (fluid phase endocytosis). The studies were initiated because the permeability characteristics of the endothelium to macromolecules in vivo is dependent in part upon the formation of endocytic vesicles and their transendothelial transport. Continuous exposure to steady shear stresses (1 to 15 dynes/cm^2) stimulated time- and amplitude-dependent increases in pinocytotic rate which returned to control levels after several hours. After 48 hr of continuous exposure to steady shear stress, removal to static conditions also resulted in a transient increase in pinocytotic rate, suggesting that temporal fluctuations in shear stress may influence endothelial cell function. Endothelial pinocytotic rates remained constant during exposure to rapidly oscillating shear stress at near-physiological frequency (1 Hz) in laminar flow. In contrast, however, a sustained elevation of pinocytotic rate occured when cells were subjected to fluctuations in shear stress amplitude

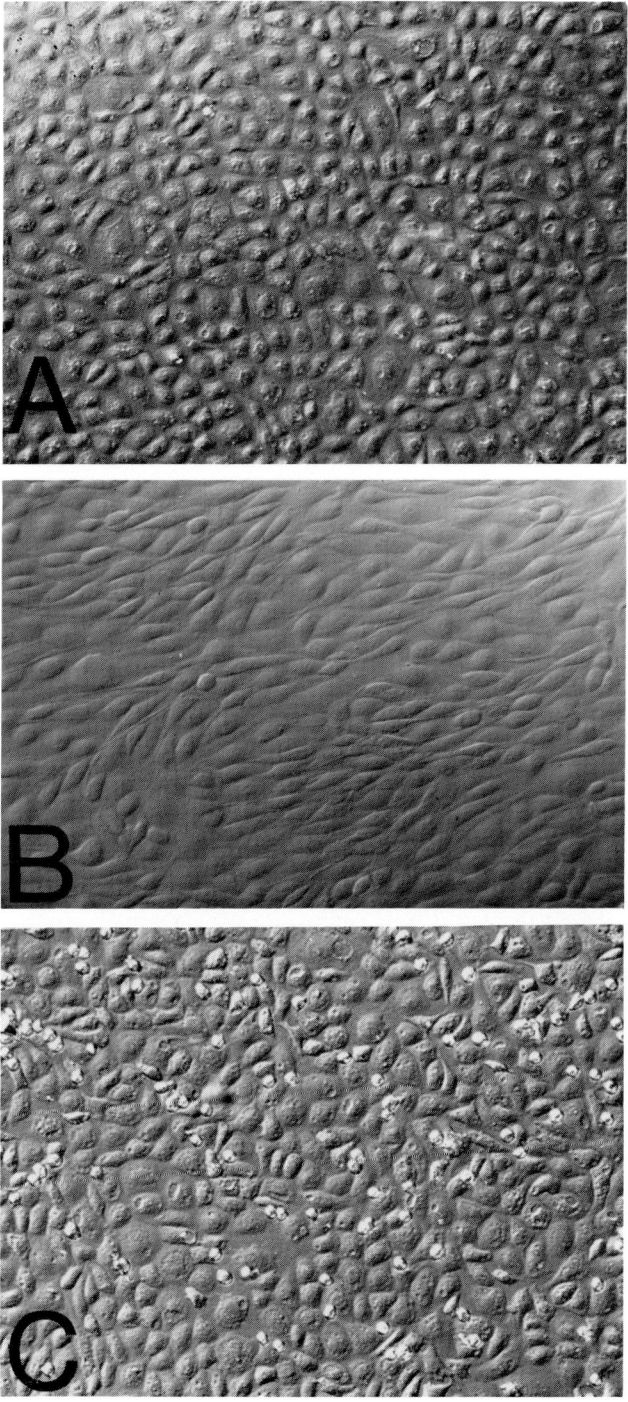

FIGURE 3. Morphological changes induced in confluent bovine aortic endothelial cells in vitro by shear stresses in laminar (B) and turbulent (C) flow. In no-flow (A), cells exhibit a polygonal configuration with no preferred orientation. After 24 hr exposure to shear stress at 8 dynes/cm² in laminar flow (B), cells elongate and align in the direction of the applied shear stress. In contrast, after 16 hr exposure to 1.5 dynes/cm² in turbulent flow (C), cell shape in the monolayer is more variable than in no-flow controls (A), no alignment occurs and a significant number of rounded cells can be seen attached to the upper surface of the monolayer. (Nomarski differential interference microscopy × 300.)

LAMINAR SHEAR STRESS EFFECTS UPON ENDOTHELIAL CELL
PINOCYTOSIS IN VITRO

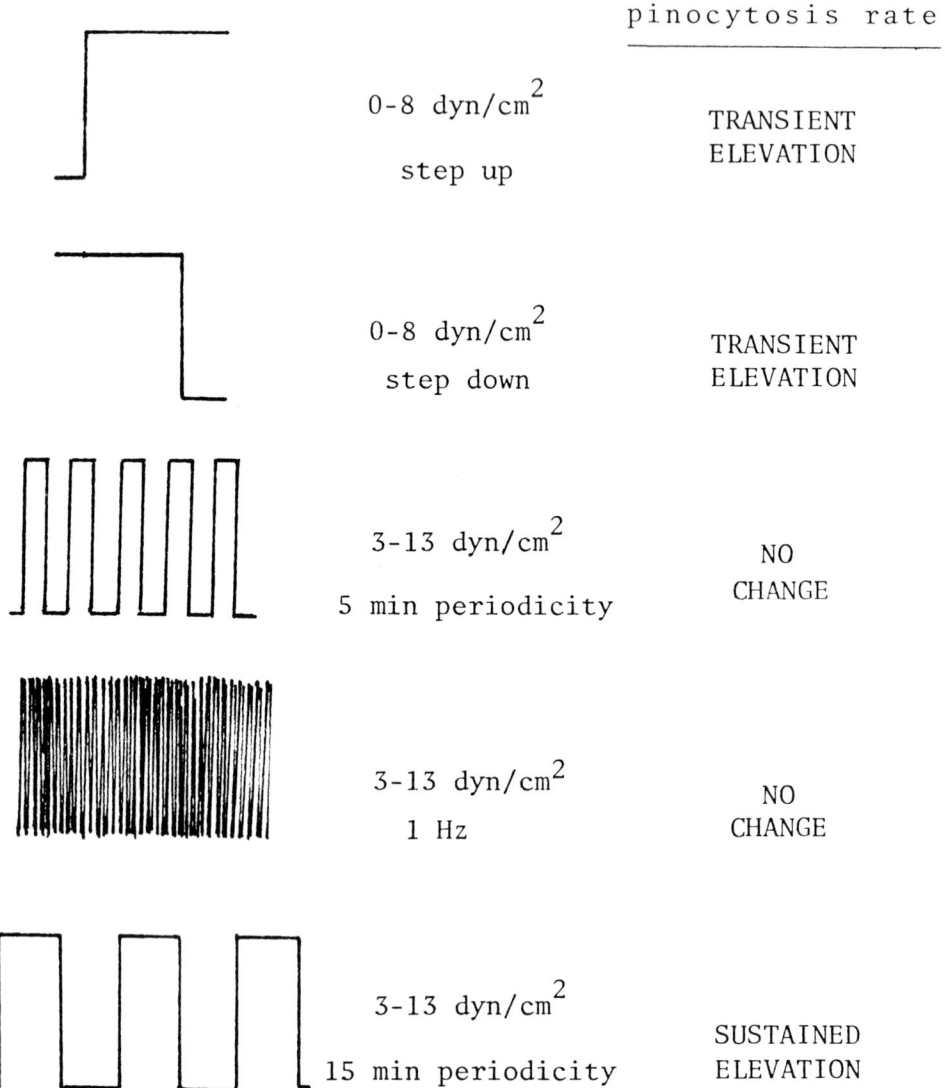

FIGURE 4.   Shear stress-induced responses of endothelial cell pinocytosis in laminar flow. Pinocytotic rate was measured by uptake of horse radish peroxidase (HRP). Following its addition to the culture medium, cells were washed and HRP enzyme activity was measured in cell lysates by a spectrophotometric assay. (Condensed from Davies, P. F. et al., *J. Clin. Invest.*, 73, 1121, 1984.)

(3 to 13 dynes/cm²) of longer cycle time (15 min), suggesting that changes in blood flow of slower periodicity may influence pinocytotic vesicle formation (Figure 4). These observations indicate that alterations in fluid shear stress can significantly influence the rate of formation of endothelial pinocytotic vesicles in a force- and time-dependent fashion, but that the cell also shows accomodation to the new conditions. If, however, the force is applied with a periodicity of long enough duration, the pinocytotic elevation remains sustained, presumably because the putative transduction system at the cell surface requires stimulation for a significant period before initiating changes of intracellular metabolism.

Prostacyclin (PGI_2) is an important endothelial metabolite which is involved in the anti-coagulant properties of the blood vessel.[63] Grabowski and colleagues[45] have reported that prostacyclin production by cultured endothelial cell monolayers is stimulated following exposure to step increases in laminar shear stress. Using a parallel plate flow device, these authors found that step increases in shear stress from 0 to 14 dynes/cm^2 increased prostacyclin production by a factor of 3 after 2 min of exposure to 0.9 dynes/cm^2; 13-fold increases were recorded after exposure to 14 dynes/cm$_2$. The peak values were followed by a decline of PGI_2 production over the subsequent several minutes. These results demonstrate that shear stress changes the *rate of production* of PGI_2; when total PGI_2 production over the 2-min period was calculated, there were no differences in *total* PGI_2 production at any value of applied shear stress. It appears therefore that endothelial cells can produce bursts of pros-tacyclin in response to suddenly imposed laminar shear stress. Similar results have recently been reported by Frangos et al.[31]

The codistribution of fibronectin and F-actin in monolayers of vascular endothelial cells subjected to laminar shear stress was reported by Wechezak et al.[90] using fluorescent labeled antibodies to fibronectin and rhodamine-phalloidin dye for actin staining. Reorientation of the cells exposed to shear stresses of 6 or 26 dynes/cm^2 was accompanied by a reorganization in cellular fibronectin and F-actin. Fibronectin became more uniformly distrubuted beneath the monolayers and was frequently organized into bands of densely packed fibrils. Despite reorganization of fibronectin, its distribution did not colocalize with F-actin. Thus, although a reticular network of fibronectin was induced by laminar shear stress, the studies did not suggest a strong association between extracellular fibronectin and intracellular F-actin filaments.

The relevance of endothelial cell turnover in vivo to mechanisms of atherogenesis has been discussed earlier in this chapter. A striking feature of the in vitro endothelial cell alignment with laminar shear stress is the maintenance of contact inhibition of cell growth. In unstressed, confluent endothelial monolayers, ^{3}H-thymidine autoradiography of DNA shows a low incidence of cell turnover, typically 2 to 6% labeled cells.[18,21,74] During and following cell alignment with flow, the labeling index remains unchanged[18-21] — a remarkable observation considering the major changes of cell-cell contact which must occur during reorientation. When confluent endothelial monolayers are disrupted by scraping with a blunt instrument[74] or individual cells are induced to retract from their neighbours by various cytoskeletal-disrupting drugs,[78] loss of contact inhibition of growth results in cell cycle entry and cell turnover. Absence of enhanced cell turnover in realigned cells in vitro has also been observed *in situ* following experimental alteration of the flow patterns in the rabbit aorta by coarctation.[59] These observations *in vitro* and *in vivo* are consistent with quiescent endothelial cell growth in regions of steady laminar shear stress, i.e., regions not normally associated with the development of atherosclerotic lesions. In contrast, endothelial cell morphology and cell turnover is affected in quite different ways in vitro by turbulent shear stress as described in the next section

B. Shear Stresses in Turbulent Flow

Recent experiments in the cone plate device[21] have explored the endothelial response to shear stresses associated with turbulent flow as a model for disturbed flow. In contrast to the induction of cell alignment by laminar shear stresses, the shape of confluent cultured endothelial cells exposed to a mean shear stress in turbulent flow as low as 1.5 dynes/cm^2 for 16 hr resulted in random orientation of the cells in the monolayer accompanied by many cells which rounded up out of the plane of the monolayer (Figure 3C). Prolonged exposure to turbulent flow resulted in cell-cell retraction and cell loss. These morphological changes were accompanied by a substantially increased endothelial cell turnover in the monolayers exposed to turbulent flow compared to laminar flow over a comparable range of shear stresses (Figure 5). While exposure to relatively high shear stresses in laminar flow for up

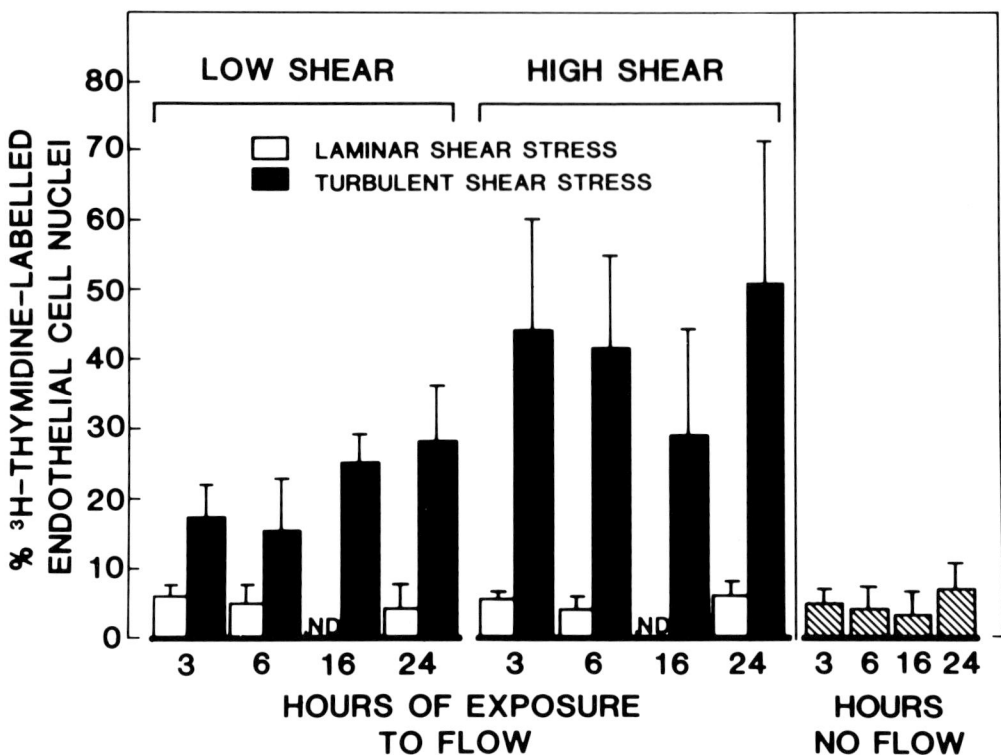

FIGURE 5.   Endothelial cell turnover in vitro in laminar and turbulent flow. Confluent endothelial cell monolayers were subjected to low or high shear stress in laminar or turbulent flow for various periods. The nuclear labeling of DNA ^{3}H-thymidine during a subsequent 24-hr period was determined by autoradiography. Low shear stress was 1.0 dynes/cm^2 in turbulent flow. High shear stress was 15 dynes/cm^2 in laminar flow and 14 dynes/cm^2 in turbulent flow. These values represent the upper and lower limits of turbulent shear stress amplitude in the con-plate apparatus. ND, not determined. (From Davies, P. F. et al., *Proc. Natl. Acad. Sci. U.S.A.*, 83, 2114, 1986. With permission.)

to 24 hr caused no significant change in the percentage of ^{3}H-thymidine-labeled cell nuclei, relatively short exposure (3 hr) to shear stresses as low as 1.5 or as high as 14 dynes/cm^2 in turbulent flow significantly stimulated DNA synthesis. These experiments demonstrate that endothelial cell turnover in vitro is considerably more sensitive to relatively low shear stresses in turbulent flow than to much higher shear stresses applied in laminar flow, implying endothelial cell susceptibility to the flow characteristics rather than to the magnitude of shear stresses alone. Turbulence is a chaotic event involving rapidly changing flow directions, frequences, and durations. Its characteristics in the cone plate apparatus have been investigated by Sdougos et al.,[76] who reported that turbulence, triggered by an instability associated with the distorted velocity profile produced by secondary flow, did not contain any discrete prominent frequences. As noted above, true turbulence occurs rarely in vivo; rather secondary laminar flow of complex frequencies appears to be the dominant pattern in regions of disturbed flow. Nevertheless, the sensitivity of endothelial cells *in vitro* to turbulent flow at very low shear stresses provides a working model for disturbed flow until we can more precisely create complex secondary laminar flow of variable frequencies and flow directions in vitro.

VIII. CONCLUSION

During the last several years, three related areas have overlapped to stimulate exciting new research concerning the mechanisms responsible for the localization of atherosclerotic lesions. First, the distribution of lesions in man and experimental animals has been well documented and provides an invaluable information base. Second, significant progress has been made in understanding the engineering aspects of blood flow in arteries, flow characteristics have been more precisely localized, and new techniques to increase the resolution of flow measurements in vivo have been developed. Third, as a structural and metabolic "connector", the endothelium stands as an interface between flowing blood and developing lesion. The morphological, biochemical, and cell and molecular biology techniques now available to probe the endothelial cell both in vitro and *in situ* is impressive. This integration of fluid dynamics with vascular biology promises to advance our knowledge at the mechanistic level concerning the regional distribution of atherosclerosis.

ACKNOWLEDGMENTS

I am indebted to many vascular biologists, pathologists, and engineers, and particularly grateful to my collaborators Professors Michael Gimbrone, of the Department of Pathology at Harvard Medical School, and Forbes Dewey, of the Massachusetts Institute of Technology, for many hours of discussion and experimentation. Supported by grants HL25536, 22602/36028, and HL24612/36049 from the National Heart, Lung, and Blood Institute of the National Institutes of Health.

REFERENCES

1. **Aschoff, L.,** *Lectures in Pathology,* Hueber, New York, 1924
2. **Baker, W. D.,** Pulsed ultrasonic doppler blood flow sensing, *IEEE Trans. Sonics Ultrason.,* 17 (Suppl.), 170, 1970.
3. **Bell, F. P., Gallus, A. S., and Schwartz, C. J.,** Focal and regional patterns of uptake and the transmural distribution of 131I-fibrinogen in the pig aorta in vivo, *Exp. Mol. Pathol.,* 20, 281, 1974.
4. **Bell, F. P., Gallus, A. S., and Schwartz, C. J.,** Aortic endothelial permeability to albumin: focal and regional patterns of uptake and transmural distribution of 131I albumin in the young pig, *Exp. Mol. Pathol.,* 20, 57, 1974.
5. **Bjorkerud, S.,** Reaction of the aortic wall of the rabbit after superficial, longitudinal and mechanical trauma, *Virchows Arch A,* 347, 197, 1969.
6. **Bjorkerud, S. and Bondjers, G.,** Endothelial integrity and viability in the aorta of the normal rabbit and rat as evaluated with dye exclusion tests and interference contrast microscopy, *Atherosclerosis,* 15, 285, 1972.
7. **Bjorkerud, S. and Bondjers, G.,** Arterial repair and atherosclerosis after mechanical injury. V. Tissue responses after the induction of a large superficial transverse injury, *Atherosclerosis,* 18, 235, 1973.
8. **Bowyer, D. E. and Reidy, M. A.,** Scanning electron microscope studies of the endothelium of aortic allografts in the rabbit: morphological observations, *J. Pathol.,* 123, 237, 1977.
9. **Bussolari, S. R., Dewey, C. F., and Gimbrone, M. A.,** Apparatus for subjecting living cells to fluid shear stress, *Rev. Sci. Instrum.,* 53, 1851, 1982.
10. **Caplan, B. A., Gerrity, R. G., and Schwartz, C. J.,** Endothelial cell morphology in focal areas of in vivo Evans Blue uptake in the young pig aorta. I. Quantitative light microscopic findings, *Exp. Mol. Pathol.,* 21, 102, 1974.
11. **Caplan, B. A. and Schwartz, C. J.,** Increased endothelial cell turnover in areas of in vivo Evans Blue uptake in the pig aorta, *Atherosclerosis,* 17, 401, 1973.
12. **Caro, C. G., Fitzgerald, J. M., and Schroter, R. C.,** Atheroma and arterial wall shear. Observation, correlation and proposal of a shear dependent mass transfer mechanism for atherogenesis, *Proc. R. Soc. London Ser. B,* 177, 109, 1971.
13. **Cheng, D.,** The effect of secondary flow on the viscosity measurement using a cone-and-plate viscometer, *Chem. Eng. Sci.,* 23, 895, 1968.
14. **Cornhill, J. F. and Roach, M. R.,** A quantitative study of the localization of atherosclerotic lesions in the rabbit aorta, *Atherosclerosis,* 23, 489, 1976.

15. **Cotran, R. S.,** The endothelium and inflammation, in *Current Topics in Inflammation and Infection,* Majno, G. and Cotran, R. S., Eds., Williams & Wilkens, Baltimore, 1982, 18.
16. **Davies, P. F.,** Biology of disease. Vascular cell interactions with special reference to the pathogenesis of atherosclerosis, *Lab. Invest.,* 55, 5, 1986.
17. **Davies, P. F. and Bowyer, D. E.,** Scanning electron microscopy of arterial endothelial cells after fixation at physiological pressure, *Atherosclerosis,* 21, 463, 1975.
18. **Davies, P. F., Dewey, C. F., Bussolari, S. R., Gordon, E. J., and Gimbrone, M. A.,** Influence of hemodynamic forces on vascular endothelial function, *J. Clin. Invest.,* 73, 1121, 1984.
19. **Davies, P. F., Reidy, M. A., Goode, T. B., and Bowyer, D. E.,** Scanning electron miscroscopy in the evaluation of endothelial integrity of the fatty streak lesion of atherosclerosis, *Atherosclerosis,* 25, 125, 1976.
20. **Davies, P. F., Remuzzi, A., Dewey, C. F., and Gimbrone, M. A.,** Responses of vascular endothelial cells to laminar and turbulent shear stresses in vitro: morphology and cell proliferation, *Atherosclerosis Rev.,* in press.
21. **Davies, P. F., Remuzzi, A., Gordon, E. J., Dewey, C. F., and Gimbrone, M. A.,** Turbulent fluid shear stress induces vascular endothelial cell turnover in vitro, *Proc. Natl. Acad. Sci. U.S.A.,* 83, 2114, 1986.
22. **Dewey, C. F., Bussolari, S. R., Gimbrone, M. A., and Davies, P. F.,** The dynamic response of vascular endothelial cells to fluid shear stress., *J. Biomech. Eng.,* 103, 177, 1981.
23. **Dewey, C. F., Gimbrone, M. A., Bussolari, S. R., White, G. E., and Davies, P. F.,** Response of vascular endothelium to unsteady fluid shear stress in vitro, in *Fluid Dynamics as a Localizing Factor for Atherosclerosis,* Schettler, G., Ed., Spring Press, Heidelberg, 1983, 182.
24. **Dewey, C. F., Jr.,** Dynamics of arterial flow, *Adv. Exp. Med. Biol.,* 115, 55, 1979.
25. **Duff, G. L. and McMillan, G. C.,** Pathology of atherosclerosis, *Am. J. Med.,* 11, 92, 1951.
26. **Eskin, S. G., Navarro, L. T., O'Bannon, W., and DeBakey, M. E.,** Behavior of endothelial cells cultured on silastic and dacron velour under flow conditions in vitro: implications for prelining vascular grafts with cells, *Artif. Organs,* 7, 31, 1983.
27. **Eskin, S. G., Sybers, H. D., O'Bannon, W., and Navarro, L. T.,** Performance of tissue cultured endothelial cells in a mock circulatory loop, *Artery,* 10, 159, 1982.
28. **Faggiotto, A., Ross, R., and Harker, L.,** Studies of hypercholesterolemia in the non-human primate. I. Changes that lead to fatty streak formation, *Arteriosclerosis,* 4, 323, 1984.
29. **Fewell, M. E. and Hellums, J. D.,** The secondary flow of Newtonian fluids in cone-and-plate viscometers, *Trans. Soc. Rheol.,* 21, 535, 1977.
30. **Flaherty, J. T., Ferrans, V. J., Pierce, J. E., Carew, T. E., III, and Fry, D. L.,** Localizing factors in experimental atherosclerosis, in *Atherosclerosis and Coronary Heart Disease,* Likoff, W., Segal, B. L., Insull, W., and Moyer, S. J., Eds., Grune & Stratton, New York, 1972.
31. **Frangos, J. A., Eskin, S. G., McIntire, L. V., and Ives, C. L.,** Flow effects on prostacyclin production by cultured human endothelial cells, *Science,* 227, 1477, 1985.
32. **Franke, R. P., Grafe, M., Schnittler, H., Seiffge, D., Mittermayer, C., and Drenckhahn, D.,** Induction of human vascular endothelial stress fibers by fluid shear stress, *Nature (London),* 307, 648, 1984.
33. **French, J.,** Atherosclerosis in relation to the structure and function of the arterial intima with special reference to the endothelium, *Int. Rev. Pathol.,* 5, 253, 1966.
34. **Friedman, M.,** *Pathogenesis of Coronary Heart Disease,* McGraw-Hill, New York, 1969.
35. **Friedman, M. and Byers, S. O.,** Endothelial permeability in atherosclerosis, *Arch. Pathol.,* 76, 99, 1963.
36. **Friedman, M. H., Bargeron, C. B., Hutchins, G. M., Mark, F. F., and Deters, O. J.,** Hemodynamic measurements in human arterial casts and their correlation with histology and luminal area, *J. Biomech. Eng.,* 102, 247, 1980.
37. **Friedman, M. H., Hutchins, G. M., Bargeron, C. B., Deters, O. J., and Mark, F. F.,** Correlation between intimal thickness and fluid shear in human arteries, *Atherosclerosis,* 39, 425, 1981.
38. **Fry, D. L.,** Response of the arterial wall to certain physical factors, *Ciba Found. Symp.,* 12, 93, 1972.
39. **Furie, M. B., Cramer, E. B., Naprotek, B. L., and Silverstein, S. C.,** Cultured endothelial cell monolayers that restrict the transendothelial passage of macromolecules and electrical current, *J. Cell Biol.,* 98, 1033, 1984.
40. **Geer, J. C. and Haust, M. D.,** Smooth muscle cells in atherosclerosis, *Monogr. Atheroscler.,* 2, 1972.
41. **Gerrity, R. G.,** The role of the monocyte in atherogenesis. I. Transition of blood borne monocytes into foam cells in fatty lesions, *Am. J. Pathol.,* 103, 181, 1981.
42. **Gerrity, R. G. and Naito, H. K.,** Alteration of endothelial cell surface morphology after experimental aortic coarctation, *Artery,* 8, 267, 1980.
43. **Gimbrone, M. A., Jr.,** Endothelial dysfunction and the pathogenesis of atherosclerosis, in *Atherosclerosis V, Proceedings of the Fifth International Symposium on Atherosclerosis* Gotto, A. M., Jr., Ed., Springer-Verlag, Berlin, 1980, 415.

44. **Gimbrone, M. A., Jr.,** Vascular endothelium and atherosclerosis, in *Vascular Injury and Atherosclerosis,* Moore, S., Ed., Marcel Dekker, New York, 1981, 25.

45. **Goode, T. B., Davies, P. F., Reidy, M. A., and Bowyer, D. E.,** Aortic endothelial cell morphology observed in situ by scanning electron microscopy during atherogenesis in the rabbit, *Atherosclerosis,* 27, 235, 1977.

46. **Grabowski, E. F., Jaffe, E. A., and Weksler, B. B.,** Prostacyclin production by cultured endothelial cell monolayers exposed to step increases in shear stress, *J. Lab. Clin. Med.,* 105, 36, 1985.

47. **Gutstein, W. H., Farrell, G. A., and Earmellini, C.,** Blood flow disturbance and endothelial cell injury in pre-atherosclerotic swine, *Lab. Invest.,* 29, 134, 1973.

48. **Harlan, J. M.,** Leukocyte-endothelial interactions, *Blood,* 65, 513, 1985.

49. **Helin, P., Lorenzen, I., Garbarsch, C., and Matthiesson, M. E.,** Atherosclerosis in the rabbit aorta induced by nor-adrenaline, *Atherosclerosis,* 12, 125, 1970.

50. **Kajiya, F., Hoki, N., Tomonago, F., and Nishihara, M.,** A laser doppler velocimeter using optical fiber and its application to local velocity measurements in the coronary artery, *Experentia,* 37, 1171, 1981.

51. **Karino, T. and Goldsmith, H. L.,** Disturbed flow in models of branching vessels, *Trans. Am. Soc. Artif. Intern. Organs,* 26, 500, 1980.

52. **Karino, T., Motomiya, M., and Goldsmith, H. L.,** Flow patterns in model and natural branching vessels, in *Fluid Dynamics as a Localizing Factor for Atherosclerosis,* Schettler, G., Ed., Springer-Verlag, Berlin, 1983, 60.

53. **Karnovsky, M. J.,** The ultrastructural basis of transcapillary change, *J. Gen. Physiol.,* 52, 645, 1970.

54. **Koslow, A., Stromberg, R., Gritsman, H., Friedman, L., and Batra, K.,** Response of endothelial cells grown on microcarrier beads to shear stress, *Fed. Proc.,* 43, 783, 1984.

55. **Ku, D. N. and Giddens, D. P.,** Pulsatile flow in a model carotid bifurcation, *Arteriosclerosis,* 3, 31, 1983.

56. **Ku, D. N., Giddens, D. P., Zarins, C. K., and Glagov, S.,** Pulsatile flow and atherosclerosis in the human carotid bifurcation. Positive correlation between plaque location and low and oscillating shear stress, *Arteriosclerosis,* 5, 293, 1985.

57. **Langille, B. L.,** Integrity of arterial endothelium following acute exposure to high shear stress, *Biorheology,* 21, 333, 1984.

58. **Langille, B. L. and Adamson, S. L.,** Relationship between blood flow direction and endothelial cell orientation at arterial branch sites in rabbits and mice, *Circ. Res.,* 48, 481, 1981.

59. **Langille, B. L., Reidy, M. A., and Kline, R. L.,** Injury and repair of endothelium at sites of flow disturbances near abdominal aortic coarctations in rabbits, *Arteriosclerosis,* 6, 146, 1986.

60. **Levesque, M. J., Liepsch, D., Moravec, S., and Nerem, R. M.,** Correlation of endothelial cell shape and wall shear stress in a stenosed dog aorta, *Arteriosclerosis,* 6, 220, 1986.

61. **Logerfo, F. W., Nowak, M. D., Quist, W. C., Crawshaw, H. M., and Bharadvaj, B. K.,** Flow studies in a model carotid bifurcation, *Arteriosclerosis,* 1, 235, 1981.

62. **Lutz, R. J., Cannon, J. N., Bischoff, K. B., Dedrick, R. L., Stiles, R. K., and Fry, D. L.,** Wall shear stress distribution in a model canine artery during steady flow, *Circ. Res.,* 41, 391, 1977.

63. **Moncada, S., Herman, A. G., Higgs, E. A., and Vane, J. R.,** Differential formation of prostacyclin by layers of the arterial wall: an explanation for the antithrombotic properties of vascular endothelium, *Thromb. Res.,* 11, 323, 1977.

64. **Motomiya, M. and Karino, T.,** Flow patterns in the human carotid artery bifurcation, *Stroke,* 15, 50, 1984.

65. **Nam, S. C., Lee, W. M., Jarmolych, J., Lee, K. T., and Thomas, W. A.,** Rapid production of advanced atherosclerosis in swine by a combination of endothelial injury and cholesterol feeding, *Exp. Mol. Pathol.,* 18, 369, 1973.

66. **Nerem, R. M., Levesque, M. J., and Cornhill, J. F.,** Vascular endothelial morphology as an indicator of blood flow, *J. Biomech. Eng.,* 103, 172, 1981.

67. **Nerem, R. M., Levesque, M. J., and Sato, M.,** Vascular dynamics and the endothelium, in *Frontiers in Biomechanics,* Schmid-Schonbein, G. W., Ed., Springer-Verlag, Berlin.

68. **Nerem, R. M., Mosberg, A. T., and Shwerin, W. D.,** Transendothelial transport of ^{131}I-albumin, *Biorheology,* 12, 31, 1975.

69. **Prior, J. T. and Hutter, R.,** Intimal repair of the aorta of the rabbit following experimental trauma, *Am. J. Pathol.,* 31, 107, 1955.

70. **Reidy, M. A. and Langille, H. L.,** The effect of local blood flow patterns on endothelial cell morphology, *Exp. Mol. Pathol.,* 32, 276, 1980.

71. **Remuzzi, A., Dewey, C. F., Davies, P. F., and Gimbrone, M. A.,** Orientation of endothelial cells in shear fields in vitro, *Biorheology,* 21, pp. 617-630, 1984.

72. **Rokitansky, C.,** The pathological anatomy of the organs of respiration and circulation, in *A Manual of Pathological Anatomy,* Day, G. E., Transl., The Sydenham Society, London, 1852.

73. **Ross, R. and Glomset, J. A.,** The pathogenesis of atherosclerosis, *N. Engl. J. Med.,* 295, 369, 420, 1976.

74. **Schwartz, S. M.,** Selection and characterization of bovine aortic endothelial cells, *In Vitro,* 14, 966, 1978.

75. **Schwartz, S. M. and Benditt, E. P.,** Clustering of replicating cells in aortic endothelium, *Proc. Natl. Acad. Sci. U.S.A.,* 73, 651, 1976.

76. **Sdougos, H. P., Bussolari, S. R., and Dewey, C. F.,** Secondary flow and turbulence in a cone-and-plate device, *J. Fluid Mech.,* 138, 379, 1984.

77. **Seed, W. A. and Wood, N. B.,** Velocity patterns in the aorta, *Cardiovasc. Res.,* 5, 137, 1971.

78. **Selden, S. C., Rabinovitch, P., and Schwartz, S. M.,** Effects of cytoskeletal disrupting agents on replication of bovine endothelium, *J. Cell. Physiol.,* 108, 195, 1981.

79. **Silkworth, J. B. and Stehbens, W. E.,** The shape of endothelial cells in en face preparations of rabbit blood vessels, *Angiology,* 26, 474, 1975.

80. **Simionescu, N.,** Enzymatic tracers in the study of vascular permeability, *J. Histochem. Cytochem.,* 27, 1120, 1979.

81. **Sinzinger, H., Silberbauer, K., and Auerswald, W.,** Quantitative investigation of sudanophilic lesions around the ostia of human fetuses, newborn and children, *Blood Vessels,* 17, 44, 1980.

82. **Smith, K. A., Colton, C. K., and Freedman, R. W.,** Shear stress measured at bifurcations, in *Fluid Dynamic Aspects of Arterial Disease,* Ohio University Press, Columbus, 1974, 12.

83. **Somer, J. B., Evans, G., and Schwartz, C. J.,** Influence of experimental aortic coarctation on the pattern of aortic Evans blue uptake in vivo, *Atherosclerosis,* 16, 127, 1972.

84. **Somer, J. B. and Schwartz, C. J.,** Focal 3H-cholesterol uptake in the pig, *Atherosclerosis,* 13, 293, 1971.

85. **Stehbens, W. E.,** Hemodynamics and atherosclerosis, *Exp. Mol. Pathol.,* 20, 412, 1974.

86. **Stewart, G. K. and Anderson, N. J.,** An ultrastructural study of endotoxin-induced damage in rabbit mesenteric arteries, *Br. J. Exp. Pathol.,* 52, 75, 1971.

87. **Taylor, C. B., Baldwin, D. and Hass, G. M.,** Localized arteriosclerotic lesions induced in the aorta of the juvenile rabbit by freezing, *Ann. N.Y. Acad. Sci.,* 174, 294, 1950.

88. **Vasile, E., Simionescu, M., and Simionescu, N.,** Visualization of the binding, endocytosis, and transcytosis of low density lipoprotein in the arterial endothelium in situ, *J. Cell Biol.,* 96, 1677, 1983.

89. **Virchow, R.,** *Cellular Pathology as Based Upon Physiological and Pathological Histology,* Chance, F., Transl., Churchill, London, 1860.

90. **Wechezak, A. R., Viggers, R. F., and Sauvage, L. R.,** Fibronectin and F-actin redistribution in cultured endothelial cells exposed to shear stress, *Lab Invest,* 53, 639, 1985.

91. **White, G. E., Fujiwara, K., Shefton, E. J., Dewey, C. F., and Gimbrone, M. A.,** Fluid shear stress influences cell shape and cytoskeletal organization in cultured vascular endothelium, *Fed. Proc.,* 41, 321, 1982.

92. **White, G. E., Gimbrone, M. A., and Fujiwara, K.,** Factors influencing the expression of stress fibers in vascular endothelial cells *in situ, J. Cell Biol.,* 97, 416, 1983.

93. **Wilkinson, P. C. and Lackie, M. J.,** The adhesion, migration and chemotaxis of leukocytes in inflammation, *Curr. Top. Pathol.,* 68, 48, 1979.

94. **Wright, H. P.,** Mitosis patterns in aortic endothelium, *Atherosclerosis,* 15, 93, 1972.

95. **Wright, H. P. and Giacometti, N. J.,** Circulating endothelial cells and arterial mitoses in anaphylactic shock, *Br. J. Exp. Path.,* 52, 1, 1971.

96. **Zarins, C. K., Giddens, D. P., Bharadvaj, B. K., Sottiurai, V. S., Mabon, R. F., and Glagov, S.,** Carotid bifurcation atherosclerosis. Quantitative correlation of plaque localization with flow velocity profiles and wall shear stress, *Circ. Res.,* 53, 502, 1983.

97. **Nerem, R. M. et al.,** personal communication.

Chapter 21

FLOW-INDUCED INTERACTIONS OF BLOOD CELLS WITH THE VESSEL WALL

Harry L. Goldsmith and Takeshi Karino

TABLE OF CONTENTS

I. INTRODUCTION

The mammalian circulation can be considered as the pulsatile flow of a concentrated suspension of deformable erythrocytes, ~8 μm diameter, in plasma (a Newtonian aqueous solution of salt and proteins) through a continuously branching network of vessels from >30 down to 0.005 mm in diameter. The mechanics of the motion of the suspensions is very complex. The flow regimes in the major arteries and veins[1,2] differ markedly from those in small vessels[3] and in the microcirculation.[4] Because of this, and the high number and volume concentration of erythrocytes which are in continuous collision with each other, it is very difficult to achieve a quantitative description of the mechanics of blood flow in the circulation.

Here, however, we are not concerned with an overall synthesis of the fluid dynamics in either the arterial or venous circulation; rather, we seek a description of events which occur on a scale of the same dimension as that of the corpuscles — local flow-induced interactions of cells with each other and the effect this motion has on their interaction with the vessel wall. For example, in thrombosis, flow-induced collisions between the corpuscles and the vessel wall in the presence of various coagulation factors lead to the formation of aggregates which adhere to the wall. An understanding of these aspects of blood flow requires a knowledge of the flow behavior of the individual corpuscles and the surrounding plasma in bulk flow, as well as in the immediate neighborhood of the vessel wall. To this end, it is necessary to observe particle and fluid motions under the microscope using the approach of microrheology,[5,6] whose aim is to predict the overall, macroscopic flow properties of a material from a detailed description of the behavior of its constituent elements. Here, the movements of the individual, isolated corpuscles are first studied, followed by observations of their collisions with each other and the vessel wall, initially at low, and finally at normal hematocrits.

Since the erythrocytes constitute 99% of the total volume of the corpuscles in mammalian blood, they effectively determine the macroscopic flow properties of blood, such as the relation between the pressure drop across a given length of a vessel and the volume flow rate. At the microscopic level, it is the erythrocytes that affect the movements of the less numerous platelets and leukocytes, and as we shall see, contribute to the wall interactions of these cells. Such interactions play an important role in the adhesion of platelets to the vessel wall in hemostasis and thrombosis,[7] and in the margination of leukocytes and their subsequent extravasation,[8] as well as in the adhesion to the endothelium and extravasation of circulating tumor cells in the microcirculatory bed.[9]

The present chapter gives an account of the microscopic, cellular events occurring in flowing blood under normal conditions, mainly derived from model studies carried out in our laboratory over the past 20 years. We begin by considering the flow and deformation of single cells in plasma subjected to flow in straight circular tubes, with special reference to their behavior near a wall. We then describe two- and multi-body collisions at progressively increasing hematocrit and how particle crowding modifies cell flow behavior. We also consider the behavior of blood cells at arterial bifurcations, T-junctions, and venous valves where flow is disturbed and secondary flows and eddies may form. Since it is known that platelet thrombi and atherosclerotic plaques form preferentially at such sites, it is quite possible that the enhanced and prolonged interactions of the formed elements and plasma with the vessel wall may dispose these regions of disturbed flow to atherogenesis and thrombogenesis. We will describe how flow patterns in channels of such geometry can enhance intercellular and cell-wall interactions.

II. FLOW THROUGH A CIRCULAR TUBE

A. Poiseuille Flow

We will begin with the simple case of Poiseuille flow, the steady laminar flow of a viscous

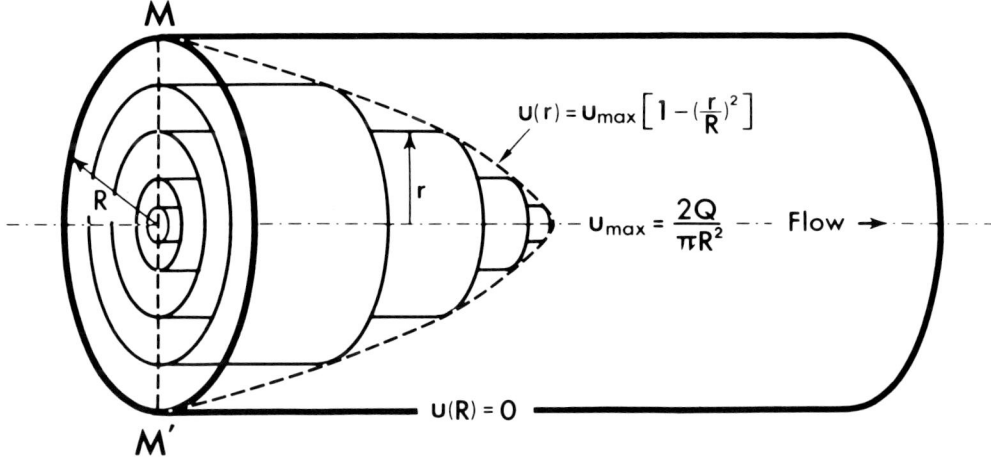

FIGURE 1.   Poiseuille or laminar viscous flow of a Newtonian liquid through a rigid circular tube. The motion of the liquid is pictured as the telescopic sliding of cylinders over each other. The layer at the wall is at rest and the velocity u(r) in the axial direction increases parabolically with increasing radial distance, r, reaching a maximum at the tube axis. (From Goldsmith, H. L., *Fed. Proc.*, 30, 1578, 1971. With permission.)

liquid through a rigid circular tube of radius R, since it provides the basis for describing the flow behavior of suspended particles and blood cells in real vessels.

As illustrated in Figure 1, Poiseuille flow may be likened to telescoping cylinders sliding over each other. The elements of the liquid at the wall are at rest since there is no slip of fluid at the interface. As one proceeds from the wall to the axis along the radial coordinate, r, adjacent layers or laminae slide over each other at progressively increasing linear velocities u, a function of r, given by:

$$u(r) = \frac{2Q}{\pi R^4} (R^2 - r^2) \tag{1}$$

where Q is the volume flow rate. Equation 1 is that of a parabola as shown by the dashed line in Figure 1. By normalizing the local velocity u(r) with the maximum velocity at the tube center, $u_{max} = u(O) = 2Q/\pi R^2$, the following simplified expression for the velocity distribution is obtained:

$$\frac{u(r)}{u(O)} = 1 - \left(\frac{r}{R}\right)^2 \tag{2}$$

While the linear fluid velocity, u(r), decreases parabolically with increasing r, the gradient in velocity or shear rate, G(r), increases linearly with increasing r, being zero at the tube center and a maximum at the tube wall:

$$G(r) = -\frac{dU}{dr} = \frac{4Q}{\pi R^4} r \tag{3}$$

Because of the gradient in velocity, work must be done to overcome the friction between adjacent layers of fluid sliding over each other. A measure of this work is given by the fluid shear stress, $\tau(r)$, defined as the tangential force per unit area exerted in the direction of flow on a layer of fluid at a radial distance r, by fluid at a radial distance r + dr. In the case of a liquid obeying Newton's Second Law, the shear stress is directly proportional to the shear rate, the constant of proportionality being the viscosity, η, or coefficient of internal friction:

$$\tau(r) = \eta G(r) \qquad (4)$$

The viscosity is thus the force per unit area required to maintain unit gradient in velocity.

B. Reynolds Number

In the larger vessels of the circulation where the velocities during the cardiac cycle are high (>100 mm/sec),[1,2] work is also required to accelerate and decelerate the fluid — so-called inertial work. The relative importance of liquid friction and liquid inertia in a given vessel is given by a dimensionless ratio known as the Reynolds number. In tube flow, the magnitude of the inertial force is proportional to the kinetic energy per unit volume, $m\overline{U}^2$/volume = $\rho\overline{U}^2$, where $\overline{U}$ is the mean velocity ($Q/\pi R^2$) and ρ the liquid density. The magnitude of the viscous force is proportional to the shear stress, ηG, where G is a representative shear rate in the tube = $\overline{U}/2R$. Hence, the tube Reynolds number, Re, is

$$Re = \frac{\rho\overline{U}^2}{\eta\overline{U}/2R} = \frac{2R\overline{U}\rho}{\eta} \qquad (5)$$

Using the above equations, valid for Poiseuille flow, representative values of Reynolds number and wall shear rate may be calculated for various parts of the vascular system from the known vessel diameters and volume flow rates. In the human circulation, maximum Reynolds numbers over one cycle decrease from ~6000 to <10^{-3}, in going from the heart to the microcirculation.[1,2]

Poiseuille flow, decribed above, is a simplified picture of blood flow in vessels, and does not strictly apply to the real circulation for various reasons. In large vessels there is not a fully developed flow with a parabolic profile. Flow is not steady but pulsatile; moreover, local flow is affected by branching and vessel curvature, where inertial effects result in the production of secondary flows having radial components, and sometimes in flow separation and eddy formation. In the smallest arteries and veins, deviations from the parabolic velocity profile are related to the effect of the red cells on the flow. Blood is not a homogeneous Newtonian fluid, and the velocity profiles are affected by flow rate and hematocrit, particularly in vessels <0.5 mm diameter. As a result of these considerations, the mean values of the wall shear rate may actually be appreciably higher than those calculated by assuming Poiseuille flow.

III. EFFECT OF SHEAR ON BLOOD CELLS AND VESSEL WALL

A. Vessel Wall

All aspects of the microscopic flow behavior of blood cells, be it the rotation, deformation, and wall migration of individual corpuscles, cell-cell collisions resulting in aggregation, or cell-wall collisions resulting in adhesion, owe their origin to the existence of a velocity gradient or shear rate in the circulating blood. The magnitude of the shear rate and the associated shear stress is of great importance both for the formed elements and the vessel wall that is in contact with blood. The exact role of shear rate and shear stress on the flow behavior of the cells is described in this and succeeding sections. With regard to the effect of shear stress on the vessel wall, it has been shown that vascular endothelial cells are affected by shear flow and exhibit morphological changes which correspond to the direction

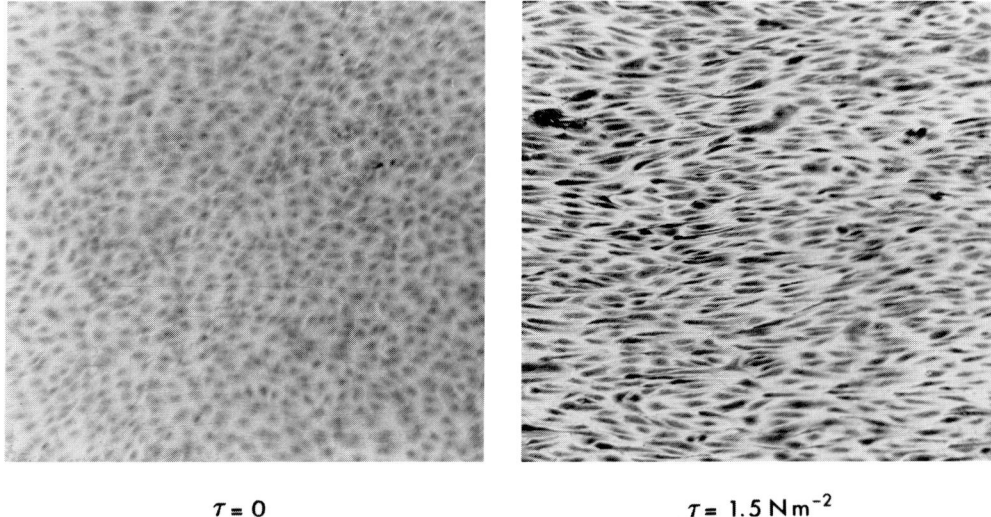

$$\tau = 0 \qquad\qquad\qquad \tau = 1.5 \, \mathrm{N \, m^{-2}}$$

FIGURE 2.   Cultured human umbilical vein endothelial cells at rest (A), and after exposure to laminar shear stress $\sim 1.5 \, N \, m^{-2}$ for 12 hr in a cone-plate shearing device (B). The sheared cells have become extremely elongated in the direction of flow (horizontally from left to right) with a pronounced tail on the downstream side.

and magnitude of the shear stress exerted on them.[10-13] Such shear stress-dependent changes in cell morphology are illustrated in Figure 2 by the photographs of cultured human endothelial cells subjected to laminar flow in a cone-plate shearing device.[14] As shown in the figure, the cells, which experienced a shear stress of $\sim 1.5 \, N \, m^{-2}$ for >12 hr, have undergone a drastic morphological change from a polygonal cobblestone shape at rest to an extremely elongated shape having an axis ratio, major:minor diameter, >10.

In man and other mammals, the shear rate, and hence shear stress, at the wall of arteries and veins varies considerably throughout the circulation. In the larger vessels, the wall shear rate also varies markedly during each cardiac cycle. The estimated mean wall shear rates over one cycle increase from $\sim 300 \, sec^{-1}$ in large arteries to $1500 \, sec^{-1}$ in arterioles and $>2000 \, sec^{-1}$ in capillaries, decrease to $\sim 70 \, sec^{-1}$ in postcapillary venules, and finally increase again to $\sim 200 \, sec^{-1}$ in large veins.[2,15] Hence, endothelial cells in different parts of the circulation are found to exhibit different morphology, depending on the magnitude of the local shear stress acting on them. In the arterial tree, the mean wall shear rates are expected to be greater than $300 \, sec,^{-1}$ exerting shear stresses $>1.0 \, N \, m^{-2}$ on endothelial cells. Thus, in the dog and rabbit aorta, most of the cells are found to be elongated and aligned in the direction of local blood flow,[10-12] in the manner shown in Figure 2. It is possible that such variation in cell morphology is linked to differences in biological and biochemical function of endothelial cells in different parts of the circulation.

B. Rotation and Deformation

1. Rigid Particles

When a particle is immersed in a liquid undergoing Poiseuille flow, it experiences stresses at its surface — there being no slip at the particle-fluid boundary. If the particle is rigid, the stresses cause it to rotate while it is carried by the flow along the tube. Thus, as shown in Figure 3, a sphere located at a fixed radial distance r rotates with uniform angular velocity, given by half the value of the shear rate at that location.[5,16] A spheroidal particle rotates with periodically varying angular velocity,[17] being maximum when its major axis is oriented across the flow, and a minimum when aligned with the flow. The rotation of a rigid disc whose angular velocity has been found to be in good agreement with theory for an oblate

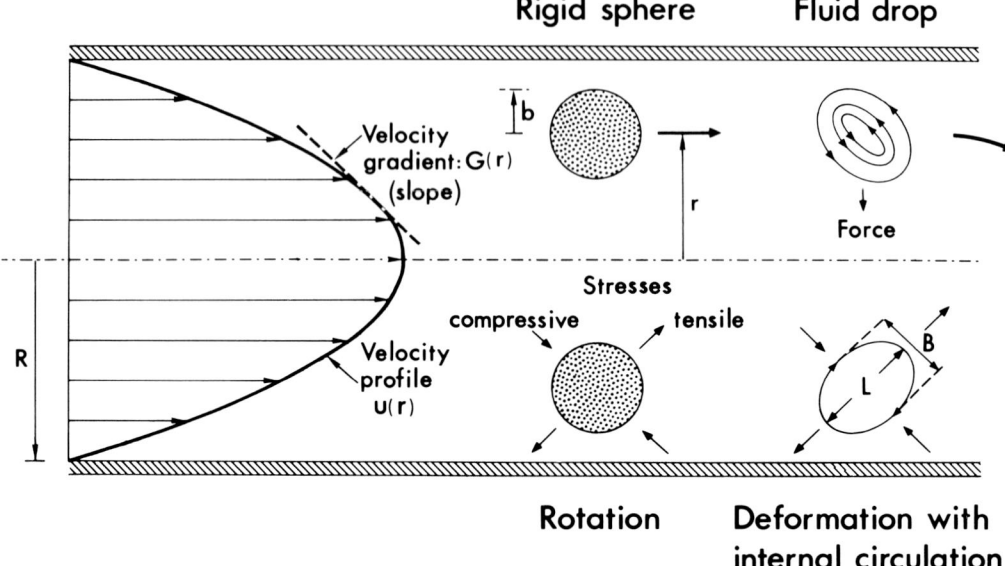

FIGURE 3.   The effect of Poiseuille flow on suspended spheres. The parabolic velocity profile and the linearly varying velocity gradient are shown in the median, MM' plane (see Figure 1) of the tube. The rigid sphere rotates with uniform angular velocity, whereas the liquid drop is distorted into an ellipsoid by the fluid stresses acting on its surface and aligns itself at a constant angle to the flow. As a result of a net inwardly directed force acting on the deformed drop, the particle migrates to the tube axis. (From Karino, T. and Goldsmith, H. L., in *Haemostasis and Thrombosis,* Bloom, A. L. and Thomas, D. P., Eds., Churchill Livingstone, 1986, chap. 42. With permission.)

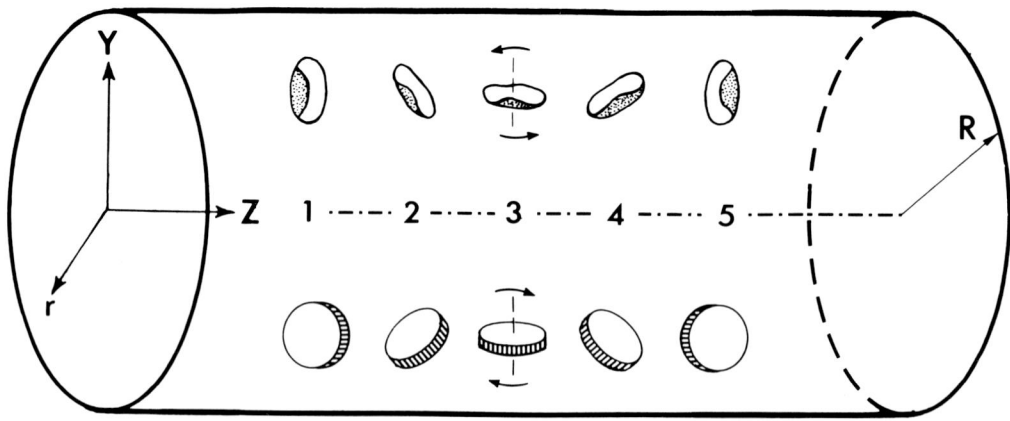

FIGURE 4.   The effect of Poiseuille flow on suspended rigid discs and on human erythrocytes at very low shear stresses. Shown are tracings from photomicrographs of the rotating particles at succeeding orientations during one half orbit, as seen in the median, MM' plane of a tube of radius R. The dashed line shows the axis of revolution of the particles whose angular velocities are greatest at positions 1 and 5. (From Goldsmith, H. L., *Fed. Proc.,* 30, 1578, 1971. With permission.)

spheroid[18,19] is shown in Figure 4. It serves as a model for the flow behavior of single human erythrocytes in plasma at hematocrits <1%,[20] provided the local shear stress is less than 0.1 N m^{-2}. The orientations of a single cell during one half orbit are also shown in Figure 4.

2. *Deformable Particles*

By contrast, if a particle is deformable, the external fluid stresses cause it to be distorted from its original shape. As shown in Figure 3, an emulsion droplet is deformed from a sphere into an ellipsoid, which is oriented at a constant angle to the direction of flow. The fluid stresses are transmitted across the droplet interface, and fluid at the surface and in the interior circulates about the particle center.

Erythrocytes can also exhibit fluid drop-like behavior at shear stresses >0.1 N m^{-2} if they are suspended in media more viscous than plasma, such as isotonic solutions of Dextran[21,22] and Ficoll.[22] The cells are then oriented at a constant angle to the flow with the membrane circulating about the interior.[23] Platelets and leukocytes, however, having structured interiors, remain undeformed in dilute suspensions and rotate as rigid discs[24] and rigid spheres, respectively.

C. Transport of Cells and Fluid

1. *Convective and Diffusive Transport*

The transport of corpuscles within the blood can occur in two ways: by convection due to bulk flow, and by diffusion due to Brownian motion. The overall transport of cells is referred to as convective-diffusive transport. In convective transport within blood vessels, both solutes and cells are carried by the plasma at velocities essentially equal to those of the fluid (providing the cell:vessel diameter ratio $<<1$)[19] given by a velocity distribution such as that in Poiseuille flow (Equation 1). The mean velocities over one cycle in the human circulation range from ~300 mm sec^{-1} in the major arteries down to 50 mm sec^{-1} in small arteries and 0.5 mm sec^{-1} in capillaries.[1,2] By contrast, diffusive transport, characterized by the translational Brownian diffusion coefficient, D_t, is relatively unimportant for the corpuscles and all but the smallest solute molecules. The diffusion coefficient is given by the Stokes Einstein equation[16]

$$D_t = \frac{kT}{F} \tag{6}$$

where k is the Boltzmann constant, T the absolute temperature, and F the friction coefficient related to the drag force opposing movement of the particle. For a sphere of radius b, F = $6\pi\eta b$, with η being the suspending medium viscosity. Values of the diffusion coefficient for various solutes of increasing molecular weight and for the corpuscles are given in Table 1.

We are concerned with transport of the corpuscles to the vessel wall. Since most flow in the body is laminar, or streamline, in the absence of cell-cell interactions, solutes or suspended cells are expected to travel along paths parallel to the vessel walls, and can only reach the endothelium through the random motion of Brownian translational diffusion. (In fact, as described below, cells actually migrate *away* from the vessel wall in flow.) An estimate of the distance which the corpuscles can travel in a direction normal to flow in a tube in a time interval Δt is given by the root mean square diffusion distance $\sqrt{\overline{\Delta x^2}}$, where:[16]

$$\overline{\Delta x^2} = 2D_t\Delta t \tag{7}$$

Table 1 gives values of the root mean square diffusion distances over a time interval Δt = 1 sec for some solute molecules in plasma and corpuscles, and compares these with the distance, Δz, over which the molecule or particle would have traveled parallel to the vessel wall had it been carried in Poiseuille flow through a large 6-mm diameter and a small 0.3-mm diameter artery with its center of rotation at a radial distance of only 2μm from the

Table 1
DIFFUSION AS A FUNCTION OF MOLECULAR AND PARTICLE SIZE

Molecule or cell	Molecular weight	Molecular or particle radius (nm)	D^a 37°C $m^2 \ sec^{-1}$	$\sqrt{\overline{\Delta x^2}}^b$ ($\Delta t = 1$ sec) μm	$\dfrac{\Delta z}{\sqrt{\overline{\Delta x^2}}}^c$ R = 0.3 mm	6.0 mm
Water	18	0.15	1.26×10^{-9}	50.2	53	13
Oxygen	32		2.75×10^{-9}	74.2	36	9
Sodium chloride	58.5	0.23	8.22×10^{-10}	40.5	65	16
Glucose	180	0.37	5.11×10^{-10}	32.0	83	21
Sucrose	342	0.48	3.94×10^{-10}	28.1	94	24
Serum albumin	67,000	3.6	5.25×10^{-11}	10.3	257	65
Fibrinogen	330,000	9.5	2.00×10^{-11}	6.3	420	106
Platelet	—	1,200	1.58×10^{-13}	0.56	4,730	1,190
Red cell	—	2,800	6.75×10^{-14}	0.37	7,160	1,800

a   Calculated from Equation 6 using molecular or particle radius, and plasma viscosity $\eta = 1.2$ mPa s at 37°C.
b   Root mean square distance moved in 1 sec, calculated from Equation 7.
c   Relative distances moved in 1 sec by a substance or cell located 2 μm from the wall of a small artery (0.3-mm diameter, $\overline{U} = 50$ mm sec^{-1}) and a large artery (6.0-mm diameter, $\overline{U} = 250$ mm sec^{-1}) *in flow* (Δz) calculated using Equation 2, and *in diffusion* ($\sqrt{\overline{\Delta x^2}}$).

wall. It should be noted that the calculated values of Δz may be overestimated by as much as 50%, since there is a retardation of both translational and rotational velocities when the cells are within two particle diameters of the wall.[25] It is evident that for the corpuscles in both vessels, $\Delta z >> \sqrt{\overline{\Delta x^2}}$, and that, therefore, one would not expect wall interactions to be significant in dilute suspensions. This becomes more apparent when one considers the observed inward radial migration of blood cells. It should be noted, however, that there are disturbed flows in the circulation in which the streamlines are curved toward the vessel wall and corpuscles are transported by convection to the endothelium. Such flow patterns are described in Section VI below.

2. Wall Migration

An important consequence of particle deformation in Poiseuille flow, such as that shown for the liquid drop in Figure 3, is lateral migration away from the wall toward the axis.[19,26] This occurs at low tube Reynolds numbers ($<<1$) when, rigid, nondeformable particles flow along paths parallel to the tube wall. The effect arises because of the interaction of the deformed particle with the wall, giving rise to an inwardly directed force which is absent in the case of the undeformed particle. As shown by Equation 8 describing the migration of deformable fluid drops in Poiseuille flow, the rate of migration, dr/dt, increases with increasing deformation, flow rate, and the third power of the particle to tube diameter ratio, and rapidly diminishes as the distance of the particle center from the wall, = (R − r), increases:

$$- \frac{dr}{dt} = \frac{D\overline{U}b^3}{R^2} \cdot \frac{r}{(R - r)^2} \cdot f(\lambda) \tag{8}$$

Here, b = undeformed drop radius, $f(\lambda)$ = function of the viscosity ratio λ = drop:suspending medium viscosity, and D is the drop deformation = L − B/L + B, where L and B are the length and breadth of the ellipsoidal drop (Figure 3).

At higher tube Reynolds numbers (>1), when inertial forces in the suspending liquid become appreciable, there is migration of rigid particles away from the tube wall *as well as* away from the tube axis toward an equilibrium radial position r*. This, the so called tubular pinch effect,[27,28] has its origin in the inertia of the fluid and the interaction of the particle with the wall,[25,29] and may be likened to the curving of a spinning tennis ball (Magnus effect), although it must be emphasized that nonrotating particles also migrate at these Reynolds numbers and that the presence of the tube wall is an essential feature of two-way migration.[29,30] As shown by Equation 9, for the case of rigid spheres, the rate of migration is found to increase with increasing Re, also rapidly increases with increasing ratio of particle to tube diameter, and decreases as the equilibrium position is approached:

$$\frac{dr}{dt} = 0.2\overline{U}Re\left(\frac{b}{R}\right)^3 \cdot \frac{r}{R}\left(1 - \frac{r}{r*}\right) \tag{9}$$

As shown in Figure 5, human erythrocytes have been observed to migrate away from the walls of small tubes <200 μm in diameter, and the results are explicable in terms of the above two migration mechanisms. Thus, at Re <10^{-2}, normal erythrocytes exhibit appreciable migration whereas glutaraldehyde fixed cells do not.[22] In viscous, low-molecular-weight Dextran solution, normal erythrocytes oriented with the flow migrate inward at markedly higher rates than cells in plasma at the same local shear rate. Thus, at a mean tube shear rate = 50 sec^{-1} in an 85-μm diameter tube, a normal erythrocyte in plasma, starting from the wall, migrates 4 μm inward while traveling 10 mm downstream, compared to a normal cell in a Dextran solution of 50 mPa sec viscosity which migrates 22 μm inward. By contrast, a rigid aldehyde-fixed cell migrates only 0.2 μm inward while traveling 10 mm downstream.

At higher Re, normal erythrocytes become subject to the tubular pinch effect, and as with rigid cells, migration out from the tube axis is also observed.[22] Presumably, this is due to the fact that the normal cell maintains a degree of rigidity.

Given the high mean wall shear rates prevailing throughout the circulation (from 300 sec^{-1} in the largest arteries to 1500 sec^{-1} in arterioles),[2] one would expect inward migration of erythrocytes due to deformation to become increasingly important as vessel diameters decrease to values not too much larger than those of the cells (R <100 μm). Migration due to inertia would not be expected to play a large role since in vessels where the mean tube Reynolds number is high, the ratio of cell to vessel diameter is too low, and where this ratio becomes appreciable, the tube Reynolds numbers are too low (cf. Equation 9).

Inward lateral migration of the relatively rigid leukocytes[31] and platelets[32,33] has also been observed, but the rate of migration is, as expected, appreciably lower than that of normal red cells. Nevertheless, in the absence of cell interactions, one would not expect to find any platelets close to the vessel wall. However, the situation in blood at hematocrits of 40 to 50% is markedly different. Here, due to the presence of large numbers of erythrocytes, there are continual interactions greatly affecting the motions of the less numerous platelets and leukocytes, as described in Section IV.

IV. CELL INTERACTIONS IN TUBE FLOW

A. Cell Collisions: Two-Body Collision Model

The velocity gradient in Poiseuille flow is responsible for bringing suspended cells, carried on adjacent streamlines, into collision with each other. A simple model of this process is the symmetrical two-body collision between equal-sized rigid spheres of radius b in a suspension containing n particles per milliliter subjected to Poiseuille flow, as shown in the upper panel of Figure 6. The number of collisions suffered by a given sphere per second, defined as the collision frequency, f_s, is given by:[5,34]

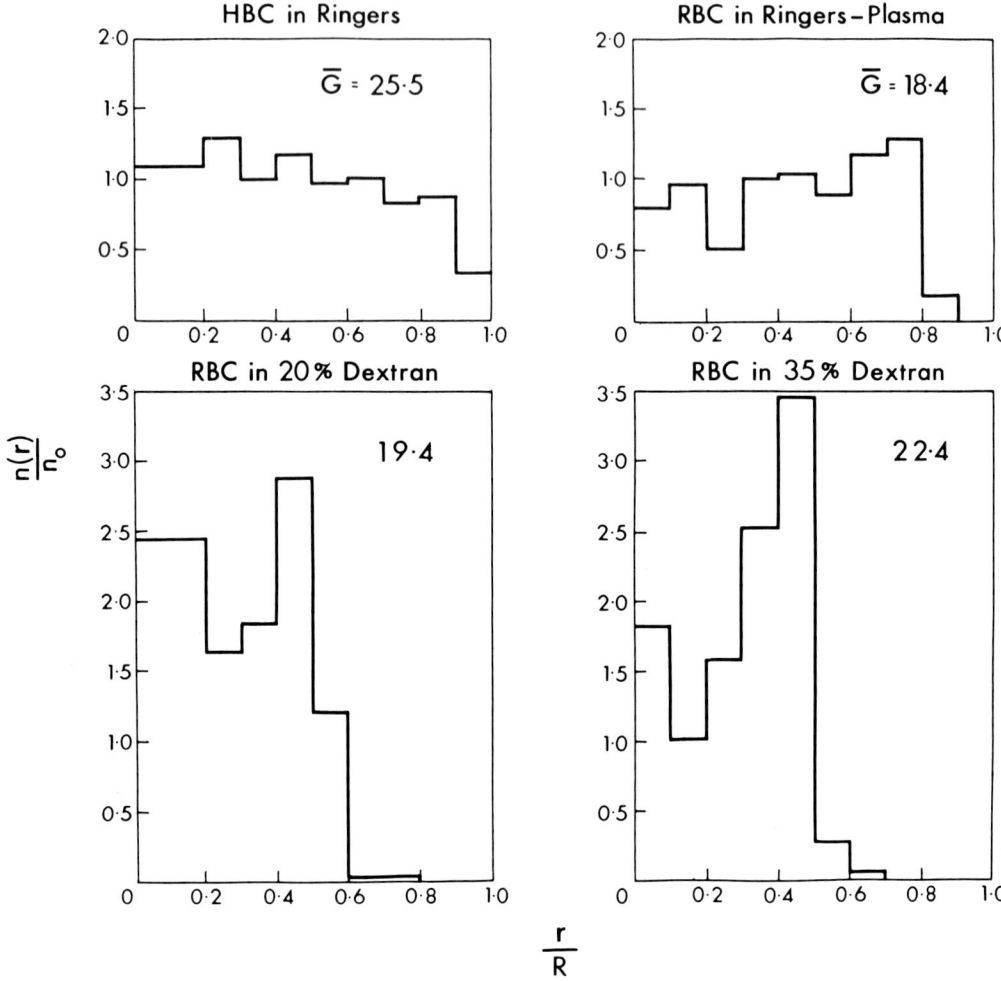

FIGURE 5. Radial migration of normal erythrocytes (RBC) and glutaraldehyde hardened rigid erythrocytes (HBC) at low Reynolds numbers in an 83-μm diameter tube measured 10 mm downstream from its entry from a reservoir. Shown are the number concentration distributions of erythrocytes, plotted as the number of cells per milliliter of suspension n(r) at intervals of $^1/_{10}$ tube radius, divided by the number per milliliter in the reservoir n_o against the dimensionless radial distance r/R. If the distribution were uniform across the tube, $n(r)/n_o = 1$ at all r/R. The figure shows that the deformable erythrocytes undergo appreciable inward migration, especially in the viscous Dextran solutions, in contrast to the rigid erythrocytes whose distribution is fairly uniform across the tube. The values of the mean shear rate $\overline{G} = 2\overline{U}/R$ are given in each panel. (From Goldsmith, H. L., *Fed. Proc.*, 30, 1578, 1971. With permission.)

$$f_s = \frac{32}{3} G(r)nb^3 \qquad (10)$$

$$= \frac{8G(r)c}{\pi} \qquad (11)$$

since the volume concentration $c = 4\pi b^3 n/3$. Thus, the total number of two-body collisions per second per milliliter of the suspension $F_s = fn/2$ is greater at the tube periphery where $G(r)$ is larger, and is greatly affected by the size of particle. In very dilute suspensions at

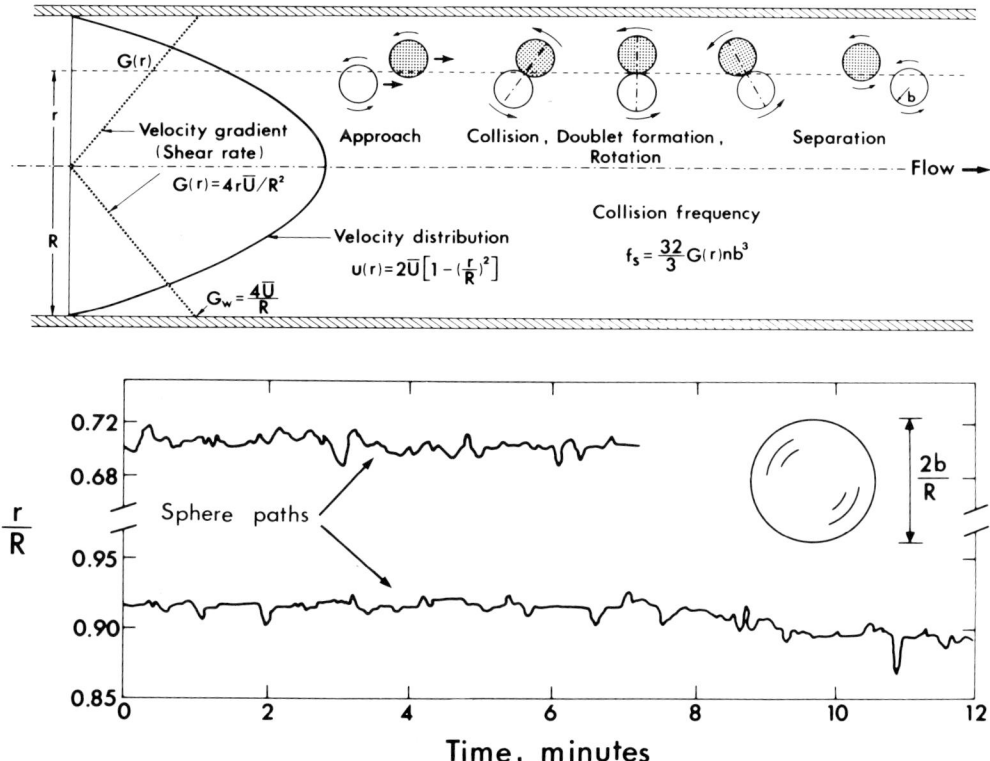

FIGURE 6. (A) A two-body, symmetrical collision between rigid spheres in Poiseuille flow. The spheres, traced from photomicrographs of 2-mm diameter polystyrene spheres in a viscous oil in an 8-mm diameter tube,[35] are shown during approach, doublet formation and rotation, and separation, with the final radial positions and velocities being the same as those before collision. (B) The variation in radial distance with time of a single sphere being tracked in flow in a suspension 2% by volume, owing to such two-body collisions.

the same number concentration and shear rate, the two-body collision frequency for red cells of equivalent sphere radius $b = 2.8$ μm (calculated from a mean cell volume $= 90$ μm^3) is more than eight times greater than that of platelets whose equivalent sphere radius $b = 1.3$ μm (mean cell volume $= 10$ μm^3). Equations 10 and 11 may be applied to platelet-rich plasma, where it can be shown that for $n = 3 \times 10^5$ cells per microliter, ($c = 0.3\%$), and at a mean wall shear rate over one cycle $= 300$ sec^{-1}, representative of those in a major artery with a diameter of 6 mm,[2] there would be 2.1 collisions per cell per sec at a distance of 2 μm from the wall, and a total of 7.1×10^6 two-body collisions per second in a 6-mm-long annulus extending 20 μm out from the wall.

B. Radial Dispersion of Particle Paths

Of significance with regard to cell-wall interactions is the fact that, as may be seen in Figure 6, the radial position of both particles are displaced during a collision. Thus, a particle flowing very close to the vessel wall may come into contact with the wall during a two-body collision. However, due to the pronounced increase in hydrodynamic resistance as the particle surface approaches very close to a wall, two-body collisions near the wall are asymmetric, with the faster moving particle being more radially displaced than the slower moving one close to the wall.[35]

As the concentration of particles or cells increases, the radial dispersion of their paths becomes more pronounced as three-, four-, and multi-body collisions occur. The multi-body collision frequency is appreciable even at hematocrits as low as 5%. Such collisions result

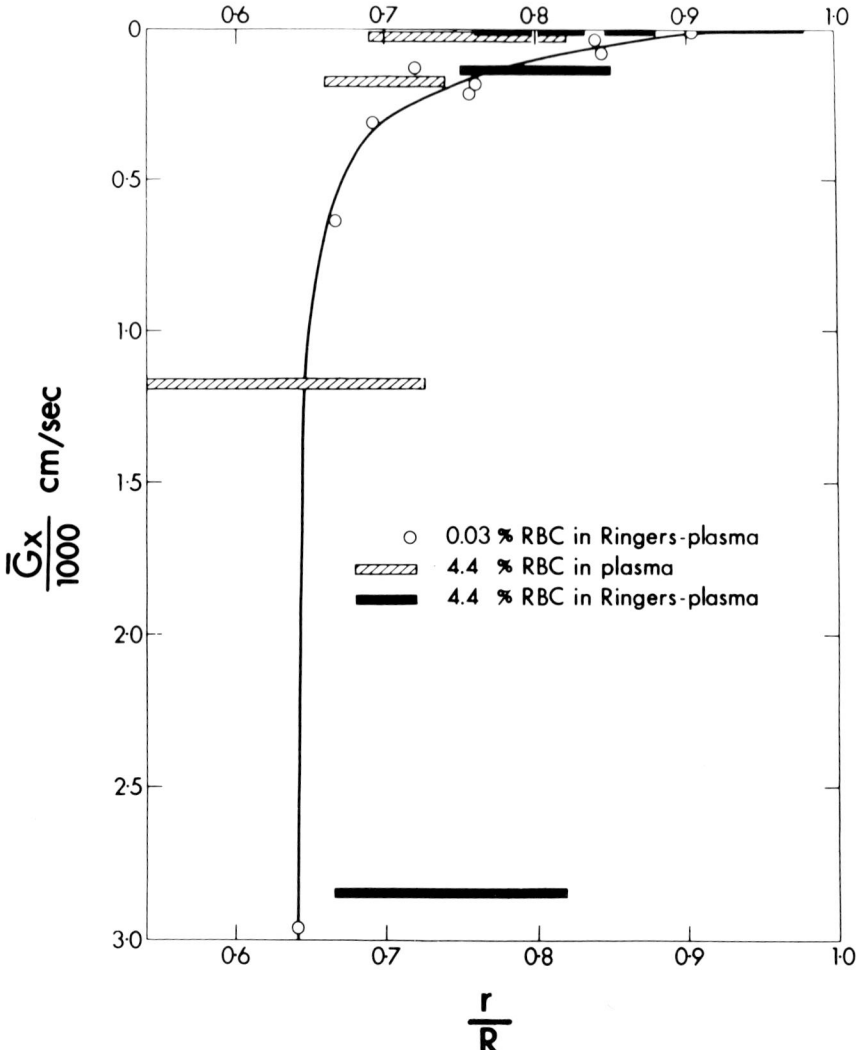

FIGURE 7.   The inward migration of human erythrocytes in a Ringers-plasma (in the absence of rouleau formation) at 0.03% hematocrit in the virtual absence of cell-cell interactions compared with that at 4.4% hematocrit when there are frequent multi-body collisions. Also shown is the migration in plasma at 4.4% hematocrit when rouleaux are present. The mean shear rate times the distance from the tube entry is plotted against the relative radial distance of the outermost single cell or aggregate. The bands represent the measured spread in the width of the particle-free layers from the wall obtained from 50 measurements taken from different frames of the cinefilm of the flowing suspensions. (From Goldsmith, H. L., *Fed. Proc.*, 30, 1578, 1971. With permission.)

in a significant radial dispersion of erythrocytes in tube flow, as illustrated by the data shown in Figure 7. The inward migration of erythrocytes in plasma, diluted with Ringers solution to prevent rouleau formation, is seen to be greater at 0.03% hematocrit when there are very few two-body collisions in the flow, than at 4.4% hematocrit when there are frequent multi-body collisions.

C. Aggregation of Cells

With blood cells, attractive forces acting while the cell membranes are in close apposition during a collision can lead to the formation of permanent doublets.[36] Rouleaux of erythrocytes in plasma readily form when suspensions of cells flow through tubes at mean shear rates $<50 \text{ sec}^{-1}$. The effect of such aggregation in enhancing wall migration is shown in Figure 7 where the rouleaux which form in the plasma, and which easily deform in the flow[20] (in a manner similar to that of flexible fibers),[37] migrate toward the tube axis more rapidly than the smaller single erythrocytes, and combine into larger stable aggregates in the low shear region.

The kinetics of platelet aggregation in Poiseuille flow has been studied by activation of platelets in plasma with adenosine diphosphate (ADP) using a double-infusion technique.[38] It was shown that the rate of aggregation increases with increasing time and shear rate.[39] In the absence of erythrocytes and at $1 \mu M$ ADP, microaggregates first form near the tube wall, and as they grow in size through collisions with other singlets and aggregates in flow along the tube, they migrate toward the axis because their rotation is physically impeded by the wall, and eventually form a large, stable but loosely bound aggregate at the tube center.[39] In the presence of erythrocytes, however, this is unlikely to happen. Even at a hematocrit of only 5%, calculations using theory applicable to rigid spheres[5] show that one third of all erythrocytes are in collision at any given instant. The resulting continuous lateral fluctuations of cell paths will affect the platelet motions and would be expected to impede the growth of the microaggregates. The effect will become more pronounced with increasing hematocrit as the platelets are surrounded by a large number of erythrocytes having a nonthrombogenic membrane. Indeed, it has been shown that the addition of erythrocytes at a hematocrit of only 4% to a flowing suspension of activated platelets results in a drastic reduction in the size of platelet aggregates.[40] In the circulation, therefore, erythrocytes are likely to play a preventative role in the growth of freely flowing platelet aggregates or thrombi. However, as described below, erythrocytes have a pronounced effect on both the distribution and flow behavior of platelets near the vessel wall.

V. CELL CROWDING AND WALL INTERACTIONS AT NORMAL HEMATOCRITS

A. Effects at Normal Hematocrits

When the hematocrit exceeds 20%, there is marked change in the flow regime, due primarily to crowding of the erythrocytes and the resulting cell-cell collisions. The following effects are obtained:

1.   The velocity profile in the tube is no longer parabolic as in Poiseuille flow but is blunted with a region of "partial plug flow" in the tube center where all the particles move with the same maximum velocity which is less than the centerline velocity u(0) in Poiseuille flow at the same volume flow rate (Figure 8).
2.   Deformation of red cells in blood occurs to a degree that is not attributable to shear alone (Figure 9).
3.   The blood cell paths exhibit erratic displacements in a direction normal to the flow. This is of particular importance with regard to the motions of platelets and their interactions with the vessel wall (Figures 10 to 12).

Effects 1 and 3 occur in many concentrated suspensions and have been documented by using photo-optical methods in which the systems are made transparent to transmitted light and a small quantity of visible particles are added to serve as tracers of the particle movements in the interior of the flowing system. In model suspensions of rigid spheres or disks, and in emulsions of oil droplets, transparency is achieved by matching refractive indices of the suspended phase to that of the particles.[41-43] In blood, transparency is achieved by use of

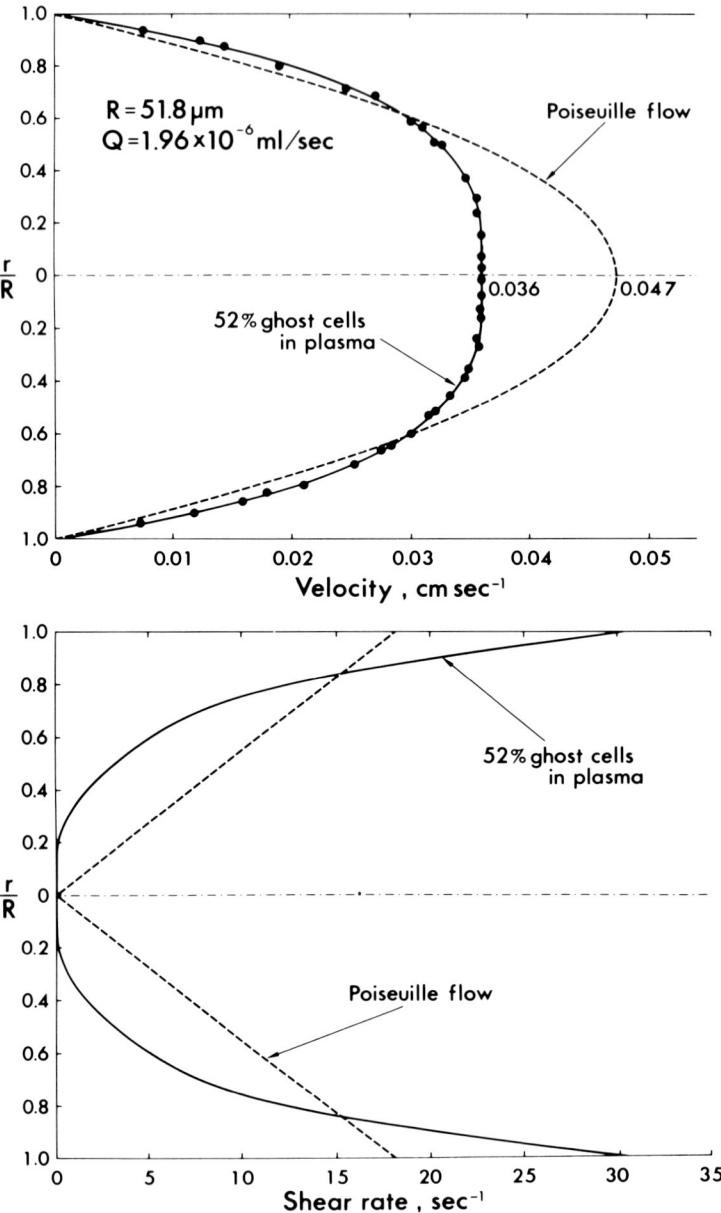

FIGURE 8.   Distribution of velocity and shear rate as a function of the relative radial position in a 52% ghost cell suspension containing tracer erythrocytes flowing through a 103.6-μm diameter tube. The points represent the measured translational velocities of the erythrocytes;[3] the line is the best fit drawn by eye. The solid line in the lower panel was obtained from measurements of the slope of the velocity distribution curve. The dashed lines represent the calculated velocity distribution and shear rate in a Poiseuille flow at the same volume flow rate as in the ghost cell suspension. The graphs illustrate the blunting of the velocity profile from the parabolic which occurs at hematocrits above 20%, with the zones of zero shear rate close to the tube axis where there is a partial plug flow. (After Karino, T. and Goldsmith, H. L., in *Haemostasis and Thrombosis,* Bloom, A. L. and Thomas, D. P., Eds., Churchill Livingstone, Edinburgh, 1986, chap. 42. With permission.)

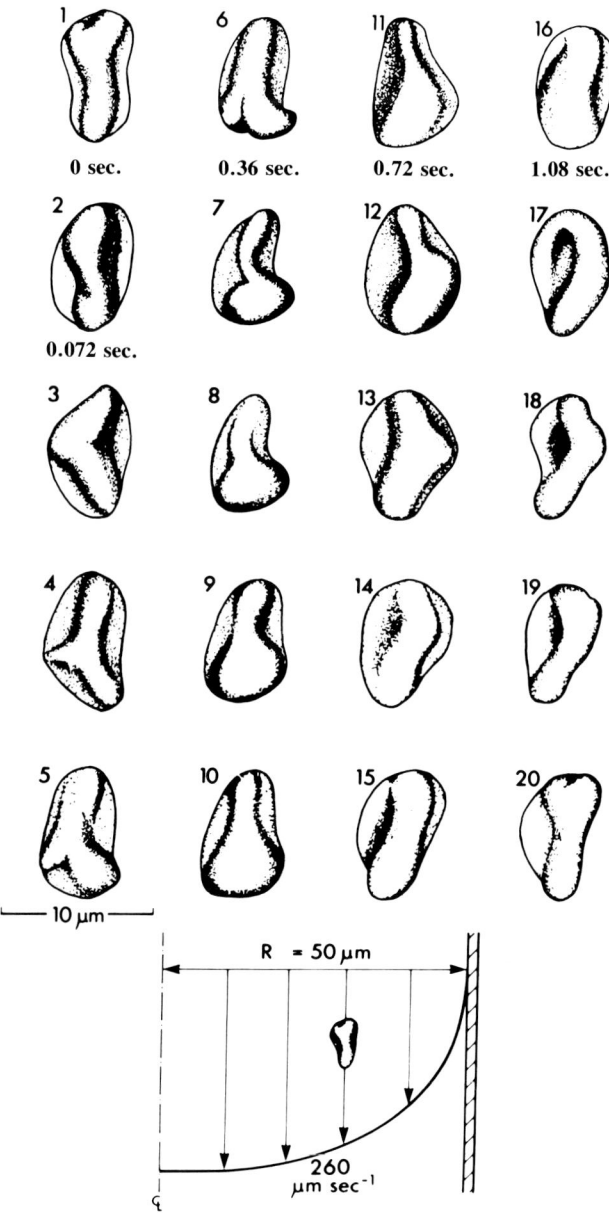

FIGURE 9. Tracings from photomicrographs of the successive deformation of a tracer erythrocyte at 0.072-sec-intervals in a 55% ghost cell suspension flowing through a 100-μm diameter tube at a mean velocity $\overline{U}$ = 250 μm sec^{-1}. The radial position of the cell (r = 0.6 R) and velocity profile are also shown. The erythrocyte is almost always aligned with the direction of flow. (From Goldsmith, H. L. and Skalak, R., *Annu. Rev. Fluid Mech.*, 7, 213, 1975. With permission.)

reconstituted biconcave ghost cells in plasma.[3,22,44] Particles of the same size and shape but of different refractive index serve as tracers in the model suspensions. Normal erythrocytes, leukocytes, platelets, or microspheres serve as tracers in the ghost cell suspensions.

Effect 2 is observed in emulsions and in ghost cell suspensions where, it appears that

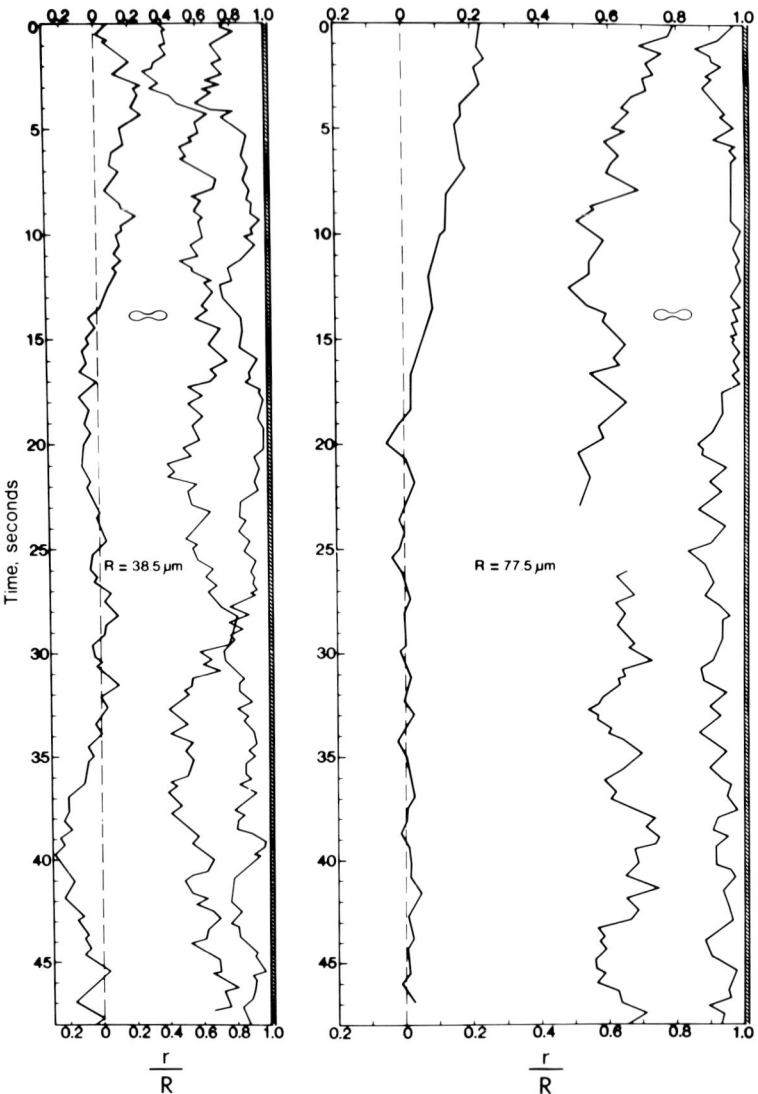

FIGURE 10. Radial displacements of tracer erythrocytes in a 40% ghost cell suspension in a 77 (left) and 155-μm (right) diameter tube. The cells situated in the region between r = 0.4 R and 0.8 R appear to undergo the largest radial displacements. In the smaller tube, for the cell near the center at a mean $\bar{r}$ = 0.092 R, the root mean square radial displacement $\sqrt{\overline{\Delta r^2}}$ = 2.00 μm; in the larger tube, for $\bar{r}$ = 0.076 R, $\sqrt{\overline{\Delta r^2}}$ = 1.62 μm. The values of $\bar{r}$/R and $\sqrt{\overline{\Delta r^2}}$ for the cells at the intermediate radial position were 0.583 and 3.32 μm, and 0.662 and 3.84 μm, in the 77- and 155-μm tubes, respectively. For the cells near the tube wall, the values were 0.830 and 2.96 μm, and 0.952 and 1.97 μm, respectively. (From Goldsmith, H. L. and Marlow, J., *J. Colloid Interface Sci.*, 71, 383, 1979. With permission.)

particle deformation is responsible for the low relative viscosities (viscosity of blood/viscosity of plasma) compared to those in rigid particle suspensions at the same concentrations.[45,46] The remarkable deformation of erythrocytes that is seen at concentrations >30%, illustrated in Figure 9, occurs at shear rates <7 sec^{-1}, corresponding to shear stresses <0.03 Nm^{-2}, i.e., values at which an isolated erythrocyte in plasma would rotate as a rigid disk (Figure 4). The erythrocytes are deformed into continually changing shapes, and rotate only irreg-

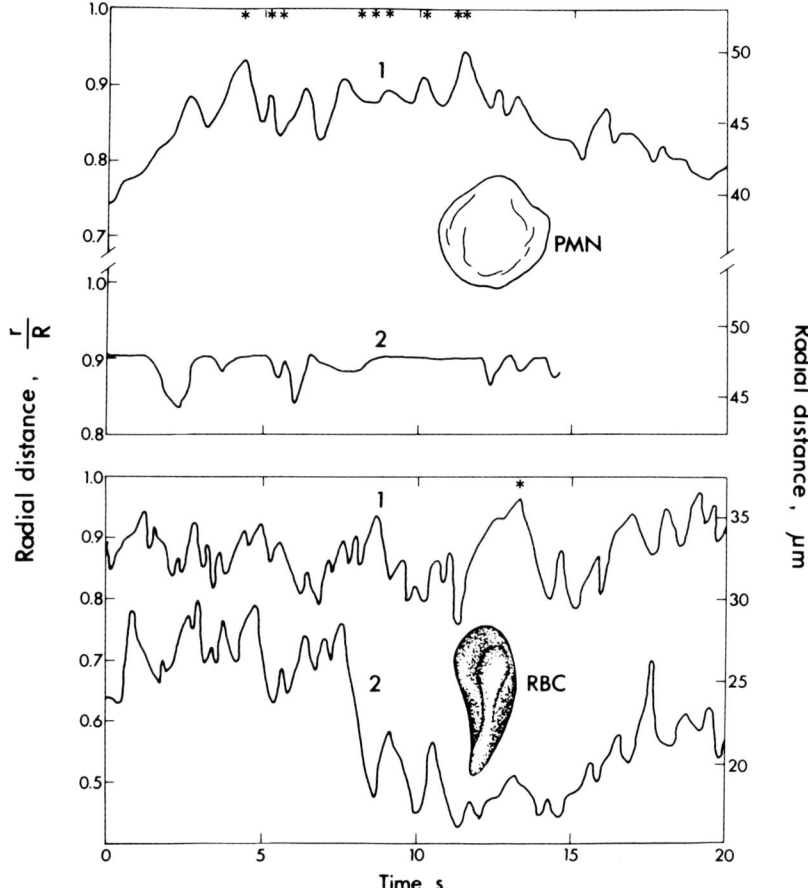

FIGURE 11.   Radial displacements of tracer blood cells in a 40% ghost cell suspension due to continual collisions between cells. Plot of the relative radial position of a polymorphon-uclear leukocyte (PMN) in a 106-μm diameter tube, and an erythrocyte in a 76-μm diameter tube. The deformed cells were drawn from actual cinemicrographs and are typical of shapes seen during flow. Cells being tracked along the paths (1) in the periphery of the tube collided with the wall at times indicated by the asterisks. (From Goldsmith, H. L. and Karino, T., *Ann. N.Y. Acad. Sci.*, 283, 241, 1977. With permission.)

ularly. At ghostcrits >50%, the cells are continuously squeezed and no longer rotate as a whole; instead, the membrane probably moves around the cell interior in an irregular fashion.

B. Cell-Wall Interactions
1. Erythrocyte-Induced Dispersion in Flowing Blood
The presence of erythrocytes at volume concentrations from 40 to 50% considerably disturbs the motion of the formed elements in the plasma. One might have expected that in such a concentrated suspension, the erythrocytes would actually impede the diffusion of other cells and solutes. This is true when blood is stationary, but the opposite is observed in flow. The continued collisions between and deformation of the erythrocytes in flowing blood actually lead to a continual radial displacement of their paths and an alternate method of solute mixing, similar on a macroscopic scale to the intermolecular collisions which result in Brownian motion. In fact, as described below, for particles as large as the erythrocytes, the movement induced by this mixing results in radial dispersion, or effective diffusion coefficients which are two to three orders of magnitude greater than those due to Brownian

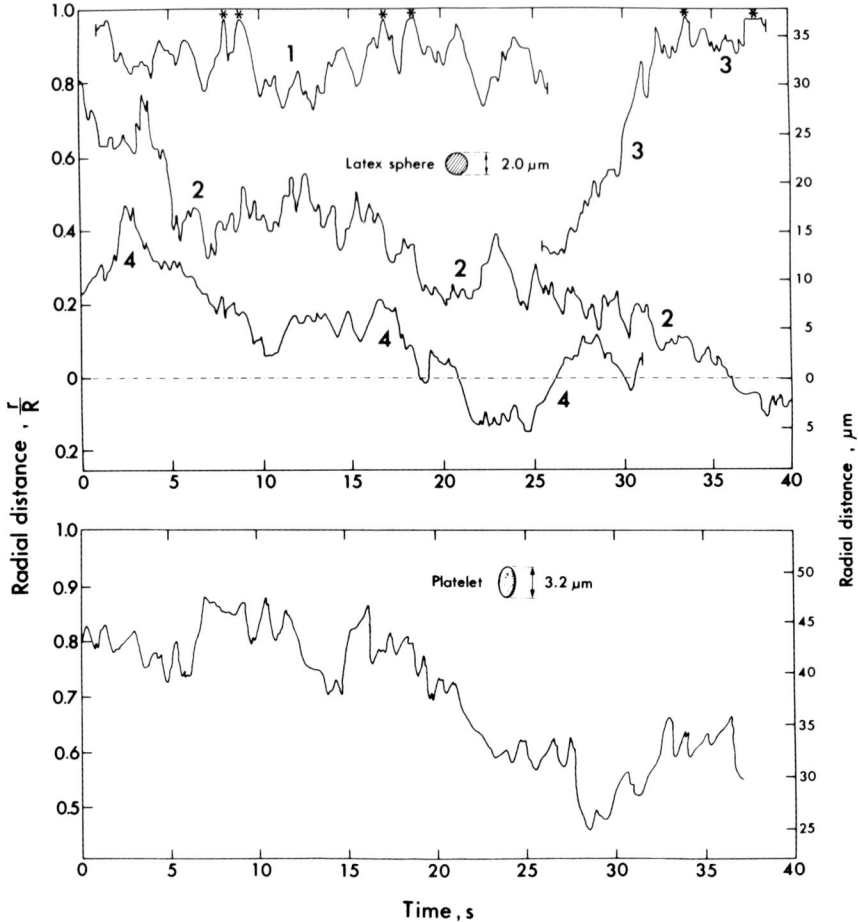

FIGURE 12.   Plot as in Figure 11 of the radial displacements of a 2-μm diameter latex sphere in a 75-μm diameter tube, and a human platelet in 104-μm diameter tube in 42% ghost cell suspensions.

motion, and this mixing motion becomes the major mechanism for permitting cells and possibly certain of the larger proteins to interact with the vascular surface.

The radial displacement of the cells in blood at normal hematocrits has been investigated using the model of the transparent ghost cell suspensions. Measurements of the paths of tracer erythrocytes, 2-μm diameter latex spheres, polymorphonuclear granulocytes, and platelets have been carried out over a range of ghostcrits, c, from 10 to 70%.[3,22,47] The magnitude of the particle radial displacements at first increases with c and then decreases at c >50%, due to the pronounced cell deformation brought about by particle crowding at the high ghostcrits. As illustrated in Figure 10, at a given c, the erythrocyte displacements are greatest at radial distances between 0.5 R and 0.8 R. Close to the wall, radial excursions of the cell paths are somewhat inhibited by the presence of the boundary. The amplitude of the displacements appears to be related both to particle deformability and size, being somewhat greater for red cells, 2-μm spheres, and platelets than for the leukocytes (Figures 11 and 12).

By analogy with Brownian translational diffusion (Section III. C), a radial dispersion coefficient, D_r, obtained from the mean square radial distance $\overline{\Delta r^2}$ traveled by the particle in a time interval Δt:

$$D_r = \frac{\overline{\Delta r^2}}{2\Delta t} \qquad (12)$$

can be calculated. This does not imply that the motions are considered to be random, as in Brownian diffusion, even though the paths in Figures 10 to 12 have the appearance of erratic random traces. The fluctuations in r are due to multibody collisions determined by the local velocity gradient, cell concentration, and deformation.

At local shear rates varying from 5 to 20 sec^{-1} and ghostcrits from 20 to 70%, the values of the calculated radial dispersion coefficients ranged between 10^{-12} to 2×10^{-11} m^2 sec^{-1} for red cells and 2-μm microspheres.[3] These values are one to two orders of magnitude greater than the translational Brownian diffusion coefficients for the blood cells (see Table 1).

As a consequence of the radial dispersion of the erythrocytes, leukocytes, and platelets, the frequency of wall encounters of the corpuscles increase with increasing hematocrit. In the case of the platelet, this has been correlated with the observed increased adhesion to subendothelium[48,49] and other surfaces[50,51] as the hematocrit is increased. The measured increase in diffusivity of platelets in blood as compared with that in plasma,[52] and the resulting increased cell-wall interactions are of considerable importance in thrombotic processes as discussed in the succeeding section. Values of the effective platelet diffusion constant in whole blood have been estimated to range from 3×10^{-11} to 8.6×10^{-10} m^2 sec^{-1}, as the shear rate is increased from 50 to 10^4 sec^{-1}.[49] An even more striking increase in platelet adhesion to subendothelium with increasing hematocrit has been noted at the highest shear rate.[53] The extent to which erythrocyte deformation or size may affect the increase in the effective diffusion constant of cells and solute has also been studied. The transport of a molecular solute was found to increase slightly with both cell size and rigidity,[54] and the wall adhesion of platelets was observed to increase substantially with erythrocyte size.[55] From the values of D_t given in Table 1, one would predict that only for large protein molecules such as fibrinogen and von Willebrand factor (mol wt = ~10^6) would the dispersion or effective diffusion constant be appreciably greater than the Brownian translational diffusion constant. Indeed, the diffusion of low-molecular-weight solutes such as oxygen and urea in blood has been shown to be unaffected by the presence of erythrocytes.[56,57] Reviews of the subject of solute transport and surface reactivity at the wall in tube flow have recently appeared.[2,53]

2. Redistribution of Leukocytes and Platelets

In addition to disturbing the motion of the other formed elements in blood, erythrocytes appear to affect the distribution of leukocytes and platelets because of their radial migration away from the wall and aggregation in the core of the vessel where shear rates are very low. To what extent there is a reduction in hematocrit near the endothelial wall and how far inward the erythrocyte-depleted zone extends is uncertain. The only in vivo measurements of the peripheral layer stem from sections of quick-frozen rabbit femoral arteries.[58] From this work, and the analysis of cine films of steady and pulsatile blood flow in glass tubes <100 μm in diameter,[59] it appears unlikely that at normal hematocrits and physiologically representative flow rates, the plasma-rich zone can be much larger than 4 μm. It should be noted, however, that inward migration of only a few microns in a vessel of 100 to 200 μm in diameter brings about a decrease in the average red cell concentration in the tube (Fahraeus effect),[60,61] resulting in a significant decrease in apparent blood viscosity.[62,63]

By contrast, the redistribution of platelets in whole blood occurs toward the vessel periphery: their concentration at a radial distance r > 0.7 R in 25-μm diameter arterioles is found to be twice that at the vessel center.[64] Similar results have been obtained in 100-μm glass tubes[33,65] and in a parallel flow chamber.[66] The effect is believed to be due in part to the differences in the inward migration rates between red cells and platelets, and in part to the aggregation of red cells near the vessel center.

It is the aggregation of red cells that is responsible for the so-called margination of white blood cells, a well documented phenomenon in the microcirculation associated with low blood-flow states such as occur in shock and in inflammation.[8] At high flow rates, leukocytes are found to be axially distributed in tube flow,[67,68] although not to the same extent as the erythrocytes.[31] However, when red cells are caused to aggregate, either by adding fibrinogen[67] or a high-molecular-weight Dextran, or by reducing the flow rate,[31,68] leukocytes accumulate at the vessel periphery. Cine films taken in 100-μm diameter tubes clearly show that the effect is due to the outward displacement of white cells by the inwardly migrating network of packed red cell rouleaux, which exclude plasma containing both white cells and platelets toward the periphery. Margination of leukocytes is no longer found when the experiments are carried out in washed erythrocyte suspensions in which rouleau formation is totally suppressed.[31]

A different mechanism has been advanced to explain leukocyte margination in postcapillary venules.[69] Due to their larger volume and lower deformability, leukocytes travel more slowly than erythrocytes through the capillaries. As they enter the postcapillary venules, they are displaced to the wall by the more rapidly moving and deformable erythrocytes. The leukocytes adhere to, or roll along the wall until they are displaced inward by the greater shear forces in the larger venules.[69,70]

VI. CELL-WALL INTERACTIONS IN REGIONS OF DISTURBED FLOW

A. Disturbed Flows

Investigators from the biomedical as well as the physical and engineering sciences have long been aware of the connection between the occurrence of localized arterial wall injury, especially at bifurcations, T-junctions, and curved segments, and the disturbed patterns of flow at such sites.[71] Rapid changes in the rate and/or direction of fluid motion in branching and curved vessels have been held responsible for bringing about alterations and injury, not only to the endothelium[11,72-74] and the media,[75] but also the corpuscles.[76,77] This, in turn, could lead to the aggregation and adhesion of the cells to the injured vessel wall. Indeed, atheromatous plaques and platelet thrombi have been observed at bifurcations and stenoses.[78-80] The question here is whether disturbed patterns of flow lead to increased, localized interactions with the vessel wall. Answers to this question have been obtained from microrheological studies in our laboratory of the flow behavior of blood cells in idealized models of stenoses[81,82] and T-junctions,[83,84] as well as in fixed transparent segments of natural arteries and veins.[85-88]

The term ''disturbed flow'' is used to distinguish the flow regimes encountered at stenoses, branches, and curved segments of vessels from laminar and turbulent flow. When viscous fluids such as plasma and blood are subjected to flow through a tube, they exhibit two distinct types of flow: laminar and turbulent flow. The first type, exemplified by Poiseuille flow (Figure 1) is characterized by the steady motion of the fluid in layers parallel to a wall; hence it is called streamline or laminar flow. The second type occurs at much higher flow rates beyond a certain critical value of the Reynolds number (Equation 5) when fluid elements begin to exhibit irregular or random motions with respect to time and space. Such flow is called turbulent.

In branching and nonuniform diameter vessels in the circulation there exists an additional flow regime that is not observed in uniform diameter straight tubes. Here, there are secondary fluid motions in directions away from that of the primary flow, and often there is separation of the streamlines from the vessel wall, with the formation of a vortex or a recirculation

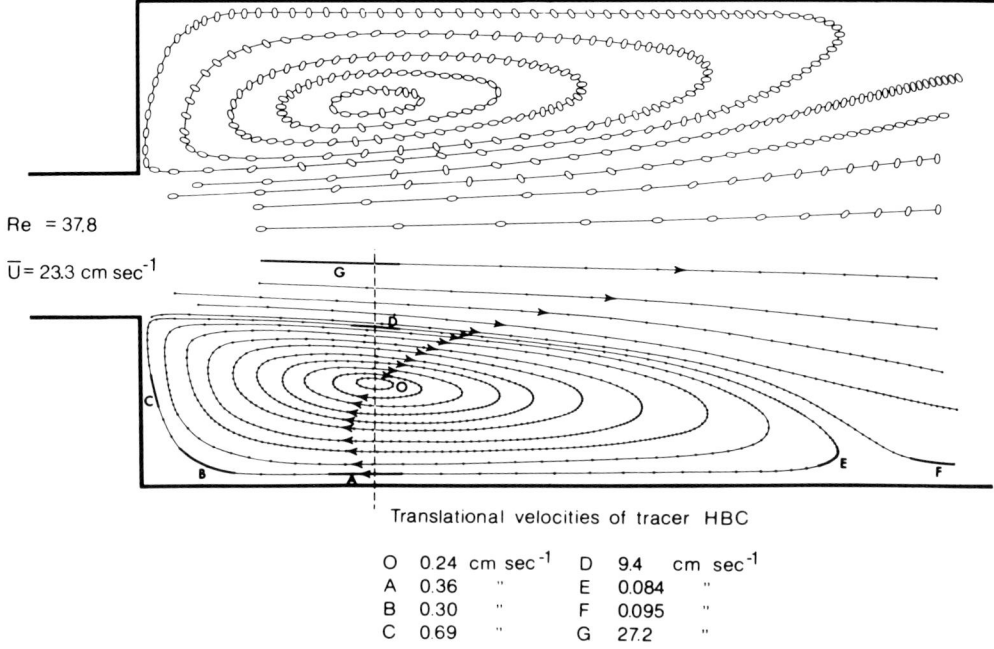

Re = 37.8

$\overline{U}$ = 23.3 cm sec^{-1}

Translational velocities of tracer HBC

O	0.24	cm sec^{-1}	D	9.4	cm sec^{-1}
A	0.36	"	E	0.084	"
B	0.30	"	F	0.095	"
C	0.69	"	G	27.2	"

FIGURE 13. Flow separation as a liquid enters a sudden concentric expansion of vessel lumen from 150 to 500 μm where it is rapidly decelerated, leading to a positive adverse gradient in pressure in the direction of flow. Liquid near the wall, unable to overcome the pressure gradient, is forced to a standstill and flows backward, finally leading to the formation of a vortex, with the mainstream being pushed away from the wall, where it is separated from the beginning of the expansion until it reattaches at the point indicated by the arrows. The region of flow separation is an annular vortex whose size increases with increasing inflow Reynolds number, Re_o, based on the mean upstream velocity, $\overline{U}$, and the degree of expansion of the lumen. The streamlines and red cell orientations in the orbits of the annular vortex are shown in the lower and upper halves, respectively. The linear velocities in mm sec^{-1} are highly variable in the region of flow separation. At the positions marked they were O, 2.4; A, 3.6; B, 3.0; C, 6.9; D, 94.1; E, 0.84; F, 0.95, and G, 272. (From Karino, T. and Goldsmith, H. L., *Phil. Trans. R. Soc. London Ser. B,* 279, 413, 1977. With permission.)

zone between the forward flowing mainstream and the wall. To describe this flow regime which is neither laminar nor turbulent, we have come to use the expression, "disturbed flow".

B. Blood Cells in Vortices and Recirculation Zones: Model Studies
1. Annular Vortex at a Sudden Tubular Expansion
a. Erythrocytes

We have studied the detailed flow behavior of human blood cells in an annular vortex formed downstream of a sudden tubular expansion of a 150-μm into a 500-μm diameter glass tube serving as a model of a stenosis.[82] As predicted by fluid mechanical theory,[89] when dilute suspensions of erythrocytes were subjected to steady flow through the model stenosis, a captive annular ring vortex was formed downstream of the expansion. Figure 13 shows paths and orientations of the erythrocytes in the median plane of the tube. During a single orbit, the measured particle paths and velocities, as well as the locations of the vortex center and reattachment point, were in good aggreement with those predicted by the theory applicable to the fluid.[89] Over longer periods, however, single cells and small aggregates <20 μm in diameter migrated outward across the closed streamlines of the vortex predicted by the theory and left the vortex after describing a series of spiral orbits of continually increasing diameter until they rejoined the mainstream. In contrast, aggregates of cells >30 µm in diameter remained trapped within the vortex, assuming equilibrium orbits or staying at the center. In pulsatile flow (a sinusoidal oscillatory flow superimposed in parallel with the steady flow), the observed phenomena were qualitatively similar to those described in steady flow. The vortex varied periodically in size and intensity; the axial location of the vortex center and reattachment point oscillated in phase with the upstream fluid velocity between maximum and minimum positions about a mean which corresponded to that measured in the absence of the component of oscillatory flow. At higher hematocrits from 15 to 45%, migration of single cells still persisted and resulted in the lowering of the hematocrit in the vortex region both in steady and pulsatile flow.[82]

The mechanism underlying the particle migration phenomenon is not a simple one, but it is likely that it is partly due to the dilution effect of the cell-poor plasma taken into the vortex from the fluid layer adjacent to the vessel wall proximal to the expansion. The mechanism for the trapping of large aggregates in the vortex was qualitatively explained[82] by using existing fluid mechanical theories[29,30] concerned with lateral particle migration near a tube wall and by the operation of a mechanical wall effect.[90,91]

b. Platelet Aggregation in the Vortex

The flow behavior and interactions of human platelets in the annular vortex formed distal to a 150- into 500-µm tubular expansion were studied at 37°C, using heparinized or citrated platelet-rich plasma (PRP) as well as washed platelets in Tyrodes-albumin solutions. It was demonstrated that the vortex provided favorable conditions for the spontaneous aggregation of normal human platelets through shear-induced collisions of particles while circulating in its orbits.[92] In a given suspension, the formation and growth of platelet aggregates could only be observed in a narrow range of Reynolds numbers (based on upstream linear velocity and tube diameter), which varied from suspension to suspension. Thus, in heparinized PRP, containing many sphered platelets with pseudopods and some microaggregates of two to six cells, the rate and extent of aggregation were the highest, with large elongated floating aggregates (>100 µm in length) being observed to form in less than 1 min and within the widest range of Re, between 4.5 and 17. In citrated PRP and washed platelet suspensions, in which no microaggregates were seen prior to flow, the degree of aggregation was much reduced. However, when platelets in these suspensions were activated with subthreshold concentrations of ADP or thrombin, the large aggregates seen in heparinized PRP were again formed. When the above suspensions were subjected to pulsatile flow in the expansion tube, there was a marked decrease in the number and size of the aggregates. Presumably, this was due to the continuously changing orbits of particles during the alternate expansion and contraction of the vortex, which shortened their residence times, as well as to the large variation in the shear rate in each cycle, beyond the range favorable for the platelet aggregation.

The above results suggest that formation of platelet aggregates in vortices will be more likely to occur in the venous circulation, where the flow is steadier and Re is lower than in the arterial circulation.

c. Wall Adhesion of Platelets in the Vortex

The effects of disturbed flow on initial platelet adhesion to the vessel wall were studied using a large-scale expansion flow tube (0.92 into 3.00 mm diameter) whose inner wall was coated with collagen fibers, and suspensions of washed human platelets containing washed erythrocytes at hematocrits from 0 to 50%.[93] As illustrated in Figure 14, it was convincingly shown that platelet adhesion was localized within the vortex and downstream on either side of the reattachment point with a local minimum at the reattachment point itself. Furthermore, platelet adhesion increased, and both adhesion peaks became more pronounced, as the hematocrit increased. Surprisingly though, the adhesion peak in the vortex decreased and flattened out as the Reynolds number increased. These results are inconsistent with a dif-

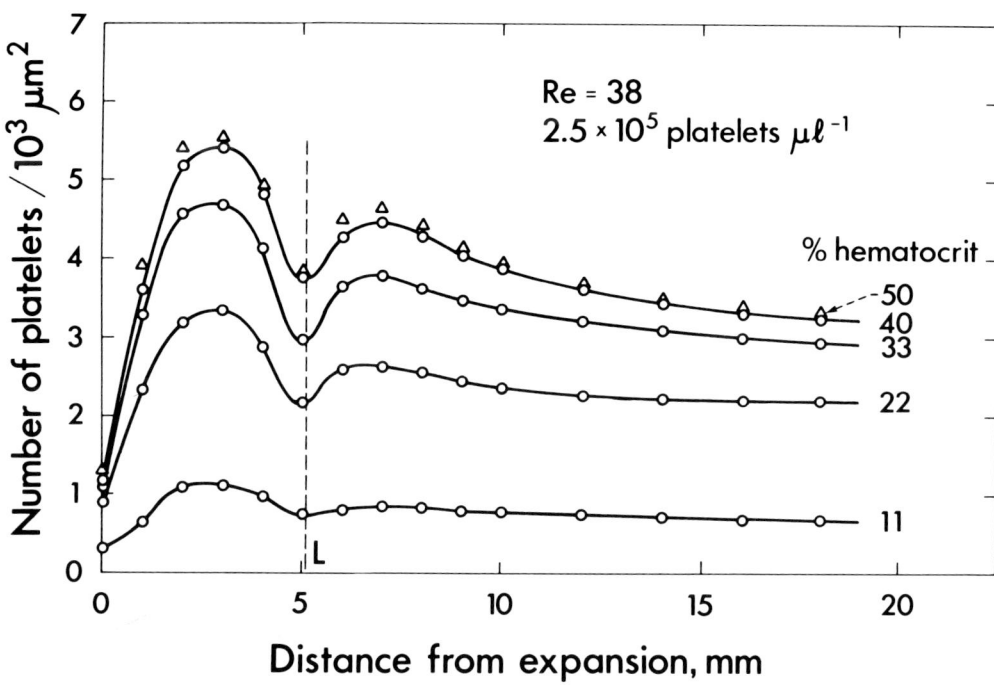

FIGURE 14.   Plot of measured number density of adhering platelets as a function of axial distance from the origin of an expansion of an 0.92- into a 3.00-mm glass tube. The figure shows the localized enhanced adhesion of platelets on either side of the reattachment point at L, indicated by the vertical dashed line, and the effect of hematocrit on the phenomenon. (From Karino, T. and Goldsmith, H. L., *Microvasc. Res.*, 17, 217, 1979. With permission.)

fusion-controlled platelet adhesion (when the rate determining step in adhesion is the rate at which cells are brought to the vessel surface), which should show an increase in adhesion number density with increasing shear rate.[53,71] It appears that the particular flow pattern within the vortex is responsible for this localization. Thus, as illustrated in the upper panel of Figure 15, only those cells carried by the curved streamlines to within one particle radius of the surface interact with the vessel wall and adhere to it on both sides of the reattachment point, which is also a stagnation point.[93,94] It follows from this that platelet adhesion onto the vessel wall, whether it be the natural endothelium or an artificial surface, will be localized wherever there is a stagnation point (or a reattachment point if it is a result of flow separation) where blood cells are carried by the flow toward the vessel wall along curved streamlines having a pronounced radial velocity component. If this mechanism operates in the circulation, a relatively higher adhesion of platelets, and hence a higher risk of thrombus formation is predictable, not only in regions of disturbed flow (adhesion peaks on either side of the reattachment point) such as downstream of aortic and venous valves, mural thrombi, and stenoses, but also in all the branching arteries at the flow divider where there is a stagnation point.

2. Secondary Flows and Recirculation Zones at T-Junctions

We, therefore, studied the flow patterns and distributions of fluid velocity and shear rate in 3-mm diameter glass models of T-junctions with branching angles from 30 to 135° and side-to-main-tube diameter ratios from 0.33 to 1.0, over a wide range of inflow Reynolds numbers Re_o and branch-to-parent tube flow ratios, Q_1/Q_o.[83,84] Figure 16 shows the flow pattern obtained by photographing the motions of 50-μm diameter latex spheres in a 90° uniform diameter T-junction when the main branch was partially occluded so that 80% of

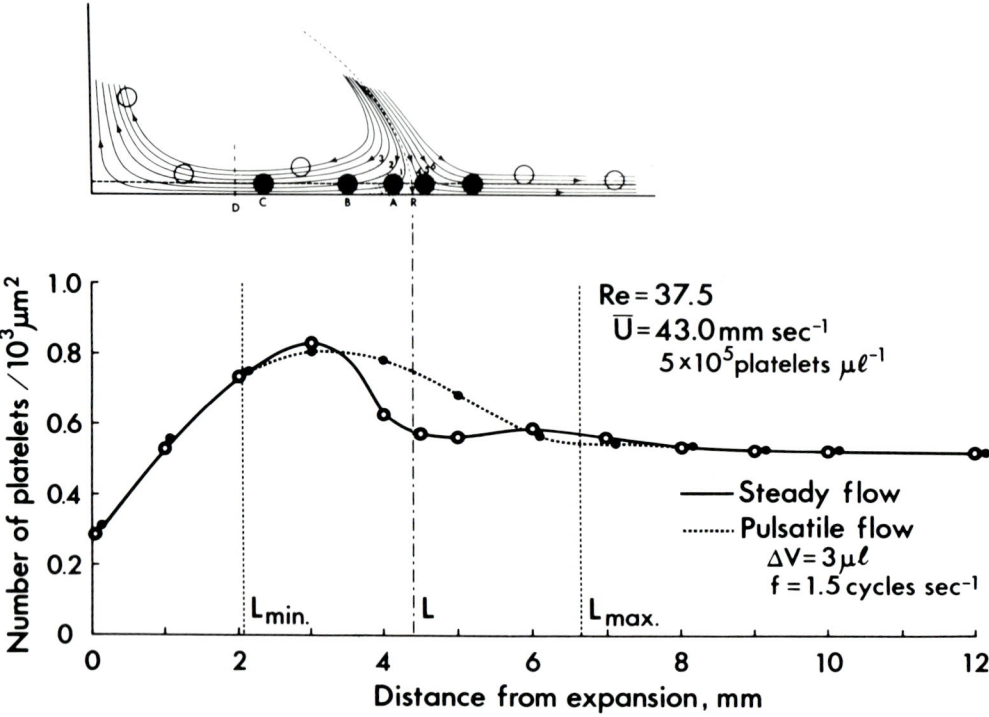

FIGURE 15. (A) Schematic representation of fluid streamlines near the tube wall downstream of the tubular expansion, showing the convective transport of particles in bulk flow to the vessel wall along the radially directed curved streamlines on either side of the reattachment point, R. The black circles represent the particles which are carried by the flow along the streamlines 1 to 6 to within the critical distance for collision with the wall at points A to C in the vortex and the corresponding points downstream. The open circles represent particles which do not come close enough to collide with the wall. (B) Plot of the measured number density of adhering platelets obtained from an experiment carried out with a suspension of washed human platelets in Tyrodes-albumin solution containing no erythrocytes. The figure shows the relationship between the flow pattern and the degree of platelet adhesion in and downstream from the vortex in steady flow. (From Karino, T. and Goldsmith, H. L., *Biorheology*, 21, 587, 1984. With permission.)

the flow left through the side branch. A large recirculation zone was formed in the main tube (due to the sudden deceleration of fluid velocity as a portion of the flow is drawn off into the side tube) and a small, secondary recirculation zone was formed in the side tube (due to the sudden change in direction of the fluid which continues to move to the outside of the 90° corner, toward the flow divider). Particles entering the large recirculation zone described complicated orbits; some of them rejoined the flow through the main tube, others entered the side branch in a paired, spiral secondary flow with pronounced radial components, and some of these circulated through the secondary recirculation zone. When the degree of occlusion of the main tube was gradually reduced, the large recirculation zone became smaller and eventually disappeared as the flow rate ratio was reversed ($Q_l/Q_o = 0.8$), while that in the side branch grew in size.

By varying the branching angle it was shown that the critical Re_o for the formation of the main recirculation zone was lowest at 90° for all Q_l/Q_o, whereas for the side recirculation zone it decreased as the branching angle increased from 45 to 135°. When the diameter of the side tube was decreased, the main recirculation zone, which actually consisted of a pair of spiral secondary flows located symmetrically on both sides of the common median plane of the T-junction, became smaller and thinner and was confined to a thin layer adjacent to the tube wall, wrapped around the mainstream.[84].

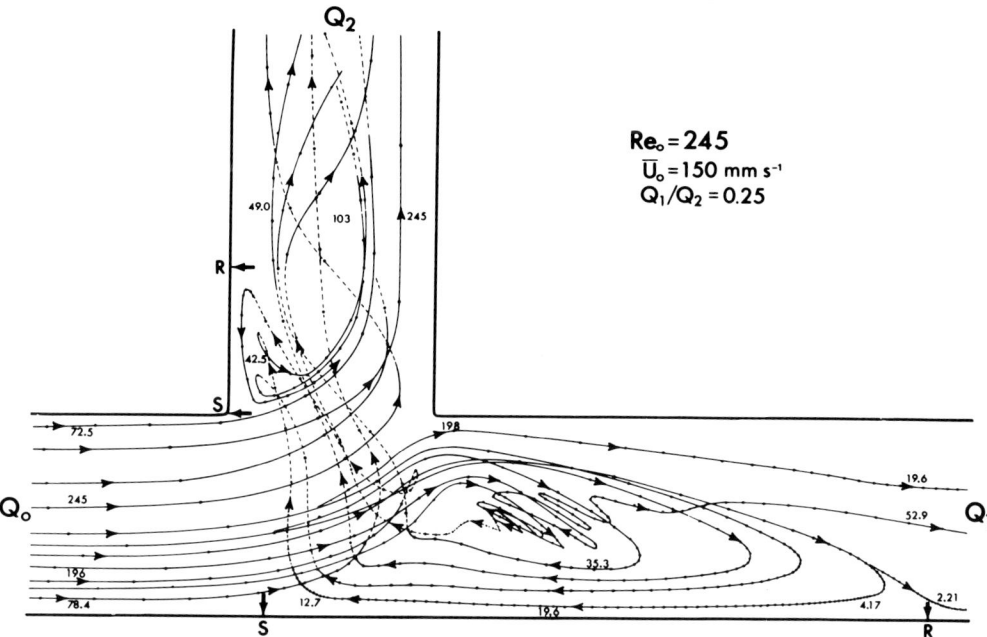

FIGURE 16.   Flow patterns in the common median plane of a model 3-mm diameter glass T-junction, as indicated by the paths of tracer 50-μm polystyrene spheres in aqueous glycerol. The suspension enters at the left with a mean velocity $\overline{U}_o$ (tube Reynolds number = Re_o), and 80% of the flow leaves through the side tube. The T-junction, made by fitting and gluing together two pieces of glass tubing, has a very low radius of curvature at the corner opposite the flow divider, as well as at the flow divider. This results in a small recirculation zone at the corner at the entry of the side tube filled with fluid from the main large recirculation zone in the parent vessel. The points are measured particle positions at intervals of 22 msec; the numbers indicate velocities in mm sec^{-1}. The solid lines represent particles traveling in or close to the median plane; the dashed lines represent those closer to the tube wall. The arrows at S and R indicate the respective points where flow first separates and then reattaches to the wall (From Karino, T., Kwong, H. M., and Goldsmith, H. L., *Biorheology*, 16, 231, 1979. With permission.)

The effect of radius of curvature of the walls at the junction was studied by comparing the critical inflow Reynolds numbers and size of the recirculation zones in the square T-junction (Figure 16; radii of curvature <2% of tube radius) with that in a rounded T-junction (radii of curvature ~ tube radius).[83] It was found that the recirculation zone in the side tube formed at a much lower Re_o in the square than the rounded junction, and that at a given Re_o and Q_1/Q_o, a larger main recirculation zone existed in the rounded junction. It appears that the formation of the side recirculation zone is largely affected by curvature of the wall at the bend opposite to the flow divider while that of the main recirculation zone is largely affected by the curvature at the flow divider.

C. Blood Cells in Vortices and Recirculation Zones: Natural Vessels

We have carried out studies of detailed flow patterns in some regions of the circulation using isolated transparent natural arteries and veins prepared from dogs and humans post-mortem by a method developed in our laboratory.[85] Here, we present results for flow at a venous valve where there is an expansion flow, and for branching flows at an aortic T-junction and at the carotid bifurcation.

1. Flow Patterns at a Venous Valve

Studies of the behavior of model particles and red cells flowing through a venous valve have been carried out using isolated transparent dog saphenous veins containing bileaflet valves.[87] Figure 17 illustrates the detailed flow patterns observed along the common median

Re = 42.1
D̄ = 2.03 mm
d = 0.81 mm
Ū = 53.3 mm s⁻¹

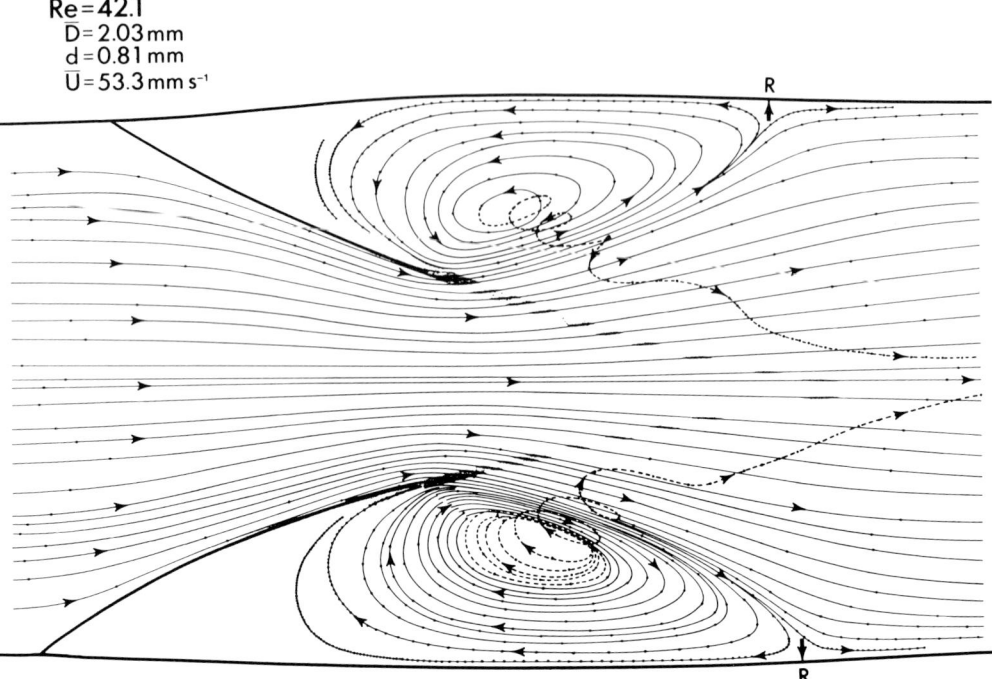

FIGURE 17.   Detailed flow patterns in the common bisector plane of the valve leaflets in a 2-mm diameter dog saphenous vein containing a bileaflet valve and showing the formation of a spiral vortex in each valve pocket. In fact, these consist of a pair of vortices located symmetrically on both sides of the bisector plane. The solid lines are the paths of particles located in or close to, and the dashed lines are those located far away from, the bisector plane (the projection of the particle paths on the bisector plane). The arrows at R indicate the location of the reattachment point. (From Karino, T. and Motomiya, M., *Thromb. Res.*, 36, 245, 1984. With permission.)

plane of a 2-mm diameter vein containing a bileaflet valve. Under physiological flow conditions, as shown in the figure, flow separation at the edge of the valve leaflet resulted in the formation of large paired vortices, located symmetrically on both sides of the bisector plane of the valve leaflets in each valve pocket. Particles continually entered the valve pockets from the mainstream, spending long periods of time describing a series of spiral orbits of decreasing diameter, while moving away from the bisector plane, and eventually left the vortex. With concentrated suspensions of hardened red cells, it was found that another smaller counter-rotating secondary vortex, driven by the large primary vortex, existed deep in each valve pocket (blank area of Figure 17) where venous thrombi are believed to originate.[95,96] Furthermore, experiments carried out with hardened erythrocyte suspensions at 25% hematocrit showed that the erythrocyte concentration in this secondary vortex remained appreciably lower than that in the primary vortex. In such stagnant regions, fluid circulated with extremely low velocities, thus creating a very low shear field which allowed erythrocytes to form aggregates. The results suggest that in some pathological states, the valve-pocket vortices could act as automatic traps and generators of thrombi in a fashion similar to that previously demonstrated in the annular vortex formed downstream from a sudden tubular expansion.[82,92]

2. Flow Patterns at Aortic T-Junctions

The flow patterns in transparent segments of a dog abdominal aorta containing branches of celiac, cranial-mesenteric, and right and left renal arteries have been studied.[85,88] The general flow patterns are similar to those observed in glass models of T-junctions, as

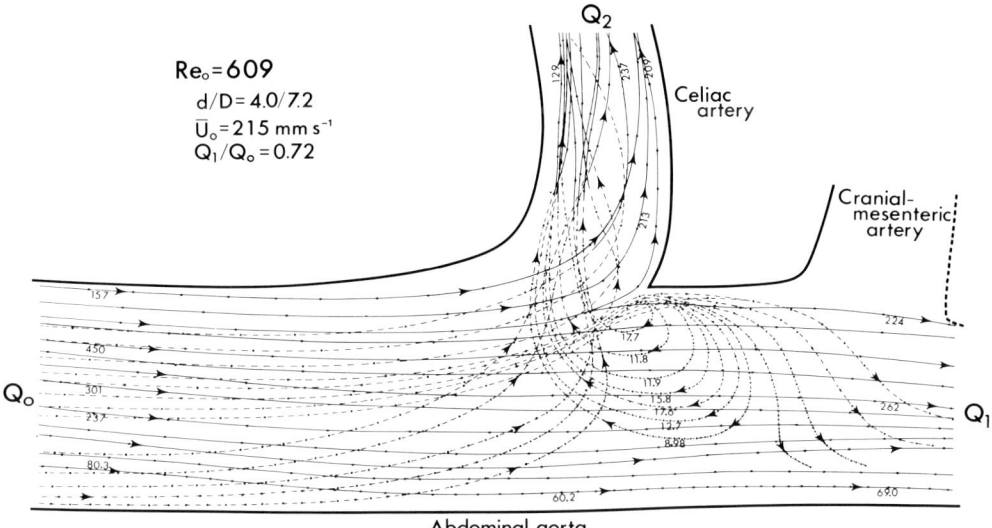

FIGURE 18. Detailed flow patterns at the aortoceliac artery junction of a dog abdominal aorta, showing the formation of paired thin-layered recirculation zones on both sides of common median plane, adjacent to vessel wall wrapped around the undisturbed mainstream. Numbers on the streamlines indicate particle velocities in mm sec^{-1}. (From Karino, T., Motomiya, M., and Goldsmith, H. L., in *Biologic and Synthetic Vascular Prosthesis*, Stanley, J. C., Ed., Grune & Stratton, New York, 1982. With permission.)

illustrated in Figure 18 for the aortoceliac artery junction. Here, at the geometrical flow rate ratio (Q_1/Q_2 = area ratio main:side tube) flow separation occurred at $Re_o \sim 300$, well below the mean physiological flow rate at $Re_o \sim 700$. However, instead of a large main standing recirculation zone as in the glass models, under physiologic flow conditions, there was only a pair of recirculation zones, confined to a thin layer close to the wall surrounding the mainstream. There was no side recirculation zone, no doubt due to the gentle curvature of the bend opposite the flow divider. This characteristic was shared by all the aortic T-junctions studied, as was the very sharp curvature of the bend at the flow divider. From the results obtained in the glass model T-junctions, this represents the optimum condition for minimizing the size of both regions of separated flow. Nevertheless, the curved streamlines of the recirculation zone and secondary flows will bring blood cells towards the vessel wall in a zone around the flow divider and in the side branch on the outer wall.

3. Flow Patterns at the Carotid Bifurcation

Flow patterns in the human carotid artery bifurcation were studied in detail using a transparent arterial segment prepared from a human subject post-mortem.[86] It was found that a standing recirculation zone consisting of a pair of complex spiral secondary flows, located symmetrically on both sides of the common median plane of the bifurcation, formed in the carotid sinus over wide ranges of inflow Reynolds numbers, Re_o, and flow ratios, Q_1/Q_o (internal/common). Figure 19 shows the detailed flow patterns in the carotid sinus. Particles were deflected at the flow divider and traveled laterally and very slowly along the wall above and below the common median plane, almost at right angles to, and encircling the mainstream. They then changed direction, moving back along the outer wall of the internal carotid artery at the site of the sinus, describing spiral orbits in the recirculation zone before rejoining the mainstream. Downstream from the reattachment point (R), a strong counter-rotating double helicoidal flow developed. The formation and the size of the recirculation zone were largely dependent on Q_1/Q_o as well as on Re_o. The size of the recirculation zone increased from ~ 4 mm at $Re_o = 300$ to a maximum of ~ 9 mm at $Re_o > 800$.

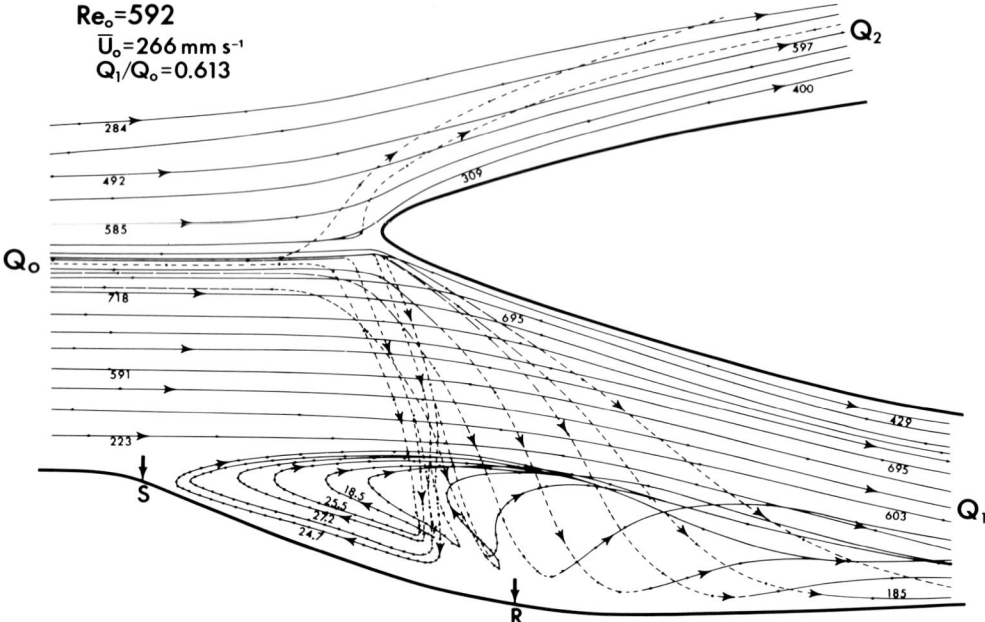

FIGURE 19.   Detailed flow patterns in the human carotid artery bifurcation in steady flow, showing the formation of a recirculation zone (paired spiral secondary flows) and a counter-rotating double helicoidal flow, both located symmetrically on either side of the common median plane in the internal carotid artery. The solid lines are the paths of particles in or close to, and the dashed lines paths which are far away from, the common median plane (the projection of the particle paths on the common median plane). The arrows at S and R denote the respective locations of the separation and stagnation points. The numbers on the streamlines (particle paths) indicate the particle translational velocities in mm sec^{-1}. (From Motomiya, M. and Karino, T., *Stroke,* 15, 50, 1984. With permission.)

Measurements of the velocity profiles showed that these were strongly skewed towards the inner walls of the bifurcation, creating a high shear field along the vessel walls downstream from the flow divider where there was a stagnation point. Due to the presence of the paired standing recirculation zones in the carotid sinus, the wall shear rate, and hence the wall shear stress, changes sign and becomes negative at the separation point (S); it becomes positive again downstream from the reattachment point (or the stagnation point, indicated by R). Thus, in the carotid sinus, there is a region where the vessel wall is stretched in opposite directions by the counter-directed wall shear stresses.

The results suggest that, under physiological conditions (mean $Re_o \sim 600$, $Q_1/Q_o \sim 0.7$),[97,98] a standing recirculation zone exists in the carotid sinus, thereby affecting local mass transfer and interactions of blood cells with the vessel wall which may lead to the incidence of thrombosis and atherosclerosis in this region.

VII. CONCLUDING REMARKS

We have described the flow behavior of human blood cells both in straight glass tubes and in regions of disturbed flow within glass models of stenoses and T-junctions, as well as in transparent segments of natural arteries and veins. We conclude that cell-wall interactions in the circulation can occur by two different mechanisms.

The first involves the dispersion or mixing motion within the blood which exists even in laminar flow through a straight vessel. Cell-wall encounters arise from the continual shear-induced collisions between the erythrocytes in the cell-crowded suspension, which leads to radial displacements of their paths. Not only erythrocytes, but also platelets and leukocytes flowing in the periphery of the vessel are frequently displaced to the wall (Figures 10 to 12). Such cell-wall collisions would not occur at low hematocrits when the individual corpuscles are free to migrate away from the vessel wall (Figure 5).

The second mechanism of cell-wall interaction involves the convective transport of corpuscles toward the vessel wall along curved streamlines having pronounced radial velocity components (Figure 15). This occurs in regions of disturbed flow which exist downstream of aortic and venous valves, mural thrombi, and stenoses, and at all bifurcations in the arterial tree.

In the case of platelets and leukocytes, wall encounters provide the means of fulfilling the normal hemostatic function of the cells. They also play a role in pathologic events in which the cells adhere to the altered vessel wall and promote the growth of thrombi. In the model symmetrical stenosis (Figure 13) we have shown that the two modes of cell-wall interaction reinforce each other. Here, localized platelet adhesion to collagen fibers on the wall of the expansion tube within the vortex is found to increase with increasing hematocrit (Figure 14). This is due to the fact that the number of cells carried by the curved streamlines to within one particle radius of the surface increases with the hematocrit as the enhanced radial motion of the platelets permits collisions of cells with the surface at distances greater than one particle radius.

We believe that the studies in glass models of straight, expansion, and branching flows, and the more recent investigations in transparent natural vessels are relevent to the understanding of flow-induced cell-wall interactions in the circulation. Although there have been no direct in vivo observations of disturbed flow patterns such as those illustrated in Figures 17 to 19 in fixed and rigid-walled segments of natural vessels, the fact that spiral secondary flows and recirculation zones were formed at Reynolds numbers much lower than those prevailing in vivo makes it plausible that such disturbed flows exist under normal physiological conditions. The regions of disturbed flow observed at flow dividers of aortic T-junctions and the outer wall of the carotid sinus generally agree with the preferred sites of thrombosis and atherosclerosis clinically found in man,[99-101] lending support to the second of the proposed mechanisms for cell-wall interaction. Clearly, in such disturbed flows, corpuscles and lipids are constantly carried to the vessel wall along the curved streamlines, and because of the much lower local translational velocities and shear stress, are able to interact with the vessel wall for longer periods than elsewhere in the circulation. One would, therefore, expect an enhanced deposition of platelets and leukocytes on the vessel wall, and an increased uptake of lipids by the endothelial cells in such regions. What is not clear at present is the extent to which uptake of atherogenic proteins by the endothelial cells is affected by the variation in local shear stress in these regions, and which in turn affects the morphology of the cells. As described earlier, endothelial cells exhibit morphological variations which reflect local flow patterns (Figure 2). Thus, in arterial bifurcations, the cells located on the inner wall distal to the flow divider (where the shear stress is high) are found to be elongated in the direction of flow, while those cells located on the outer walls opposite the flow divider (where spiral secondary flows and recirculation zones likely form, and the shear stress is much lower) are not elongated but oriented in a random fashion.[11,12,75] It is, therefore, important to ascertain whether there are differences in biological, especially metabolic function in these morphologically different cells. It is also necessary to further investigate the effect of the local flow pattern on the transport of oxygen and uptake of plasma proteins and lipids by the endothelial cells. Only then can one arrive at a full explanation for the localization of thrombosis and atherosclerosis in the circulation and the role of the formed elements in these diseases.

Continued studies of the hemodynamics of the circulation with particular reference to the effects of flow and fluid mechanical stress on the corpuscles and the endothelium will no doubt prove useful in determining the role of fluid dynamic factors in the maintenance of vascular homeostasis as well as in the genesis of vascular diseases.

REFERENCES

1. **McDonald, D. A.,** *Blood Flow in Arteries,* 2nd. ed., Williams & Wilkins, Baltimore, 1974.
2. **Goldsmith, H. L. and Turitto, V. T.,** Rheological aspects of thrombosis and haemostasis: basic principles and applications, *Thromb. Haemost.,* 55, 415, 1986.
3. **Goldsmith, H. L. and Marlow, J.,** Flow behavior of erythrocytes. II. Particle motions in concentrated suspensions of ghost cells, *J. Colloid Interface Sci.,* 71, 383, 1979.
4. **Goldsmith, H. L. and Skalak, R.,** Hemodynamics, *Annu. Rev. Fluid Mech.,* 7, 213, 1975.
5. **Goldsmith, H. L. and Mason, S. G.,** The microrheology of dispersions, in *Rheology: Theory and Applications,* Vol. 4, Eirich, F. R., Ed., Academic Press, New York, 1967, chap. 4.
6. **Goldsmith, H. L. and Mason, S. G.,** Some model experiments in hemodynamics. V. Microrheological techniques, *Biorheology,* 12, 181, 1975.
7. **Karino, T. and Goldsmith, H. L.,** Rheological factors in haemostasis and thrombosis, in *Haemostasis and Thrombosis,* Bloom, A. L. and Thomas, D. P., Eds., Churchill Livingstone, Edinburgh, 1986, chap. 42.
8. **Grant L.,** The sticking and emigration of white blood cells in inflammation, in *The Inflammatory Process,* Vol. II, Zweifach, B. W., Grant, L., and McCluskey, R. T., Eds., Academic Press, New York, 1973, chap. 7.
9. **Weiss, L. and Glaves, D.,** Cancer cell damage at the vascular endothelium, *Ann. N. Y. Acad. Sci.,* 416, 681, 1983.
10. **Flaherty, J. T., Pierce, J. E., Ferrans, V. J., Patel, D. J., Tucker, W. K., and Fry, D. L.,** Endothelial nuclear patterns in the canine arterial tree with particular reference to hemodynamic events, *Circ. Res.,* 30, 23, 1972.
11. **Reidy, M. A. and Bowyer, D. E.,** Scanning electron microscopy of arteries. The morphology of aortic endothelium in hemodynamically stressed areas associated with branches, *Atherosclerosis,* 26, 181, 1977.
12. **Reidy, M. A.,** Arterial endothelium around rabbit aortic ostia: a SEM study using vascular casts, *Exp. Mol. Pathol.,* 30, 327, 1979.
13. **Dewey, C. F., Bussolari, S. R., Gimbrone, M. A., and Davies, P. F.,** The dynamic response of vascular endothelial cells to fluid shear stress, *J. Biomech. Eng.,* 103, 177, 1981.
14. **Karino, T.,** unpublished results.
15. **Chien, S.,** Present state of blood rheology, in *Hemodilution, Theoretical Basis and Clinical Application,* Messmer, K. and Schmid-Schönbein, H., Eds., S. Karger, Basel, 1972, 1.
16. **Einstein, A.,** *The Theory of the Brownian Movement,* Dover, New York, 1956.
17. **Jeffery, G. B.,** On the motion of ellipsoidal particles immersed in a viscous fluid, *Proc. R. Soc. London Ser. A,* 102, 161, 1922.
18. **Goldsmith, H. L. and Mason, S. G.,** Particle motions in sheared suspensions. XIII. The spin and rotation of disks, *J. Fluid Mech.,* 12, 88, 1962.
19. **Goldsmith, H. L. and Mason, S. G.,** The flow of suspensions through tubes. I. Single spheres, rods and discs, *J. Colloid Sci.,* 17, 448, 1962.
20. **Goldsmith, H. L. and Marlow, J.,** Flow behaviour of erythrocytes. I. Rotation and deformation in dilute suspensions, *Proc. R. Soc. London Ser. B,* 182, 351, 1972.
21. **Schmid-Schönbein, H. and Wells, R.,** Fluid drop-like transition of erythrocytes under shear, *Science,* 165, 228, 1969.
22. **Goldsmith, H. L.,** Red cell motions and wall interactions in tube flow, *Fed. Proc.,* 30, 1578, 1971.
23. **Fischer, T. M., Stöhr, M., and Schmid-Schönbein, H.,** Red blood cell (RBC) microrheology: comparison of the behavior of single RBC and liquid droplets in shear flow, *A.I.Ch.E. Symp. Ser.,* 74, 38, 1978.
24. **Frojmovic, M. M., Newton, M., and Goldsmith, H. L.,** The microrheology of mammalian platelets: studies of rheological transients and flow in tubes, *Microvasc. Res.,* 11, 203, 1976.
25. **Cox, R. G. and Mason, S. G.,** Suspended particles in fluid flow through tubes, *Annu. Rev. Fluid Mech.,* 3, 291, 1971.
26. **Karnis, A. and Mason, S. G.,** Particle motions in sheared suspensions. XXIII. Wall migration of fluid drops, *J. Colloid Interface Sci.,* 24, 164, 1967.
27. **Segré, S. and Silberberg, A.,** Behaviour of macroscopic rigid spheres in Poiseuille flow. II. Experimental results and interpretation, *J. Fluid Mech.,* 14, 136, 1962.
28. **Karnis, A., Goldsmith, H. L., and Mason, S. G.,** The flow of suspensions through tubes. V. Inertial effects, *Can. J. Chem. Eng.,* 44, 181, 1966.
29. **Brenner, H.,** Hydrodynamic resistance of particles at small Reynolds numbers, in *Advances in Chemical Engineering,* Vol. 6, Drew, T. B., Hooper, J. W., and Vermeulen, T., Eds., Acedemic Press, New York, 1966, 287.
30. **Cox, R. G., and Hsu, S. K.,** The lateral migration of solid particles in a laminar flow near a plane wall, *Int. J. Multiphase Flow,* 3, 201, 1977.

31. **Goldsmith, H. L. and Spain, S.,** Margination of leucocytes in blood flow through small tubes, *Microvasc. Res.,* 28, 204, 1984.
32. **Palmer, A. A.,** Platelet and leukocyte skimming, *Bibl. Anat.,* 9, 300, 1967.
33. **Beck, M. R. and Eckstein, E. C.,** Preliminary report on platelet concentration in capillary tube flows of whole blood, *Biorheology,* 17, 465, 1980.
34. **Smoluchowski, M., von,** Versuch einer Mathematischen Theorie der Koagulationskinetik Kolloider Lösungen, *Z. Phys. Chem.,* 92, 129, 1917.
35. **Goldsmith, H. L. and Mason, S. G.,** The flow of suspensions through tubes. III. Collisions of small uniform spheres, *Proc. Roy. Soc. London Ser. A,* 282, 569, 1964.
36. **Goldsmith, H. L., Lichtarge, O., Tessier-Lavigne, M., and Spain, S.,** Some model experiments in hemodynamics. VI. Two-body collisions between blood cells, *Biorheology,* 18, 531, 1981.
37. **Forgacs, O. L. and Mason, S. G.,** Particle motions in sheared suspensions. X. Orbits of flexible thread-like particles, *J. Colloid Sci.,* 14, 473, 1959.
38. **Bell, D. N., Teirlinck, H. C., and Goldsmith, H. L.,** Platelet aggregation in Poiseuille flow. I. A double infusion technique, *Microvasc. Res.,* 27, 297, 1984.
39. **Bell, D. N. and Goldsmith, H. L.,** Platelet aggregation in Poiseuille flow. II. Effect of shear rate, *Microvasc. Res.,* 27, 316, 1984.
40. **Machi, J., Sigel, B., Ramos, J. R. et al.,** Role of red cells in preventing the growth of platelet aggregation, *Thromb. Res.,* 36, 53, 1984.
41. **Karnis, A., Goldsmith, H. L., and Mason, S. G.,** The kinetics of flowing dispersions. I. Concentrated suspensions of rigid particles, *J. Colloid Interface Sci.,* 22, 531, 1966.
42. **Gauthier, F. J., Goldsmith, H. L., and Mason, S. G.,** The flow of suspensions through tubes. X. Liquid drops as models of erythrocytes, *Biorheology,* 9, 205, 1972.
43. **Vadas, E., Goldsmith, H. L., and Mason, S. G.,** The microrheology of colloid dispersions. III. Concentrated emulsions, *Trans. Soc. Rheol.,* 20, 373, 1976.
44. **Goldsmith, H. L.,** Microscopic flow properties of red cells, *Fed. Proc.,* 26, 1813, 1967.
45. **Goldsmith, H. L.,** The microrheology of red blood cell suspensions, *J. Gen. Physiol.,* 52, 5s, 1968.
46. **Chien, S.,** Shear dependence of effective cell volume as a determinant of blood viscosity, *Science,* 168, 977, 1970.
47. **Goldsmith, H. L. and Karino, T.,** Microscopic considerations: the motions of individual particles, *Ann. N.Y. Acad. Sci.,* 283, 241, 1977.
48. **Turitto, V. T. and Baumgärtner, H.,** Platelet interaction with subendothelium in a perfusion system: physical role of red cells, *Microvasc. Res.* 9, 335, 1975.
49. **Turitto, V. T. and Weiss, H. J.,** Red Cells: their dual role in thrombus formation, *Science,* 207, 541, 1980.
50. **Grabowski, E. F., Friedman, L. I., and Leonard, E. F.,** Effects of shear rate on the diffusion and adhesion of blood platelets to a foreign surface, *Ind. Eng. Chem. Fund.,* 11, 224, 1972.
51. **Feuerstein, I. A., Brophy, B. M., and Brash, J. L.,** Platelet transport and adhesion to reconstituted collagen and artificial surfaces, *Trans. Am. Soc. Artif. Intern. Organs,* 21, 427, 1975.
52. **Turitto, V. T., Benis, A. M., and Leonard, E. F.,** Platelet diffusion in flowing blood, *Ind. Eng. Chem. Fund.,* 11, 216, 1972.
53. **Turitto, V. T.,** Viscosity, transport and thrombogenesis, in *Progress in Hemostasis and Thrombosis,* Vol. 6, Spaet, T. H., Ed.. Grune & Stratton, New York, 1982, 139.
54. **Wang, N. H. and Keller, K. H.,** Solute transport induced by erythrocyte motions in shear flow, *Trans. Am. Soc. Artif. Intern. Organs,* 25, 14, 1979.
55. **Aarts, P. A., Bolhuis, P. A., Sakariassen, K. S., Heethar, R. M., and Sixma, J. J.,** Red blood cell size is important for adhesion of blood platelets to artery subendothelium, *Blood,* 62, 214, 1983.
56. **Collingham, R. E.,** Mass Transfer in Flowing Suspensions, Ph.D. thesis, University of Minnesota, Minneapolis, 1968.
57. **Walton, C. K., Smith, K. A., Merrill, E. R., and Friedman, S.,** Diffusion of urea in flowing blood, *Am. Inst. Chem. Eng. J.,* 17, 800, 1971.
58. **Phibbs, R. H.,** Orientation and distribution of erythrocytes in blood flowing through medium-sized arteries, in *Hemorheology: Proceedings of the 1st International Conference,* Copley, A. L., Ed., Pergamon Press, New York, 1967, 617.
59. **Bugliarello, G. and Sevilla, J.,** Velocity distribution and other characteristics of steady and pulsatile flow in glass tubes, *Biorheology,* 7, 85, 1971.
60. **Fahraeus, R.,** The suspension stability of blood, *Physiol. Rev.,* 9, 241, 1929.
61. **Cokelet, G. R.,** Macroscopic rheology and the tube flow of human blood, in *Microcirculation,* Vol. I, Grayson, J. and Zingg, W., Eds., Plenum Press, New York, 1976, 9.
62. **Thomas, H. W.,** The wall effect in capillaries: an improved analysis suitable for application to blood and other particulate suspensions, *Biorheology,* 1, 41, 1962.

63. **Barbee, J. H.,** The Flow of Human Blood Through Capillary Tubes with Inside Diameters Between 8.7 and 221 Microns, Ph.D. thesis, California Institute of Technology, Pasadena, 1970.

64. **Tangelder, G. J., Slaaf, D. W., Teirlinck, H. C., Alweijnse, R., and Reneman, R. S.,** Localization within thin optical sections of fluorescent blood platelets flowing in a microvessel, *Microvasc. Res.,* 23, 214, 1982.

65. **Corattiyl, V. and Eckstein, E. C.,** Regional platelet concentration in blood flow through capillary tubes, *Microvasc. Res.,* 32, 261, 1986.

66. **Blackshear, P. H., Bartelt, K. W., and Forstrom, R. J.,** Fluid dynamic factors affecting particle capture and retention, *Ann. N.Y. Acad. Sci.,* 283, 270, 1977.

67. **Vejlens, G.,** The distribution of leukocytes in the vascular system, *Acta Pathol. Microbiol. Scand. Suppl.,* 33, 11, 1938.

68. **Nobis, V., Fries, A. R., and Gähtgens, P.,** Rheological mechanisms contributing to WBC-margination, in *White Blood Cells: Morphology and Rheology Related to Function,* Bagge, U., Born, G. V. R., and Gähtgens, P., Eds., Martinus Nijhoff, The Hague, 1982, 57.

69. **Schmid-Schönbein, G. W., Usami, S., Skalak, R., and Chien, S.,** The interaction of leukocytes and erythrocytes in capillary and postcapillary vessels, *Microvasc. Res.,* 19, 45, 1980.

70. **Schmid-Schönbein, G. W., Fung, Y.-C., and Zweifach, B. W.,** Vascular endothelium-leukocyte interaction: sticking force in venules, *Circ. Res.,* 36, 173, 1975.

71. **Goldsmith, H. L. and Karino, T.,** Mechanically induced thromboemboli, in *Quantitative Cardiovascular Studies: Clinical and Research Applications,* Hwang, N. H. C., Gross, D. R., and Patel, D. J., Eds., University Park Press, Baltimore, 1978, 289.

72. **Caro, C. G., Fitz-Gerald, J. M., and Schroter, R. C.,** Arterial wall shear, observation, correlation and proposal of a shear-dependent mass transfer mechanism for atherogenesis, *Proc. Roy. Soc. London Ser. B,* 177, 109, 1971.

73. **Glagov, S.,** Hemodynamic risk factors: mechanical stress, mural architecture, medial nutrition and the vulnerability of arteries to atherosclerosis, in *The Pathogenesis of Atherosclerosis,* Wissler, R. W. and Geer, J. C., Eds., Williams & Wilkins, Baltimore, 1972, chap. 6.

74. **Fry, D. L.,** Hemodynamic forces in atherogenesis, in *Cardiovascular Diseases,* Scheinberg, P., Ed., Raven Press, New York, 1976, 77.

75. **Roach, M. R.,** The effect of bifurcations and stenoses on arterial disease, in *Cardiovascular Flow Dynamics and Measurements,* Hwang, N. H. C. and Normann, N. A., Eds., University Park Press, Baltimore, 1977, 489.

76. **Mustard, J. F., Murphy, E. A., Rowsell, H. C., and Downie, H. G.,** Factors influencing thrombus formation *in vivo, Am. J. Med.,* 33, 621, 1962.

77. **Mustard, J. F. and Packham, M. A.,** The role of blood and platelets in atherosclerosis and the complications of atherosclerosis, *Thromb. Diath. Haemorrh.,* 33, 444, 1975.

78. **Geissinger, H. D., Mustard, J. F., and Rowsell, H. C.,** The occurrence of microthrombi on the aortic endothelium of swine, *Can. Med. Assoc. J.,* 87, 405, 1962.

79. **Mitchell, J. R. A. and Schwartz, C. J.,** The relationship between myocardial lesions and coronary artery disease. II. A select group of patients with massive cardiac necrosis or scarring, *Br. Heart J.,* 25, 1, 1963.

80. **Packham, M. A., Rowsell, H. C., Jörgensen, L., and Mustard, J. F.,** Localized protein accumulation in the wall of the aorta, *Exp. Mol. Pathol.,* 7, 214, 1967.

81. **Yu, S. K. and Goldsmith, H. L.,** Behavior of model particles and blood cells at spherical obstructions in tube flow, *Microvasc. Res.,* 6, 5, 1973.

82. **Karino, T. and Goldsmith, H. L.,** Flow behaviour of blood cells and rigid spheres in an annular vortex, *Phil. Trans. Roy. Soc. London Ser. B,* 279, 413, 1977.

83. **Karino, T., Kwong, H. M., and Goldsmith, H. L.,** Particle flow behavior in models of branching vessels. I. Vortices in 90° T-junctions, *Biorheology,* 16, 231, 1979.

84. **Karino, T. and Goldsmith, H. L.,** Particle flow behavior in models of branching vessels. II. Effects of branching angle and diameter ratio on flow patterns, *Biorheology,* 20, 119, 1983.

85. **Karino, T. and Motomiya, M.,** Flow visualization in isolated transparent natural blood vessels, *Biorheology,* 22, 87, 1985.

86. **Motomiya, M. and Karino, T.,** Flow patterns in the human carotid artery bifurcation, *Stroke,* 15, 50, 1984.

87. **Karino, T. and Motomiya, M.,** Flow through a venous valve and its implication in thrombogenesis, *Thromb. Res.,* 36, 245, 1984.

88. **Karino, T., Motomiya, M., and Goldsmith, H. L.,** Flow patterns in model and natural vessels, in *Biologic and Synthetic Vascular Prostheses,* Stanley, J. C., Ed., Grune & Stratton, New York, 1982, 153.

89. **Macagno, E. O. and Hung, T. K.,** Computational and experimental study of a captive annular eddy, *J. Fluid Mech.,* 28, 43, 1967.

90. **Whitmore, R. L.,** The viscous flow of disperse suspensions in tubes, in *Rheology of Disperse Systems,* Mill, C. C., Ed., Pergamon Press, New York, 1959, 49.

91. **Karnis, A. and Mason, S. G.,** The flow of suspensions through tubes. VI. Meniscus effects, *J. Colloid Interface Sci.,* 23, 120, 1967.
92. **Karino, T. and Goldsmith, H. L.,** Aggregation of human platelets in an annular vortex distal to a tubular expansion, *Microvasc. Res.,* 17, 217, 1979.
93. **Karino, T. and Goldsmith, H. L.,** Adhesion of human platelets to collagen on the wall distal to a tubular expansion, *Microvasc. Res.,* 17, 238, 1979.
94. **Karino, T. and Goldsmith, H. L.,** Role of blood cell-wall interactions in thrombogenesis and atherogenesis: a microrheological study, *Biorheology,* 21, 587, 1984.
95. **Diener, L., Ericsson, J. L. E., and Lund, F.,** The role of venous valve pockets in thrombogenesis. A postmortem study in a geriatric unit, in *Atherogenesis,* Shimamoto, T. and Numano, F., Eds., Excerpta Medica, Amsterdam, 1969, 125.
96. **Sevitt, S.,** Pathology and pathogenesis of deep vein thrombi, in *Venous Problem,* Bergan, J. J. and Yao, J. S. T., Eds., Year Book Medical Publishing, Chicago, 1978, 257.
97. **Kristiansen, K. and Krog, J.,** Electromagnetic studies on the blood flow through the carotid system in man, *Neurology,* 12, 20, 1962.
98. **Uematsu, S., Yang, A., Preziosi, T. J., Kouba, R., and Toung, T. J. K.,** Measurement of carotid blood flow in man and its clinical application, *Stroke,* 14, 256, 1983.
99. **Flaherty, J. T., Ferrans, V. J., Pierce, J. E., Caren, T. E., and Fry, D. L.,** Localization factors in experimental atherosclerosis, in *Atherosclerosis and Coronary Heart Disease,* Likoff, W., Segal, B. E., Insule, W., and Moyer, J. H., Eds., Grune & Stratton, New York, 1972, 40.
100. **Cornhill, J. F. and Roach, M. R.,** A quantitative study of the localization of atherosclerotic lesions in the rabbit aorta, *Atherosclerosis,* 23, 498, 1976.
101. **Zarins, C. K., Giddens, D. P., Bharadvaj, B. K., Scottiurai, V. S., Mabon, R. F., and Glagov, S.,** Carotid bifurcation atherosclerosis. Quantitative correlation of plaque localization with flow velocity profiles and shear stresses, *Circ. Res.,* 53, 502, 1983.

Chapter 22

ENDOTHELIAL CELL INJURY BY NEUTROPHILS

David E. Gannon, James Varani, and Peter A. Ward

TABLE OF CONTENTS

I. INTRODUCTION

Circulating polymorphonuclear leukocytes normally exist in a quiescent state. Following activation by soluble or particulate stimuli, these cells undergo changes in their surface properties, motility characteristics, and metabolism. In addition, they produce and secrete (internally into phagosomes or externally into the extracellular milieu) various proteolytic enzymes, toxic products, and inflammatory mediators. In this manner, neutrophils can phagocytize and kill microorganisms, injure eukaryotic cells, alter extracellular matrices, and amplify and modulate inflammatory reactions.

There is little interaction between inactive circulating neutrophils and healthy vascular endothelial cells. Even when present as "marginated" cells, neutrophils do not damage the vascular endothelium. However, when neutrophils are activated, they interact extensively with the endothelial layer lining the vessel walls. These interactions include adhesion to the endothelial cells, migration through the endothelium and into the extravascular space, disruption of the endothelial cell layer, and cytotoxicity of the endothelial cells. Thus, normal vascular endothelium can be injured indiscriminately by neutrophils, especially endothelium that is proximate to a nidus of inflammation that causes chemotaxis of neutrophils into the area and activation of the neutrophils.

Many experimental models have been developed to explore the changes in activated neutrophils and the interactions of activated neutrophils with endothelial cells. These include in vivo models in which circulating neutrophils are activated in animals and in vitro models in which neutrophils isolated from blood are activated while in contact with endothelial cells in culture. We will discuss some of these models and the information they yield about the interactions between neutrophils and endothelial cells.

II. NEUTROPHIL ACTIVATION

Phagocytic cells can become activated in response to a number of stimuli. These include both soluble factors (i.e., complement system components, n-formyl chemotactic peptides, platelet-activating factor, metabolites of arachidonic acid, phorbol esters, various lymphokines) and particulate stimuli (i.e., immune complexes, opsinized particles). The initial interaction between the stimulus and the cells is mediated through ligand binding to receptors present on leukocyte surfaces. Following this, activation of the g-binding protein complex occurs in association with a rapid membrane depolarization followed by (1) change in the cytosolic levels of monovalent and divalent cations; (2) activation of phospholipases resulting in generation of diacyl glycerol and phosphoinositol (which is subsequently converted to inositol bis and triphosphates) in the case of phospholipase C activation, or arachidonic acid and lysophosphoglycerate in the case of phopholipase A_2 activation; and (3) activation of cellular kinases. Structurally, rapid polymerization of cytoplasmic actin and formation of microfilament bundles occur in submembraneous areas. The cell becomes polarized, increasingly adhesive (to other cells and to extracellular matrix components), and actively motile. Activation of the neutrophil with chemotactic peptide results in an increase in the amount of CDW_{18} complex, which is a surface glycoprotein responsible for the increased adherence to endothelial cells via a surface moiety on the endothelial cell (which is not yet further defined). Simultaneously, lysozomal granules may fuse with cell surface membranes and release granule contents into the extracellular environment. In addition, membrane NADPH oxidase may be activated and give rise to the production of superoxide and, in turn, a whole series of toxic oxygen metabolites. The cellular and molecular events underlying leukocyte activation have been extensively studied. It is beyond the scope of the present work to review this area, but several good monographs have been written on this topic.[1-9]

We will concentrate here on the possible consequences of lysozomal enzyme release and oxygen radical production on target cells (specifically endothelial cells) present where leukocyte activation occurs.

III. NONLETHAL INJURY OF ENDOTHELIAL CELLS BY NEUTROPHILS

Polymorphonuclear leukocytes contain several types of hydrolytic enzymes within their granules. These include elastase, cathepsin G, collagenase, plasminogen activator, and a class of enzymes referred to as proteoglycon-degrading neutral proteases.[10] These enzymes are stored in lysozomal granules and are released into developing phagosomes when the lysozomes fuse with phagocytic vacuoles. When the cell attempts to phagocytose a particle which is too large to be completely engulfed, the lysozomal contents may be released into the extracellular environment. Cells activated by stimuli other than those leading to phagocytosis may also release their lysozomal enzymes into the extracellular space. When large amounts of these enzymes are released extracellularly at sites of inflammation, they may overwhelm the antiprotease defenses of the host[11] and damage host tissues, especially extracellular matrix material. The potent proteolytic enzymes of leukocytes are capable of degrading a number of extracellular matrix components. These enzymes may also be injurious to intact cells. It is thought that the release of these enzymes at sites of chronic inflammation may be related to the etiology of tissue-destructive diseases such as emphysema and rheumatoid arthritis.[12,13]

Several studies have examined the effects of activated polymorphonuclear leukocytes on components of the extracellular matrix using in vitro model systems.[14-16] These studies have attempted to analyze the mechanisms by which leukocytic proteases may function in the presence of high concentrations of antiproteases. The results have shown that upon activation, neutrophilic leukocytes rapidly attach and spread on substrates of extracellular matrix molecules and hydrolyze these substrates. Hydrolysis is mediated primarily by cells in direct contact with the substrate. This contact facilities the release of the hydrolytic enzymes in a concentrated form at the target site, and facilitates the exclusion of antiproteases from the active site as well. The direct contact between the cells and the substrate probably accounts, therefore, for the functioning of these enzymes in a milieu of antiproteases (Table 1). In this regard, the functioning of leukocytic proteases is similar to the activation of several other protease cascades (i.e., complement-activations, blood coagulation, and fibrinolysis) which occur in the presence of inhibitors by solid-support sequestration of the active sites from the inhibitors. In addition to physical isolation from the antiprotease defense mechanisms of the host, the α-1-proteinase inhibitor of the host is susceptible to inactivation by leukocyte oxidants.[15,16] In this way, leukocytic protease action is enhanced by chemical inactivation of host antiprotease defenses in the immediate vicinity of an inflammatory nidus.

Although leukocytes can adhere tenaciously to extracellular matrices and degrade matrix components, circulating neutrophils do not come into direct contact with these substances under normal circumstances. Intact endothelium serves as an effective barrier to the interaction between leukocytes and the subendothelial cell extracellular matrix. Activated leukocytes must have a mechanism for facilitating initial adherence to and subsequent migration through an intact endothelial cell layer. Recent studies suggest that this may involve sublethal injury to the endothelium mediated by leukocyte proteases. Harlan et al.[17,18] showed that exposure of an intact monolayer of endothelial cells to activated neutrophils led to disruption of the monolayer with destruction of the target cells. This action was not attenuated by inhibitors of oxygen radicals, but could be blocked with a variety of protease inhibitors. Although the authors quantitated injury in the in vitro system by counting the cells actually released from the monolayer, only cell retraction would be required to occur under physiological conditions. This would expose the subendothelial cell basement membrane to

Table 1
**ROLE OF DIRECT CONTACT
BETWEEN POLYMORPHONUCLEAR
LEUKOCYTES AND THE
SUBSTRATUM FOR HYDROLYSIS IN
THE PRESENCE OF ANTIPROTEASES**

Test group	Radioactivity released (%)
No activation	7
Immune complexes in suspension	33
+ α-1-Anti-protease	13
+ α-2-Macroglobulin	10
+ Soybean trypsin inhibitor	8
Immune complexes bound to substrate	25
+ α-1-Anti-protease	23
+ α-2-Macroglobulin	27
+ Soybean trypsin inhibitor	20

The hydrolysis of radiolabeled hemoglobin was assessed by immune complex-stimulated neutrophils. When the neutrophils were activated in suspension, the activity of the proteolytic enzymes was effectively blocked by antiproteases. When the cells were activated in direct contact with the substratum, the antiproteases were much less effective.(From Johnson, K. J. and Varani, J., *J. Immunol.*, 127, 1875, 1981. With permission.)

activated neutrophils and allow injury of the matrix. It should be pointed out that the in vitro system developed by Harlan et al.[17,18] uses a suspension of cells activated in a buffer devoid of the normal protease inhibitors. Under physiological conditions, the direct contact between the neutrophils and the target endothelial cells may serve to inhibit the effects of the plasma inhibitors. While this type of injury to endothelial cells may not be directly cytotoxic, the retraction of the individual cells from each other and from the basement membrane could still have two important consequences. These are (1) discontinuity of the endothelial cell layer facilitating the extravasation of activated neutrophils into the extravascular space, and (2) exposure of the subendothelial basement membrane to neutrophils, facilitating the destruction of the subendothelial cell basement membrane which itself may ultimately lead to endothelial cell death.

There is recent evidence that human microvascular endothelial cells are susceptible to the cytotoxic effects of human neutrophils stimulated with chemotactic peptide or zymosan particles. The killing process does not appear to be related to the generation of toxic oxygen products by the neutrophils, but does appear to be related to a lysosomal constituent which has been previously identified as elastase.[19] Thus, neutrophils may damage or kill endothelial cells by a nonoxidative mechanism.

IV. CYTOTOXIC INJURY OF ENDOTHELIAL CELLS BY NEUTROPHILS IN VIVO

The mechanism of neutrophil-mediated cytotoxicity of eukaryotic cells is under active investigation, and the mechanism of their cytotoxic injury of endothelial cells is of particular interest. Normal vascular endothelium is disrupted during a number of acute and chronic inflammatory processes, and this can result in loss of integrity of the vasculature. Subsequent extravasation of fluid and cells from the vasculature can contribute to further damage of the surrounding tissue. Thus, the injury of vascular endothelium can play a significant role in the manifestation of an inflammatory process.

A. Dermal Vascular Injury

Neutrophils have long been associated with experimental models of vascular injury. Stetson[20] first showed in 1951 that an arthus reaction produced by intradermal injection of antigen in sensitized rabbits was associated with intravascular and perivascular accumulation of neutrophils. Using neutrophil-depleted animals, Humphrey[21,22] later showed that neutrophils were essential for both the edema formation and the vascular damage that occurred during the reverse passive arthus reaction in rabbits and guinea pigs.

In a model of dermal arthus in rabbits, Cochrane et al.[23] also demonstrated that neutrophils were essential to produce vasculitis. In similar models of dermal arthus in rats and guinea pigs, Cochrane and Ward[24] demonstrated that in addition to neutrophils, complement was required for the production of vasculitis. Further studies of immune complex-induced vasculitis in rats by Fligiel et al.[25] confirmed the dependence of the injury on the complement system and on neutrophils. In addition, when injury was assessed by an index of vascular leak and morphologic appearance of vessels and endothelial cells, protection from injury was seen when animals were given the antioxidant catalase, the hydroxyl radical scavenger dimethyl sulfoxide, or the iron chelators apolactoferrin or deferoxamine mesylate. These observations implicated neutrophil-derived oxygen radicals in the injury of endothelial cells during dermal vasculitis.

Models of nephrotoxic glomerulonephritis in rats and rabbits have shown a similar dependence on neutrophils.[26] The dependence of experimental glomerulonephritis on complement and neutrophils has been recently reviewed by Cochrane.[27]

B. Pulmonary Vascular Injury

Injury of dermal vascular endothelium can result in exudation of vascular contents with local edema and swelling. Injury of pulmonary vasculature, however, can have profound systemic consequences. Loss of integrity of the pulmonary vascular endothelium can result in leak of vascular contents into the lung with commensurate pulmonary dysfunction and, if extensive enough, respiratory failure. Because of the importance of the pulmonary vasculature in maintaining viability of the organism, the lungs contain one of the most extensively studied vascular endotheliums.

In 1910, Andrewes[28] reviewed work by himself and others dating from before 1894 in which rabbits were injected with bacteria or bacterial protein. In all cases, a transient leukopenia resulted during which accumulation of neutrophils in lung was demonstrated. Nearly a century later, Hosea et al.[29] showed that injection of guinea pigs with bacteria resulted in complement activation and pulmonary alveolar capillary leak. The apparent alteration of the capillary endothelium was both complement- and neutrophil-dependent. Since then, much work has been done to define the mechanism of neutrophil-mediated injury of pulmonary vascular endothelial cells.

The possible mechanisms involved in neutrophil-dependent lung injury have been extensively reviewed.[30-37] We will concentrate here on the mechanisms underlying pulmonary vascular endothelial cell injury.

1. Clinically Observed Lung Injury in Patients
a. Extracorporeal Circulation of Blood

Observations of the white blood cell counts of patients undergoing dialysis led to the first demonstration of neutrophil interaction with pulmonary vascular endothelial cells in humans.

A universal marked temporary neutropenia was noted independently on both sides of the Atlantic Ocean in uremic patients undergoing hemodialysis with cellophane membrane equipment.[38-40] The same effect was noted in patients and volunteers undergoing continuous-flow filtration leukapheresis with nylon fiber filter equipment,[41,42] and could be reproduced in normal volunteers reinfused with heparinized blood that had been stagnated briefly in a hemodialysis coil.[43] Parallel studies in dogs revealed that a transient sequestration of granulocytes in the pulmonary vascular bed was the cause of the peripheral neutropenia.[44]

Studies of blood during filtration leukapheresis revealed an apparent mediator of the altered neutrophil kinetics that was plasma-derived and -borne.[42]. It had been independently determined that activation of the alternative pathway of the complement cascade by either cobra venom factor, inulin, zymosan, or trypsinized human C5 resulted in a marked neutropenia in rabbits.[45,46] In addition, it was shown that complement activation in rabbits and humans played a central role in the increased granulocyte adherence observed in activated granulocytes.[47] Since activation of complement was found to cause both increased adherence of neutrophils and neutropenia, the possibility of complement involvement in the pulmonary vascular sequestration of neutrophils and the peripheral neutropenia induced by hemodialysis and filtration leukapheresis was investigated.

Analysis of human plasma incubated with nylon fibers in vitro and of plasma returning from the filtration equipment during filtration leukapheresis of patients demonstrated activation of the alternative pathway of the complement cascade.[48] This suggested that the transient neutropenia that occurs during filtration leukapheresis with nylon fiber filter equipment is mediated by complement. In addition, the same study noted that the neutropenia was more pronounced than could be explained by trapping of neutrophils in the filter, and sequestration was inferred. Moreover, some pulmonary dysfunction was measured in patients undergoing filtration leukapheresis.

A similar study using dialyzer cellophane also showed activation of the alternative pathway of complement in patients' plasma returning from the dialyzer and in human plasma incubated with dialyzer cellophane in vitro.[49] In a parallel animal study, reinfusion of autologous plasma activated by cellophane dialysis membrane into rabbits resulted in transient leukopenia identical to that seen in hemodialysis patients, and lungs from the animals revealed sequestration of granulocytes in the pulmonary vasculature.[49] These studies showed that the activation of the alternative pathway of the complement cascade induces pulmonary vascular sequestration of neutrophils which results in a transient leukopenia as observed during hemodialysis with cellophane membrane equipment.

Patients undergoing cardiopulmonary bypass with nylon mesh bubble oxygenators exhibited neutrophil kinetics similar to those exhibited by patients and volunteers undergoing hemodialysis with cellophane membrane equipment and patients and volunteers undergoing filtration leukapheresis with nylon fiber filter equipment. Patients on cardiopulmonary bypass were found to have pulmonary vascular sequestration of neutrophils with plugging of the pulmonary capillaries, interstitial and intra-alveolar edema, and damage of endothelial cells.[50-51]Analysis of plasma before and after cardiopulmonary bypass as well as plasma incubated with the nylon mesh liner of bubble oxygenators demonstrated activation of complement, suggesting that these effects were also mediated by complement.[51]

Studies in sheep were undertaken to investigate the physiologic effects of neutrophil aggregation in the lung.[52] Complement activation by exposure of plasma to zymosan or cellophane dialysis membrane resulted in pulmonary vascular leukostasis. The activation of complement also resulted in pulmonary dysfunction which was manifested by hypoxemia, hypercapnia, and impaired diffusing capacity. Pulmonary artery pressures doubled, presumably secondary to plugging of the pulmonary vasculature with aggregating granulocytes. In addition, pulmonary edema was demonstrated histologically. Collected lymph fluid was not a simple transudate, but was found to be rich in protein, and the protein leak suggested endothelial layer disruption. The data from this study demonstrated that complement-activation results in plugging of the pulmonary vasculature with activated granulocytes, pulmonary dysfunction, and endothelial cell injury.

b. Adult Respiratory Distress Syndrome

The observations that activation of complement in patients and animals resulted in aggregation of neutrophils in the pulmonary vasculature, pulmonary edema, and pulmonary dysfunction led to speculation by 1980 that a similar mechanism may be involved in the pathogenesis of the adult respiratory distress syndrome (ARDS).[29] Bacterial endotoxin during sepsis or bacteremia was thought to be the most likely activator of serum complement in patients who developed ARDS. Since then, the involvement of complement and neutrophils in ARDS has been a subject of much investigation in patients and in animal models of ARDS.

Numerous studies in sheep have demonstrated that activation of serum complement by injection of bacteria or bacterial endotoxin results in accumulation of neutrophils in the pulmonary vasculature and characteristic patterns of lung injury. The injury is neutrophil-dependent and includes pulmonary hypertension, increased vascular permeability manifested by two phases of increased flow of lung lymph (with protein-rich lymph in the late phase indicating loss of integrity of the vascular endothelium), and pulmonary dysfunction as measured by physiologic parameters of airway function, compliance, and oxygenation of blood.[53-56] Morphologic evidence of endothelial cell injury and pulmonary edema has been found to correspond to these injury patterns.[33,34]

Studies in humans with ARDS have been limited by the difficulties of investigating critically ill patients. Despite this, however, some evidence for the involvement of complement and neutrophils has been accumulated. A prospective study of patients at risk for ARDS found a correlation between neutrophil-aggregating activity (reflecting C5a levels) and the development of ARDS.[57] Studies of bronchoalveolar lavage fluid have found oxidant activity in the lungs of patients with ARDS, suggesting that neutrophil-derived oxygen metabolites may be involved.[58]

Although it is not yet certain whether all the experimental models of ARDS in animals[59] and ARDS in humans[60-63] depend, in fact, on neutrophils, most of the evidence suggests that neutrophils and toxic oxygen matabolites derived from neutrophils may be related to the lung damage in ARDS. The evidence for the involvement of neutrophils in the pathogenesis of ARDS has been extensively reviewed.[30,32,35-37]

c. Oxygen Toxicity

Early studies of hyperoxic lung injury in rats and monkeys demonstrated that the primary cellular damage was of the pulmonary capillary endothelial cells with subsequent accumulation of edema in the interstitial space and destruction of capillaries.[64-66] The massive endothelial damage coincided with increased numbers of neutrophils in the lung.[64,66] A study using rats exposed to hyperoxia found a close temporal relationship between the appearance of chemoattractants in bronchoalveolar lavage fluid and death of the rats.[67] However, a study of bronchoalveolar lavage in humans exposed to increased concentrations of oxygen found alterations in alveolar capillary permeability without influx of neutrophils into lavage fluid.[68] The dependence of the lung injury on neutrophils is not certain, and hyperoxic lung injury may involve oxygen radicals not derived from neutrophils.[69,70]

2. Experimentally Induced Lung Injury in Animals

A number of animal models of acute lung injury have been developed. Although they differ in the agent used to precipitate the injury and the species of animal used, most involve activation of complement and require neutrophils for full expression of injury.

a. Phorbol Ester-Induced Lung Injury

Phorbol myristate acetate (PMA) is a soluble nonspecific activator of neutrophils that causes neutrophils to adhere, aggregate, release enzymes from granules, and produce oxygen radicals. Intravenous injection of PMA into rabbits has been shown to result in a striking neutropenia and accumulation of neutrophils in the pulmonary vasculature.[71,72] The animals manifested lung injury characterized by increased permeability of the pulmonary vasculature,[71,72] and accumulation of a protein-rich edema in the lungs,[73] which suggested loss of integrity of pulmonary vascular endothelium. Granulocytopenic rabbits did not manifest the injury, which indicated that the injury was mediated by neutrophils.[73]

In a different model, PMA was given to rabbits and rats by intratracheal instillation instead of intravenous injection.[72,74] This also produced an acute lung injury. The injury was found to be dependent on the presence of neutrophils in rabbits,[72] but not in rats.[74]

Intrabronchial instillation of PMA or the formylated peptide norleu-leu-phe resulted in acute lung injury in a primate model, characterized by accumulation of neutrophils in the lung and pulmonary edema.[75] Bronchoalveolar lavage fluid contained an increased protein content, indicating loss of integrity of the pulmonary vascular endothelium. In addition, oxidants and proteases appeared in the lavage fluid concomitant with the lung injury, suggesting the involvement of neutrophil-derived oxidants and proteases in the pathogenesis of the injury.

b. Immune Complex-Induced Lung Injury

In studies by Johnson et al.,[74,76,77] deposition of IgG-containing immune complexes in lung after airway instillation of antibody and intravenous injection of antigen resulted in an acute lung injury in rats as measured by lung vascular permeability. Morphologically, the lungs manifested edema and injury of pulmonary vascular endothelial cells under microscopic examination. The injury was dependent on the presence of neutrophils and appeared to be mediated by complement. Catalase did not inhibit the accumulation of neutrophils in the lung vasculature, but did markedly suppress the lung injury, suggesting that toxic products of neutrophil oxygen metabolism such as hydrogen peroxide or its derivatives were involved in the injury.

Deposition of IgA-containing immune complexes in rat lung resulted in a similar acute lung injury as measured by lung vascular permeability and as seen morphologically.[77] The lung injury was complement-dependent, but in contrast to lung injury induced by IgG-containing immune complexes, the lung injury induced by IgA-containing immune complexes was not dependent on the presence of neutrophils.

c. Cobra Venom Factor-Induced Lung Injury

Cobra venom factor (CVF) activates complement and produces lung injury in animal models. The lung injury produced by injection of CVF in rats has been studied primarily by Ward's group.[74,78-82] Injection of CVF caused intravascular activation of the complement system and resulted in a profound peripheral neutropenia and acute lung injury as measured by lung permeability.[74,78,82] The lung injury manifested morphologically by accumulation of neutrophils in the lung, plugging of pulmonary capillaries with neutrophils, disruption of the pulmonary capillary endothelium, and damage and destruction of endothelial cells. The injury was dramatically attenuated in animals that had been previously depleted of complement or neutrophils, indicating that the injury was complement- and neutrophil-dependent. The injury produced by complement-activated neutrophils resembled the injury produced by enzymatically produced oxygen metabolites,[83] and catalase markedly attenuated the injured in both models. This suggested that the lung injury produced by intravenous CVF injection may involve the production of toxic oxygen metabolites by complement-activated neutrophils. Similar results were seen in parallel studies done in mice.[80]

Further studies in rats demonstrated that pretreatment of animals with the hydroxyl radical scavenger dimethyl sulfoxide or the iron chelators apolactoferrin or deferoxamine mesylate protected the animals from lung injury induced by injection of CVF.[79,80,82] This suggested that the hydroxyl radical, a product possibly generated via the iron-catalyzed Haber-Weiss reaction,[84] may be involved in the lung injury induced by complement which had been activated by CVF. Recent studies have found a quantitative relationship between intensity of lung injury and levels of products of lipid peroxidation, further supporting the involvement of the hydroxyl radical and suggesting that products of lipid peroxidation may be used as markers of oxygen-radical-mediated lung injury.[81]

d. Thermal Skin Burn-Induced Lung Injury

Acute thermal skin injury in rats has been shown to produce systemic activation of the complement system.[85] In studies assessing lung injury, thermal skin injury resulted in peripheral neutropenia and lung injury as evidenced by increases in lung permeability in rats.[80,82,85] Morphologically, the lung injury resembled the injury seen after systemic complement activation with CVF, and was manifested by leukoaggregates within pulmonary capillaries and injury of pulmonary capillary endothelial cells. The lung injury was shown to be complement- and neutrophil-dependent. In addition, it was attenuated by the antioxidant catalase, the hydroxyl radical scavengers dimethylsulfoxide and dimethyl thiourea, and the iron chelator deferoxamine mesylate. These studies indivated that the injury was mediated by neutrophil-generated toxic oxygen metabolites, including perhaps the hydroxyl radical. In addition, a similar relationship between lung injury and the appearance of products of lipid peroxidation was observed after thermal skin burns[86] as was observed after injection of CVF.

3. Experimentally Induced Injury of Isolated Perfused Animal Lungs

Isolated perfused animal lungs have become helpful tools in investigating lung injury as they allow close control and manipulation of lung perfusate as well as close measurement of lung effluent. The results of these studies have strengthened the argument for the involvement of neutrophil-derived oxygen radicals in the pathogenesis of pulmonary vascular injury in animal models.

In studies by Shasby et al.[73,87,88] granulocytes activated by PMA or arachidonate, but not granulocytes, PMA, or arachidonate alone, caused lung injury in isolated perfused rabbit lungs. The injury was manifested by lung edema and leak of vascular protein into airways, which suggested loss of integrity of pulmonary vascular endothelium. Pretreatment of the granulocytes with agents that reduce granulocyte adherence (cytochalasin B or 2% dextran) markedly reduced edema formation, which suggested that PMA-stimulated granulocytes must be closely apposed to the endothclial cells to induce injury. In addition, neutrophils from patients with chronic granulomatous disease did not mediate injury, which indicated that the injury depended on granulocyte-derived oxygen radicals.

Morganroth et al.[89] found that CVF infusion into isolated perfused rat lungs resulted in sequestration of leukocytes in the lung vasculature and in lung injury manifested by leak of vascular proteins into lung parenchyma and airways. Histologically, neutrophil accumulation, pulmonary capillary endothelial cell injury, and interstitial and intra-alveolar edema were seen. The injury was dependent on complement and neutrophils, and was attenuated by catalase. These experiments indicated that the injury involved toxic oxygen metabolites derived from activated neutrophils.

V. CYTOTOXIC INJURY OF ENDOTHELIAL CELLS BY NEUTROPHILS IN VITRO

In the previous section we reviewed a number of in vivo models in which pulmonary vascular endothelial cells are injured after systemic activation of neutrophils. It is unlikely that previously healthy endothelial cells are specific targets in these injuries. It is more likely that the endothelial cells are "innocent bystanders" injured nonspecifically by activated neutrophils aggregating in the pulmonary vasculature. Similar nonspecific damage of healthy pulmonary vascular endothelium by circulating neutrophils may represent the mechanism of endothelial cell damage that occurs during inflammatory processes in humans. Although these injuries have been studied in a variety of animal models, the specific cellular mechanisms involved in the injury of endothelial cells have been difficult to define. Animal models are limited by the extent to which the cellular microenvironment can be controlled and the extent to which cellular products, especially local mediators and short-lived reactive products, can be measured. In vitro systems have been developed to study cellular aspects of the injury of endothelial cells by non specifically activated mammalian neutrophils.

The first in vitro system for studying neutrophil-mediated cytotoxicity of endothelial cells was reported by Sacks et al.[90] Monolayers of human umbilical vein endothelial cells in culture were exposed to human neutrophils which were stimulated by C5a-rich fractions of zymosan-treated plasma. Injury was measured by assaying the amount of radioactive chromium (^{51}Cr) released from damaged endothelial cells which had been prelabeled with the isotope, and the method was validated by measuring fluorochromasia of viable endothelial cells remaining after exposure to activated neutrophils in parallel experiments. These investigators found that complement-activated human neutrophils were able to kill human endothelial cells. The injury depended on the presence of neutrophils and required the neutrophils to be stimulated by activated complement. Injury was prevented by the presence of cytochalasin B, which inhibits complement-induced spreading and adherence of neutrophils to endothelial cells. This suggested that close proximity between effector cells and target cells may be required for injury, although changes in other surface properties may have been responsible for the observed effect. In parallel experiments, injury similar to that seen with activated neutrophils was seen with oxygen radicals generated enzymatically by the xanthine/xanthine oxidase system. In both cases, injury of the endothelial cells was prevented by the addition of a mixture of superoxide dismutase and catalase, which enhances the dissipation of superoxide and hydrogen peroxide. This indicated that cytotoxicity of endothelial cells by complement-activated neutrophils depends on toxic metabolites of oxygen generated by the neutrophils.

Neutrophil-mediated cytotoxic injury of endothelial cells was further characterized by Weiss et al.[91] Monolayers and suspensions of human umbilical vein endothelial cells were exposed to human neutrophils which were activated by the phorbol ester PMA. Injury of endothelial cells was measured by a ^{51}Cr-release assay, and the method was validated by counting intact endothelial cells remaining after exposure to activated neutrophils in parallel experiments. These investigators found that PMA-activated human neutrophils were able to destroy human endothelial cells. Neutrophils from patients with chronic granulomatous disease were not able to destroy endothelial cells, however, which indicated the importance of neutrophil-generated toxic oxygen metabolites in mediating the injury. In parallel experiments, hydrogen peroxide generated enzymatically by the glucose/glucose oxidase system caused similar destruction of human endothelial cells. In both systems, cytolysis of endothelial cells was prevented by the addition of the antioxidant enzyme catalase. These experiments indicated that hydrogen peroxide plays a key role in the events leading to cytotoxicity. In contrast, myeloperoxidase-deficient neutrophils successfully mediated cytotoxicity, and addition of exogenous myeloperoxidase to the experiments did not augment the cytotoxicity mediated by the myeloperoxidase-deficient neutrophils. These experiments indicated that products of the myeloperoxidase system did not play a strong role in the neutrophil-mediated injury. Moreover, catalase prevented cytotoxicity of endothelial cells by myeloperoxidase-deficient neutrophils, confirming the important role that hydrogen peroxide plays in the injury.

The identities of the oxygen metabolites involved in neutrophil-mediated cytotoxicity of endothelial cells were further investigated by Martin.[92] In this study, monolayers of bovine pulmonary artery endothelial cells in culture were exposed to PMA-activated neutrophils. The injury was measured by a ^{51}Cr-release assay, and the method was validated by assaying LDH released by endothelial cells after exposure to activated neutrophils in parallel experiments. This study confirmed that PMA-activated human neutrophils can cause cytotoxic injury of endothelial cells. Neutrophils from a subject with chronic granulomatous disease were not able to induce significant cytotoxic injury, however, which confirmed the importance of toxic oxygen products in the injury. Exogenous hydrogen peroxide or oxygen radicals generated enzymatically by the xanthine/xanthine oxidase system caused cytotoxic injury of endothelial cells which was similar to injury caused by activated neutrophils. Catalase protected endothelial cells from cytotoxicity in both of these systems as well as in the neutrophil-mediated system, which confirmed that hydrogen peroxide is of pivotal importance in mediating neutrophil toxicity. In addition, injury of endothelial cells was significantly reduced by the presence of the antioxidant dimethyl sulfoxide, a recognized scavenger of the hydroxyl radicals, and a role for the hydroxyl radical in neutrophil-mediated cytotoxic injury of endothelial cells was suggested.

Using bovine pulmonary artery endothelial cells in culture, we have further investigated the species of oxygen metabolites involved in neutrophil-mediated cytotoxicity of endothelial cells.[93,94] In our studies, injury was measured by a ^{51}Cr-release assay and visualized by phase and electron microscopy. We found that human neutrophils activated by immune complexes, opsonized zymosan, or PMA caused significant injury of endothelial cells (Table 2). Figure 1 shows the sequence of changes which the endothelial cells undergo morphologically when subjected to activated neutrophils. Figure 2 shows the dependence of the injury on neutrophils and the time course of injury as indicated by the amount of ^{51}Cr released from prelabeled target cells.

Immune complexes, opsonized zymosan, and PMA all caused neutrophils to both release proteolytic enzymes and produce oxygen metabolites. However, the extent of injury correlated with the quantity of hydrogen peroxide produced and not with the quantity of proteases released. Moreover, the injury was inhibited by the antioxidant catalase and not by soybean trypsin inhibitor. In addition, when neutrophils were activated by N-formyl-methionyl-leucyl-phenylalanine or platelet-activating factor, they released proteolytic enzymes but produced little or no superoxide or hydrogen peroxide and did not injure endothelial cells. These experiments indicated that cytotoxic injury of endothelial cells by human neutrophils involves toxic oxygen metabolites and not proteolytic enzymes.

In order to identify which species of oxygen metabolites are involved in neutrophil-mediated cytotoxicity of endothelial cells, we evaluated a number of potential inhibitors (Table 3). Superoxide dismutase did not protect endothelial cells from injury. Since this enzyme dismutates superoxide to hydrogen peroxide, this obeservation is consistent with the involvement of hydrogen peroxide or its metabolites. In fact, catalase did protect endothelial cells from injury, which indicated that hydrogen peroxide is necessary for cytotoxic injury. Products of the myeloperoxidase pathway did not seem to be involved in the injury of endothelial cells by neutrophils, as evidenced by the lack of protection afforded by sodium azide, an inhibitor of myeloperoxidase. The hydroxyl radical scavengers *N,N*-dimethylthiourea and D-mannitol each inhibited killing of endothelial cells by neutrophils, which im-

Table 2
ENDOTHELIAL CELL KILLING BY UNSTIMULATED AND STIMULATED HUMAN NEUTROPHILS: CORRELATION WITH OXYGEN METABOLITE PRODUCTION AND PROTEOLYTIC ENZYME RELEASE

Stimulating agent	O_2^- production (nmol/ 5×10^5 cells/hr)	H_2O_2 production (nmol/ 5×10^5 cells/hr)	Protease release (μg of hemoglobin hydrolyzed/5×10^5 cells/18 hr)	Cytotoxicity (% specific ^{51}Cr release)
None	10 ± 1	<1	20 ± 1	3 ± 1
PMA				
1.6 × 10⁻⁸ M	37 ± 0	20 ± 1	51 ± 10	40 ± 7
Immune complexes				
25 μg	46 ± 1	14 ± 1	59 ± 1	32 ± 8
10 μg	45 ± 1	13 ± 2	56 ± 1	30 ± 12
1 μg	22 ± 1	2 ± 1	ND[a]	5 ± 1
Zymosan				
100 μg	16 ± 1	8 ± 0	32 ± 1	10 ± 2
50 μg	19 ± 0	6 ± 0	25 ± 1	8 ± 1
FMLP				
10⁻⁶ M	19 ± 1	1 ± 0	63 ± 5	0
Platelet-activating factor				
10⁻⁵ M	16 ± 2	1 ± 0	45 ± 2	3 ± 1

[a] ND = not determined.

From Varani, J. et al., *Lab. Invest.*, 53, 656, 1985.

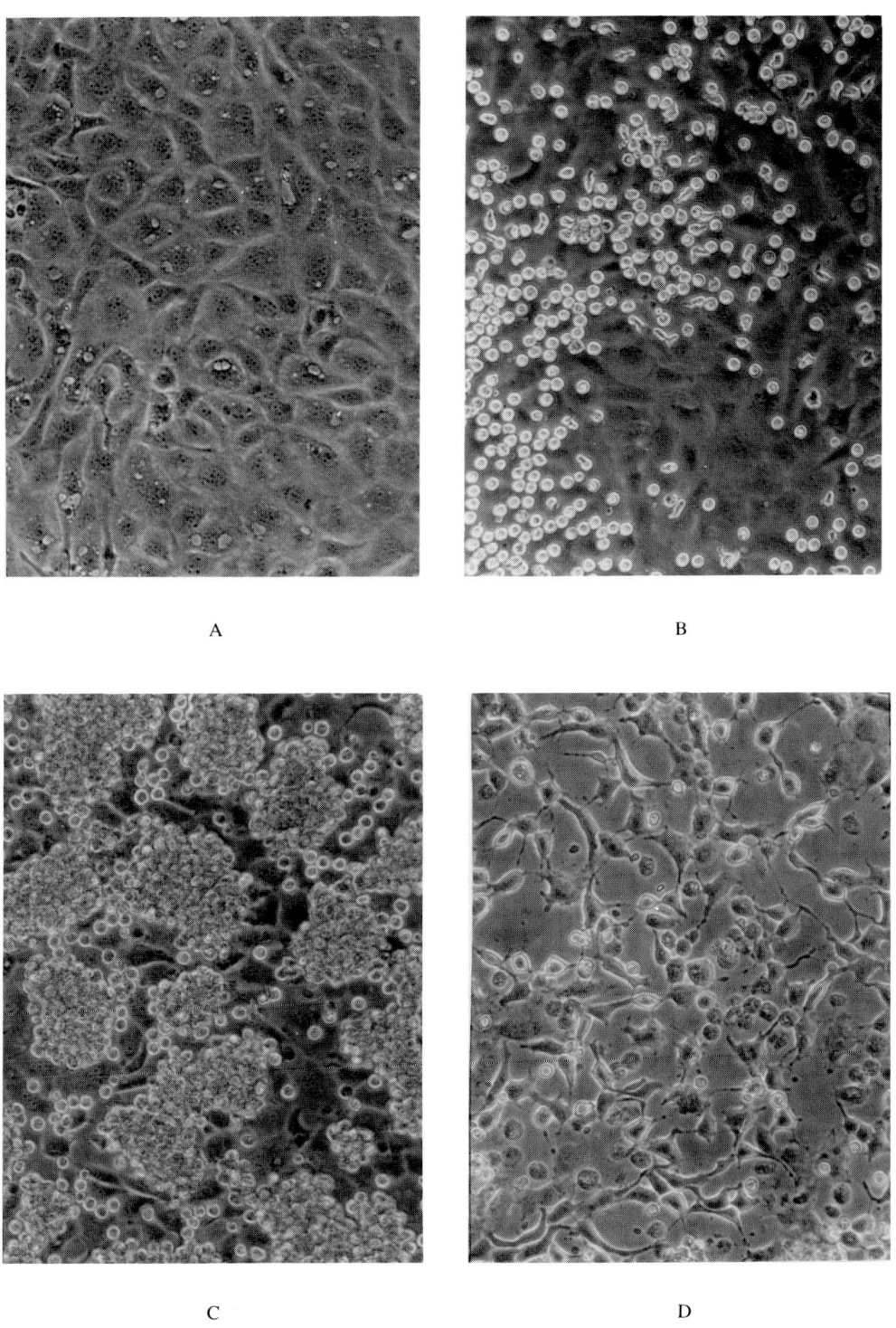

FIGURE 1.   Phase contrast photomicrographs of bovine pulmonary artery endothelial cells in culture. (A) Endothelial cells alone; (B) endothelial cells and unstimulated neutrophils; (C) endothelial cells and neutrophils 5 min after stimulation with PMA, neutrophils are clumped; (D) endothelial cells and neutrophils 30 min after stimulation with PMA, endothelial cells are retracted; (E) endothelial cells and neutrophils 1 hr after stimulation with PMA, endothelial cells are vacuolated; (F) endothelial cells and neutrophils 2 hr after stimulation with PMA, endothelial cells are rounded up.

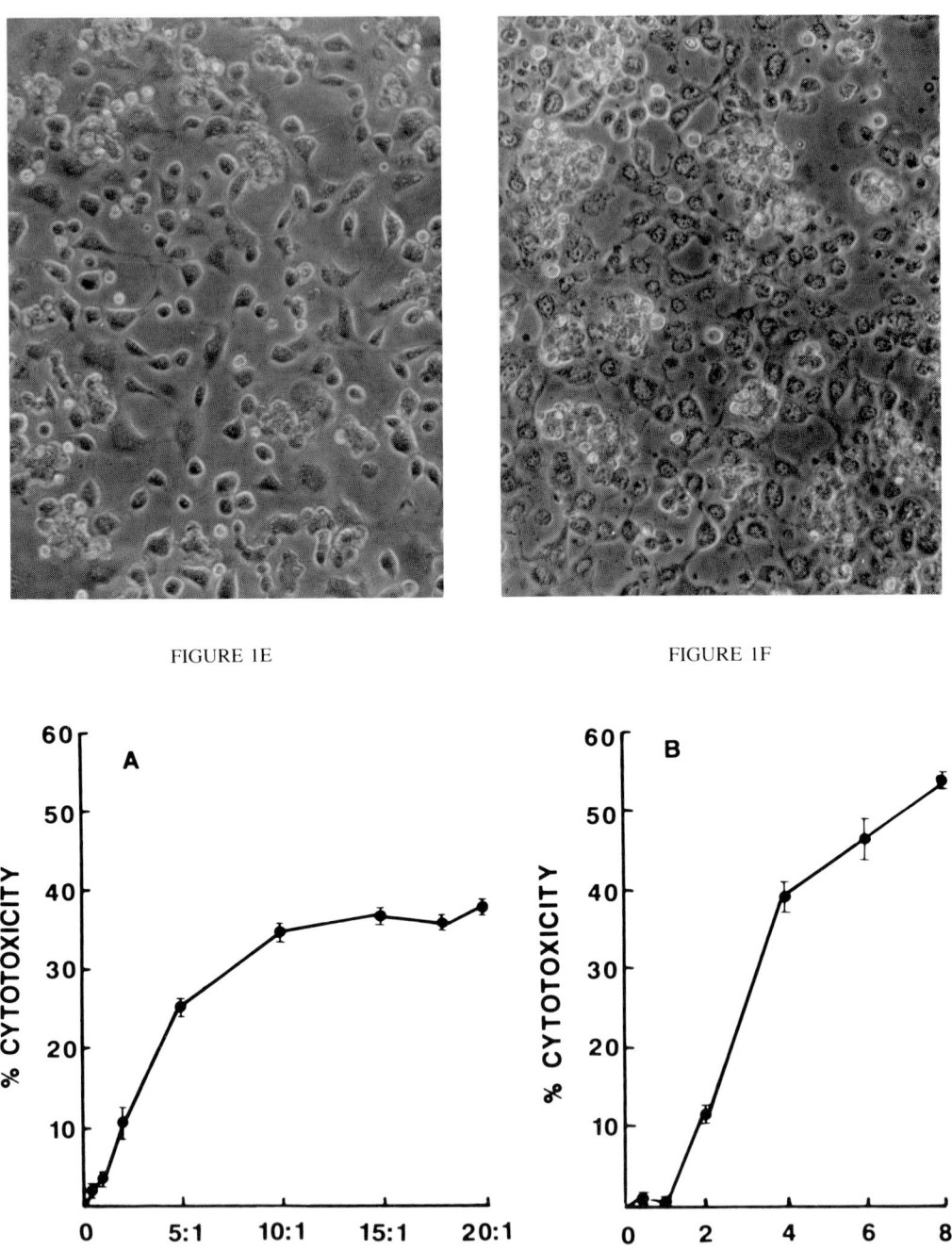

FIGURE 1E   FIGURE 1F

FIGURE 2.   Cytolysis of bovine pulmonary artery endothelial cells in culture by PMA-stimulated human neutrophils. (A) Effect of neutrophil number, in terms of neutrophil to (target) endothelial cell ratio; (B) time course of cytotoxicity as measured by ^{51}Cr released from lysed endothelial cells. (From Gannon, D. E. et al., *Lab. Invest.,* 53, 37, 1987. With permission.)

Table 3
ENDOTHELIAL CELL KILLING BY HUMAN NEUTROPHILS IN THE PRESENCE OF VARIOUS INHIBITORS

Inhibitor	O_2^- production (nmol/5 × 10⁵ cells/ hr)	H_2O_2 production (nmol/5 × 10⁵ cells/ hr)	Cytotoxicity (% specific ⁵¹Cr release)
None	48 ± 3	20 ± 1	27 ± 3
100 μg SBTI	48 ± 1	ND[a]	24 ± 4
280 units SOD	ND	ND	20 ± 7
1850 units catalase	41 ± 1	ND	9 ± 1
10 mM DMTU	59 ± 1	19 ± 1	8 ± 3
55 nM D-mannitol	58 ± 3	25 ± 1	13 ± 2
75 μM deferoxamine mesylate	47 ± 1	20 ± 1	5 ± 2
75 μM deferoxamine mesylate (iron saturated)	ND	ND	17 ± 4
1 μM sodium azide	49 ± 1	ND	30 ± 4

[a] ND = not determined.

From Varani, J. et al., *Lab. Invest.*, 53, 656, 1985.

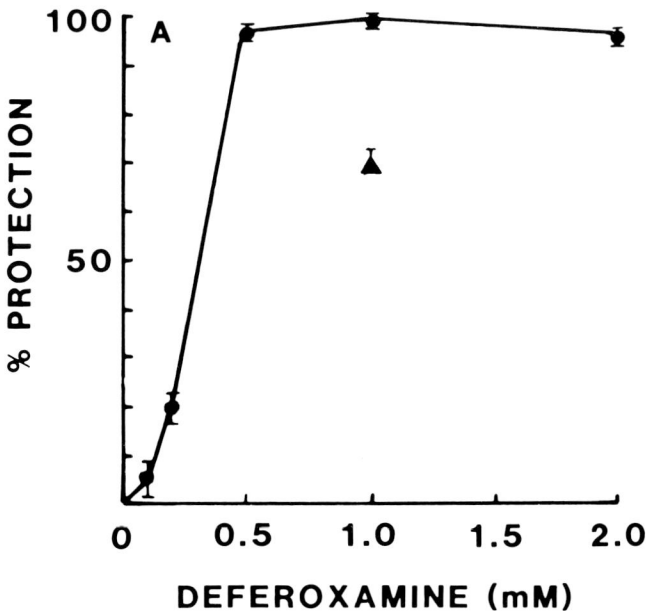

FIGURE 3. Protection of bovine pulmonary artery endothelial cells in culture from cytotoxic injury by PMA-stimulated human neutrophils in the presence of deferoxamine mesylate (●). Protection is inhibited when the deferoxamine mesylate is saturated with ferric iron (▲). (From Gannon, D. E. et al., *Lab. Invest.*, 53, 37, 1987. With permission.)

plicated the hydroxyl radical in the mechanism of neutrophil-mediated cytotoxicity of endothelial cells. In addition, the iron chelator deferoxamine mesylate inhibited endothelial cell injury (Figure 3), which indicated the importance of iron in the mechanism of cytotoxicity and thereby suggested the involvement of the iron-catalyzed Haber-Weiss reaction and its product, the hydroxyl radical.[84] From this study, we concluded that neutrophil-mediated cytotoxic injury of endothelial cells does not depend on proteolytic enzymes, but does depend on hydrogen peroxide and iron, and probably involves formation of hydroxyl radical.

VI. CONCLUSIONS AND FUTURE INVESTIGATIONS

An extensive literature now exists showing that vascular endothelial cells can be injured by activated neutrophils. Injury may be either noncytotoxic or cytotoxic, depending on the mechanism of activation and the physical proximity of the effector cell to the target. Cytotoxic injury is mediated through the generation of superoxide anion by activated neutrophils followed by conversion of the superoxide to hydrogen peroxide and subsequent metabolites. Protection afforded by hydroxyl radical scavengers and iron chelators suggest that the highly-reactive hydroxyl radical may be the ultimate metabolite producing the lethal injury to the endothelial cell. This is difficult to confirm, however, since the hydroxyl radical is a very short-lived species[95] and its formation can not be readily measured in the presence of viable cells. Other iron-dependent mechanisms may also play critical roles.

Subsequent investigations need to precisely define the role of various oxygen metabolites in the injury process, including their sites of generation, their mechanisms of generation, and the ways in which they interact with the target cells to bring about the lethal injury.

Based on an improved understanding of the injury of eukaryotic cells caused by neutrophil-generated oxygen metabolites, therapeutic modalities could perhaps be formulated to interfere with the production or activity of these toxic metabolites and their damage of normal host tissue.

REFERENCES

1. **Babior, B. M.,** Oxygen-dependent microbial killing by phaogyctes, *N. Engl. J. Med.,* 298, 659, and 721, 1978.
2. **Klebanoff, S. J.,** Oxygen metabolism and the toxic properties of phagocytes, *Ann. Intern. Med.,* 93, 480, 1980.
3. **Roos, D.,** The metabolic response to phagocytosis, in *The Cell Biology of Inflammation,* Weissman, G., Ed., Elsevier, Amsterdam, 1980, 337.
4. **Ward, P. A.,** Chemotaxis, in *Textbook of Immunology,* Vol. 1, Parker, C. W., Ed., W. B. Saunders, Philadelphia, 1980, 272.
5. **Snyderman, R. and Goetzl, E. J.,** Molecular and cellular mechanisms of leukocyte chemotaxis, *Science,* 213, 830, 1981.
6. **Fantone, J. C. and Ward, P. A.,** The role of oxygen-derived free radicals and metabolites in leukocyte-dependent inflammatory reactions, *Am. J. Pathol.,* 107, 397, 1982.
7. **Weiss, S. J.,** Oxygen as a weapon in the phagocyte armamentarium, in *Handbook of Inflammation,* Vol. 4, Ward, P. A., Ed., Elsevier, New York, 1983, chap. 2.
8. **Segal, A. W.,** How do phagocytic cells kill bacteria?, *Med. Biol.,* 62, 81, 1984.
9. **Fantone, J. C., and Ward, P. A.,** *Oxygen-Derived Radicals and Their Metabolites: Relationship to Tissue Injury. Current Concepts,* Upjohn, Kalamazoo, Mich., 1985.
10. **Kieser, H. D.,** The effect of lysozomal enzymes on extracellular substrates, in *The Cell Biology of Inflammation,* Weissman, G., Ed., Elsevier, Amsterdam, 1980, 461.
11. **Ohlsson, K.,** Interrelation of granulocytic neutral protease with alpha-1-antitrypsin, alpha-2-macroglobulin and alpha-1-antichymotrypsin, in *Neutral proteases of Human Polymorphonuclear Leukocytes,* Urban and Schwarzenberg, Baltimore, 1978, 167.
12. **Senior, R. M., Tegner, H., Kuhn, C., Ohlsson, K., Stascher, B. C., and Pierce, J. A.,** The induction of pulmonary emphysema with human leukocyte elastase, *Am. Rev. Respir. Dis.,* 116, 469, 1977.

13. **Starkey, P. M., Barrett, A. J., and Burleigh, M. C.,** The degradation of articular collagen by neutrophil proteinases, *Biochim. Biophys. Acta,* 483, 367, 1977.
14. **Johnson, K. J., and Varani, J.,** Substrate hydrolysis by immune complex-activated neutrophils: effect of physical presentation of complexes and protease inhibitors, *J. Immunol.,* 127, 1875, 1981.
15. **Campbell, E. J., Senior, R. M., McDonald, J. A., and Cox, D. L.,** Proteolysis by neutrophils. Relative importance of cell-substrate contact and oxidative inactivation of proteinase inhibitors in vitro, *J. Clin. Invest.,* 70, 845, 1982.
16. **Weiss, S. J. and Regiani, S.,** Neutrophils degrade subendothelial matrices in the presence of alpha-1-proteinase inhibitor: cooperative use of lysosomal proteinases and oxygen metabolites, *J. Clin. Invest.,* 73, 1297, 1984.
17. **Harlan, J. M., Killen, P. D., Harker, L. A., and Striker, G. F.,** Neutrophil-mediated endothelial injury in vitro: mechanisms of cell detachment, *J. Clin. Invest.,* 68, 1394, 1981.
18. **Harlan, J. M., Schwartz, B. R., Reidy, M. A., Schwartz, S. M., Ochs, H. D., and Harker, L. A.,** Activated neutrophils disrupt endothelial monolayer integrity by an oxygen radical-independent mechanism, *Lab. Invest.,* 52, 141, 1985.
19. **Smedly, L. A., Tonnesen, M. G., Sandhaus, R. A., Haslett, C., Guthrie, L. A., Johnston, R. B., Henson, P. M., and Worthen, G. S.,** Neutrophil-mediated injury to endothelial cells. Enhancement by endotoxin and essential role of neutrophil elastase, *J. Clin. Invest.,* 77, 1233, 1986.
20. **Stetson, C. A.,** Similarities in the mechanisms determining the arthus and Shwartzman phenomena, *J. Exp. Med.,* 94, 347, 1951.
21. **Humphrey, J. H.,** The mechanism of arthus reactions. I. The role of polymorphonuclear leukocytes and other factors in reversed passive arthus reactions in rabbits, *Br. J. Exp. Pathol.,* 36, 268, 1955.
22. **Humphrey, J. H.,** The mechanism of arthus reactions. II. The role of polymorphonuclear leucocytes and platelets in reversed passive reactions in the guinea-pig, *Br. J. Exp. Pathol.,* 36, 283, 1955.
23. **Cochrane, C. G., Weigle, W. O., and Dixon, F. J.,** The role of polymorphonuclear leukocytes in the initiation and cessation of the arthus vasculitis, *J. Exp. Med.,* 110, 481, 1959.
24. **Ward, P. A. and Cochrane, C. G.,** Bound complement and immunologic injury of blood vessels, *J. Exp. Med.,* 121, 215, 1965.
25. **Fligiel, S. E. G., Ward, P. A., Johnson, K. J., and Till, G. O.,** Evidence for a role of hydroxyl radical in immune-complex-induced vasculitis, *Am. J. Pathol.,* 115, 375, 1984.
26. **Cochrane, C. G., Unanue, E. R., and Dixon, F. J.,** A role of polymorphonuclear leukocytes and complement in nephrotoxic nephritis, *J. Exp. Med.,* 122, 99, 1965.
27. **Cochrane, C. G.,** The role of complement in experimental disease models, *Springer Semin. Immunopathol.,* 7, 263, 1984.
28. **Andrewes, F. W.,** The Croonian lectures on the behaviour of the leucocytes in infection and immunity. Lecture II, *Lancet,* 2, 8, 1910.
29. **Hosea, S., Brown, E., Hammer, C., and Frank, M.,** Role of complement activation in a model of adult respiratory distress syndrome, *J. Clin. Invest.,* 66, 375, 1980.
30. **Brigham, K. L.,** Mechanisms of lung injury, *Clin. Chest Med.,* 3, 9, 1982.
31. **Fantone, J. C. and Ward, P. A.,** Mechanisms of neutrophil-dependent lung injury, in *Handbook of Inflammation, Vol. 4, Immunology of Inflammation,* Ward, P. A., Ed., Elsevier, New York, 1983, chap. 3.
32. **Tate, R. M. and Repine, J. E.,** Neutrophils and the adult respiratory distress syndrome, *Am. Rev. Respir. Dis.,* 125, 552, 1983.
33. **Brigham, K. L. and Meyrick, B.,** Granulocyte-dependent injury of pulmonary endothelium: a case of miscommunication?, *Tissue Cell,* 16, 137, 1984.
34. **Brigham, K. L. and Meyrick, B.,** Interactions of granulocytes with the lungs, *Circ. Res.,* 54, 623, 1984.
35. **Malik, A. B., Selig, W. M., and Burhop, K. E.,** Cellular and humoral mediators of pulmonary edema, *Lung,* 163, 193, 1985.
36. **Hyers, T. M. and Fowler, A. A.,** Adult respiratory distress syndrome: causes, morbidity, and mortality, *Fed. Proc.,* 45, 25, 1986.
37. **Worthen, G. S., Haslett, C., Smedly, L. A., Rees, A. J., Gumbay, R. S., Henson, J. E., and Henson, P. M.,** Lung vascular injury induced by chemotactic factors: enhancement by bacterial endotoxins, *Fed. Proc.,* 45, 7, 1986.
38. **Kaplow, L. S. and Goffinet, J. A.,** Profound neutropenia during early phase of hemodialysis, *JAMA,* 203, 1135, 1968.
39. **Papadimitriou, M., Baker, L. R. I., Seitanidis, B., Sevitt, L. H., and Kulatilake, A. E.,** White blood count in patients on regular hemodialysis, *Br. Med. J.,* 4, 67, 1969.
40. **Smith, E. K. M. and Jobbins, K.,** Observations on neutropenia associated with haemodialysis, *Br. Med. J.,* 4, 70, 1969.
41. **Schiffer, C. A., Aisner, J., and Wiernik, P. H.,** Transient neutropenia induced by transfusion of blood exposed to nylon fiber filters, *Blood,* 45, 141, 1975.

42. **Rubins, J. M., MacPherson, J. L., Nusbacher, J., and Wiltbank, T.,** Granulocyte kinetics in donors undergoing filtration leukapheresis, *Transfusion,* 16, 56, 1976.

43. **Jensen, D. P., Brubaker, L. H., Nolph, K. D., Johnson, C. A., and Nothum, R. J.,** Hemodialysis coil-induced transient neutropenia and overshoot neutrophils in normal man, *Blood,* 41, 399, 1973.

44. **Toren, M., Goffinet, J. A., and Kaplow, L. S.,** Pulmonary bed sequestration of neutrophils during hemodialysis, *Blood,* 36, 337, 1970.

45. **McCall, C. E., DeChatelet, L. R., Brown, D., and Lackman, P.,** New biologic activity following intravascular activation of the complement cascade, *Nature (London),* 249, 841, 1974.

46. **O'Flaherty, J. T., Showell, H. J., and Ward, P. A.,** Neutropenia induced by systemic infusion of chemotactic factors, *J. Immunol.,* 118, 1586, 1977.

47. **Fehr, J., and Jacob, H. S.,** In vitro granulocyte adherence and in vivo margination: two associated complement-dependent functions. Studies based on the acute neutropenia of filtration leukophoresis, *J. Exp. Med.,* 146, 641, 1977.

48. **Hammerschmidt, D. E., Craddock, P. R., McCullough, J., Kronenberg, R. S., Dalmasso, A. P., and Jacob, H. S.,** Complement activation and pulmonary leukostasis during nylon fiber filtration leukapheresis, *Blood,* 51, 721, 1978.

49. **Craddock, P. R., Fehr, J., Dalmasso, A. P., Brigham, K. L., and Jacob, H. S.,** Hemodialysis leukopenia: pulmonary vascular leukostasis resulting from complement activation by dialyzer cellophane membranes, *J. Clin. Invest.,* 59, 879, 1977.

50. **Pennock, J. L., Pierce, W. S., and Waldhausen, J. A.,** The management of the lungs during cardiopulmonary bypass, *Surg. Gynecol. Obstet.,* 145, 917, 1977.

51. **Chenoweth, D. E., Cooper, S. W., Hugli, T. E., Stewart, R. W., Blackstone, E. H., and Kirklin, J. W.,** Complement activation during cardiopulmonary bypass. Evidence for generation of C3a and C5a anaphylatoxins, *N. Engl. J. Med.,* 304, 497, 1981.

52. **Craddock, P. R., Fehr, J., Brigham, K. L., Kronenberg, R. S., and Jacob, H. S.,** Complement and leukocyte-mediated pulmonary dysfunction in hemodialysis, *N. Engl. J. Med.,* 296, 769, 1977.

53. **Brigham, K. L., Woolverton, W. C., Blake, L. H., and Staub, N. C.,** Increased sheep lung vascular permeability caused by pseudomonas bacteremia, *J. Clin. Invest.,* 54, 792, 1974.

54. **Heflin, A. C., and Brigham, K. L.,** Granulocyte depletion prevents increased lung vascular permeability after endotoxemia in sheep, *Clin. Res.,* 27 (Abstr.), 399, 1979.

55. **Demling, R. H., Smith, M., Gunther, R., Flynn, J. T., and Maryls, H. G.,** Pulmonary injury and prostaglandin production during endotoxemia in conscious sheep, *Am. J. Physiol.,* 240, H348, 1981.

56. **Ogletree, M. L. and Brigham, K. L.,** Effects of cyclooxygenase inhibitors on pulmonary vascular responses to endotoxin in unanesthetized sheep, *Prostagl. Leukotr. Med.,* 8, 489, 1982.

57. **Hammerschmidt, D. E., Weaver, L. J., Hudson, L. D., Craddock, P. R., and Jacob, H. S.,** Association of complement activation and elevated plasma-C5a with adult respiratory distress syndrome. Pathophysiologic relevance and possible prognostic value, *Lancet,* 1, 947, 1980.

58. **Cochrane, C. G., Spragg, R., and Revak, S. D.,** Pathogenesis of the adult respiratory distress syndrome. Evidence of oxidant activity in bronchoalveolar lavage fluid, *J. Clin. Invest.,* 71, 754, 1983.

59. **Meyrick, B. O.,** Endotoxin-mediated pulmonary endothelial cell injury, *Fed. Proc.,* 45, 19, 1986.

60. **Glauser, F. L. and Fairman, R. P.,** The uncertain role of the neutrophil in increased permeability pulmonary edema, *Chest,* 88, 601, 1985.

61. **Ognibene, F. P., Martin, S. E., Parker, M. M., Schlesinger, T., Roach, P., Burch, C., Shelhamer, J. H., and Parrillo, J. E.,** Adult respiratory distress syndrome in patients with severe neutropenia, *N. Engl. J. Med.,* 315, 547, 1986.

62. **Rinaldo, J. E.,** Mediation of ARDS by leukocytes. Clinical evidence and implications for therapy, *Chest,* 89, 590, 1986.

63. **Rinaldo, J. E. and Rogers, R. M.,** Adult respiratory distress syndrome (editorial), *N. Engl. J. Med.,* 315, 578, 1986.

64. **Kistler, G. S., Caldwell, P. B., and Weibel, E. R.,** Development of fine structural damage to alveolar and capillary lining cells in oxygen-poisoned rat lungs, *J. Cell Biol.,* 32, 605, 1967.

65. **Kapanci, Y., Weibel, E. R., Kaplan, H. P., and Robinson, F. R.,** Pathogenesis and reversibility of the pulmonary lesions of oxygen toxicity in monkeys. II. Ultrastructural and morphometric studies, *Lab. Invest.,* 20, 101, 1969.

66. **Crapo, J. D., Barry, B. E., Foscue, H. A., and Shelburne, J.,** Structural and biochemical changes in rat lungs occurring during exposures to lethal and adaptive doses of oxygen, *Am. Rev. Respir. Dis.,* 122, 123, 1980.

67. **Fox, R. B., Hoidal, J. R., Brown, D. M., Repine, J. E.,** Pulmonary inflammation due to oxygen toxicity: involvement of chemotactic factors and polymorphonuclear leukocytes, *Am. Rev. Respir. Dis.,* 123, 521, 1981.

68. **Davis, W. B., Rennard, S. I., Bitterman, P. B., Gadek, J. E., Sun, X. H., Wewers, M., Keogh, B. A., and Crystal, R. G.**, Pulmonary oxygen toxicity. Bronchoalveolar lavage demonstration of early parameters of alveolitis, *Chest*, 83 (Suppl.), 35S, 1983.
69. **Frank, L. and Massaro, D.**, Oxygen toxicity, *Am. J. Med.*, 69, 117, 1980.
70. **McCord, J. M.**, Oxygen radicals and lung injury, *Chest*, 83 (Suppl.), 35S, 1983.
71. **O'Flaherty, J. T., Cousart, S., Lineberger, A. S., Bond, E., Bass, D. A., DeChatelet, L. R., Leake, E. S., and McCall, C. E.**, Phorbol myristate acetate. *In vivo* effects upon neutrophils, platelets, and lung, *Am. J. Pathol.*, 101, 79, 1980.
72. **Schraufstätter, I. U., Revak, S. D., and Cochrane, C. G.**, Proteases and oxidants in experimental pulmonary inflammatory injury, *J. Clin. Invest.*, 73, 1175, 1984.
73. **Shasby, D. M., VanBenthuysen, K. M., Tate, R. M., Shasby, S. S., McMurtry, I., and Repine, J. E.**, Granulocytes mediate acute edematous lung injury in rabbits and in isolated rabbit lungs perfused with phorbol myristate acetate: role of oxygen radicals, *Am. Rev. Respir. Dis.*, 125, 443, 1982.
74. **Fantone, J. C., Johnson, K. J., Till, G. O., and Ward, P. A.**, Acute and progressive lung injury secondary to toxic oxygen products from leukocytes, *Chest*, 83 (Suppl.), 46S, 1983.
75. **Revak, S. D., Rice, C. L., Schraufstätter, I. U., Halsey, W. A., Bohl, B. P., Clancy, R. M., and Cochrane, C. G.**, Experimental pulmonary inflammatory injury in the monkey, *J. Clin. Invest.*, 76, 1182, 1985.
76. **Johnson, K. J. and Ward, P. A.**, Role of oxygen metabolites in immune complex injury of lung, *J. Immunol.*, 126, 2365, 1981.
77. **Johnson, K. J., Wilson, B. S., Till, G. O., and Ward, P. A.**, Acute lung injury in rat caused by immunoglobulin A immune complexes, *J. Clin. Invest.*, 74, 358, 1984.
78. **Till, G. O., Johnson, K. J., Kunkel, R., and Ward, P. A.**, Intravascular activation of complement and acute lung injury. Dependency on neutrophils and toxic oxygen metabolites, *J. Clin. Invest.*, 69, 1126, 1982.
79. **Ward, P. A., Till, G. O., Kunkel, R., and Beauchamp, C.**, Evidence for role of hydroxyl radical in complement and neutrophil-dependent tissue injury, *J. Clin Invest.*, 72, 789, 1983.
80. **Till, G. O. and Ward, P. A.**, Oxygen radicals in complement and neutrophil-mediated acute lung injury, *J. Free Rad. Biol. Med.*, 1, 163, 1985.
81. **Ward, P. A., Till, G. O., Hatherill, J. R., Annesley, T. M., and Kunkel, R. G.**, Systemic complement activation, lung injury, and products of lipid peroxidation, *J. Clin. Invest.*, 76, 517, 1985.
82. **Till, G. O. and Ward, P. A.**, Systemic complement activation and acute lung injury, *Fed. Proc.*, 45, 13, 1986.
83. **Johnson, K. J., Fantone, J. C., Kaplan, J., and Ward, P. A.**, In vivo damage of rat lungs by oxygen metabolites, *J. Clin. Invest.*, 67, 983, 1981.
84. **Halliwell, B. and Gutteridge, J. M. C.**, Oxygen toxicity, oxygen radicals, transition metals and disease, *Biochem. J.*, 219, 1, 1984.
85. **Till, G. O., Beauchamp, C., Menapace, D., Tourtellotte, W., Kunkel, R., Johnson, K. J., and Ward, P. A.**, Oxygen radical dependent lung damage following thermal injury of rat skin, *J. Trauma*, 23, 269, 1983.
86. **Till, G. O., Hatherill, J. R., Tourtellotte, W., Lutz, M. J., and Ward, P. A.**, Lipid peroxidation and acute lung injury after thermal trauma to skin. Evidence for a role of hydroxyl radical, *Am. J. Pathol.*, 119, 376, 1985.
87. **Shasby, D. M., Shasby, S. S., and Peach, M. J.**, Polymorphonuclear leukocyte: arachidonate edema, *J. Appl. Physiol.*, 59, 47, 1985.
88. **Shasby, D. M., Shasby, S. S., and Peach, M. J.**, Granulocytes and phorbol myristate acetate increase permeability to albumin of cultured endothelial monolayers and isolated perfused lungs. Role of oxygen radicals and granulocyte adherence, *Am. Rev. Respir. Dis.*, 127, 72, 1983.
89. **Morganroth, M. L., Till, G. O., Kunkel, R. G., and Ward, P. A.**, Complement and neutrophil-mediated injury of perfused rat lungs, *Lab. Invest.*, 54, 507, 1986.
90. **Sacks, T., Moldow, C. F., Craddock, P. R., Bowers, T. K., and Jacob, H. S.**, Oxygen radicals mediate endothelial cell damage by complement-stimulated granulocytes. An in vitro model of immune vascular damage, *J. Clin. Invest.*, 61, 1161, 1978.
91. **Weiss, S. J., Young, J., LoBuglio, A. F., Slivka, A., and Nimeh, N. F.**, Role of hydrogen peroxide in neutrophil-mediated destruction of cultured endothelial cells, *J. Clin. Invest.*, 68, 714, 1981.
92. **Martin, W. J.**, Neutrophils kill pulmonary endothelial cells by a hydrogen peroxide-dependent pathway. An in vitro model of neutrophil-mediated lung injury, *Am. Rev. Respir. Dis.*, 130, 209, 1984.
93. **Varani, J., Fligiel, S. E. G., Till, G. O., Kunkel, R. G., Ryan, U. S., and Ward, P. A.**, Pulmonary endothelial cell killing by human neutrophils. Possible involvement of hydroxyl radical, *Lab. Invest.*, 53, 656, 1985.

94. **Gannon, D. E., Varani, J., Phan, S. H., Ward, J. H., Till, G. O., Ryan, U. S., Simon, R. H., and Ward, P. A.,** Source of iron in neutrophil-mediated killing of endothelial cells, *Lab. Invest.,* 57, 37, 1987.
95. **Dorfman, L. M. and Adams, G. E.,** Reactivity of the hydroxyl radical in aqueous solutions, in *Natl. Stand. Ref. Data Ser.,* Natl. Bur. Stand., Washington, D.C., 1973, 46.

Chapter 23

INTERACTION OF LEUKOCYTES WITH THE VASCULAR ENDOTHELIUM

Marcia G. Tonnesen, G. Scott Worthen, Dale C. Lien, and Peter M. Henson

TABLE OF CONTENTS

I. INTRODUCTION

Because of its unique location lining all blood vessels in the body, the vascular endothelium serves as the interface between circulating leukocytes and underlying tissues. Interaction of neutrophils with the endothelium can be considered to be an ongoing normal process which results in a marginating pool of mature neutrophils, sequestered largely within the pulmonary vasculature and capable of being mobilized by such physiological events as exercise and stress. Neutrophil interaction with the endothelium is greatly enhanced, however, in the setting of the acute inflammatory response. Mediators released during inflammation directly stimulate neutrophil and endothelial adhesiveness and neutrophil migration. Local hemodynamic changes which accompany the inflammatory process may modulate the dynamic equilibrium between adherent and circulating neutrophils with potent inflammatory potential. Thus, these events contribute to leukocyte adherence and subsequent diapedesis and accumulation at sites of inflammation. However, the nature of the leukocyte-endothelial interaction and the mechanisms involved still have not been completely elucidated despite over a century of intense interest and investigation.[1,2]

This chapter will focus on the neutrophil because of its crucial role as an initiator of the inflammatory process and our better understanding of its biology. However, at this point in our knowledge, it seems reasonable to suggest that similar mechanisms of interaction with the endothelium are likely operative for eosinophils and monocytes, and perhaps also basophils. In vitro and in vivo systems designed to study neutrophil adherence to endothelium will be considered. Despite their limitations, they have provided useful information regarding the action of inflammatory mediators, the relative contribution of the neutrophil and of the endothelial cell to the adhesion process, the site of interaction in the pulmonary vasculature, and the importance of hemodynamic effects. Nevertheless, a great deal still remains to be learned about the complex, critical, and fascinating interaction between the circulating leukocyte and the vascular endothelial cell.

II. NEUTROPHIL ADHERENCE TO ENDOTHELIUM

A. Mediators and Mechanisms In Vitro

Early in vivo studies by Clark and Clark[3] involving direct visualization of the microvasculature in the transparent tadpole tail clearly demonstrated that adhesion of leukocytes to the endothelium regularly precedes diapedesis, the migration of leukocytes through the vessel wall. The widely held presumption that diapedesis is the result of the chemotactic attraction of leukocytes from outside the vessel wall dates back to the pioneering work of Metchnikoff[4] in 1893. This concept gained credence from the observation of Allison et al.[5] that adherence of leukocytes to the endothelium induced by thermal injury in the rabbit ear chamber first occurred on the side of the vessel closest to the site of injury, indicating that the leukocyte-vascular interaction was influenced by products of tissue damage that had diffused to the vessel wall from the adjacent site of injury. These classic investigations emphasized the need for further elucidation of the role of inflammatory mediators in the process of leukocyte diapedesis. Since then, substantial in vivo and in vitro evidence has accumulated to implicate chemotactic peptides and lipid mediators as key participants in neutrophil adherence and emigration. In vitro studies will be considered here, and in vivo work will be addressed in Section II.C.

In vitro investigations of leukocyte-endothelial interactions were made possible by the relatively recent advances in tissue culture technology which facilitated the isolation and culture of vascular endothelial cells from a variety of species, including man.[6-8] As a result, significant progress has been made in our understanding of mechanisms of human neutrophil adherence to endothelium and the role of inflammatory mediators. In vitro studies have shown that neutrophils adhere preferentially to cultured endothelial cells, compared with other cell types, including smooth muscle cells and fibroblasts.[9-13] Neutrophils move freely over the surface of endothelium in culture,[9,13,14] and eventually migrate between the endothelial cells.[12-14] In a static monolayer-adhesion assay using glutaraldehyde to capture the motile, loosely adherent cells presumably capable of subsequent disadhesion or migration beneath the monolayer, human neutrophils demonstrate a baseline level of spontaneous adhesion to human umbilical vein or omental microvascular endothelial cells ranging from 15 to 55% (Reference 13 and our unpublished observations). Such spontaneous adhesion in vitro may be a manifestation of the in vivo process of leukocyte margination (Section II.C).

Human neutrophil adherence to cultured endothelial cells is markedly enhanced by chemotactic peptides (the biologically active fragments of the fifth component of complement [C5fr] and formyl-methionyl-leucyl-phenylalanine [FMLP]), and lipid mediators (platelet activating factor [PAF] and leukotriene B4 [LTB4]) in a dose-dependent manner[11,13,15-18] As many as 70 to 80% of the neutrophils adhered in some experiments (Reference 13 and our unpublished observations).

Controversy as to whether the initial leukocyte-endothelial adhesive interaction is the result of increased adhesiveness of the endothelial cell, the leukocyte, or both has persisted since the early in vivo observations cited above. Clark and Clark[3] postulated an alteration in endothelial ''consistency'' since localized leukocyte adherence could be observed without subsequent emigration. Metchnikoff[4] claimed that chemoattractants acted on the leukocytes. Electron-microscopic studies by Marchesi and Florey of adherent leukocytes during inflammation induced in the rat mesentery and rabbit ear chamber failed to resolve this issue as they revealed neither morphologic alterations in leukocytes or endothelial cells nor the presence of an intercellular ''cement'' substance.[19,20] Recent investigations, however, clearly demonstrate that chemotactic peptides and lipid mediators act primarily on the neutrophil to enhance its adherence to endothelium.

An effect on the neutrophil was initially suspected based on comparative adherence studies in which neutrophils, when stimulated, adhered to protein-coated surfaces and to a variety of cultured cell types to the same degree as they adhered to endothelial cells,[13] and on pretreatment studies in which pretreatment of the neutrophil, but not the endothelial cell, with chemotactic peptides produced increased adhesion.[13] More conclusive evidence derives from the recent elucidation of the critical role of a family of leukocyte surface adhesive glycoproteins in neutrophil adherence-dependent functions.[21] The family, termed CDw18 by the Second International Workshop on Leukocyte Differentiation Antigens[22] or Mac-1 glycoprotein family by Springer and coworkers,[23,24] consists of three structurally and functionally related glycoproteins: Mac-1 (the C3bi complement receptor), LFA-1 (lymphocyte function associated antigen), and p150,95. Each of the three glycoproteins is composed of an alpha and a beta subunit which are noncovalently associated.[23] The three alpha subunits are immunologically distinct, while the beta subunits are immunologically identical.[23] Only white blood cells and hematopoietic precursor cells have thus far been shown to express this glycoprotein family.[25] Mac-1 and p150,95 are present in unstimulated human neutrophils in an intracellular vesicular compartment as well as on the cell surface.[24,26,27] Chemotactic mediators including FMLP, C5fr, LTB4, and PAF, as well as the calcium ionophore A23187 and phorbol myristate acetate (PMA), increase the cell surface expression two- to tenfold.[18,24,26-30]

This increased expression of the glycoprotein family on the surface of stimulated neutrophils appears to be essential to the development of enhanced adhesiveness for endothelium since neutrophils from patients with leukocyte adhesion deficiency,[25] a genetic deficiency of the Mac-1 glycoprotein family, show defective adhesion to cultured human endothelial cells compared to normal control neutrophils when stimulated with chemotactic peptides,[30] lipid mediators,[18] or PMA.[29] In addition, chemotactic peptides,[30] lipid mediators,[18] and

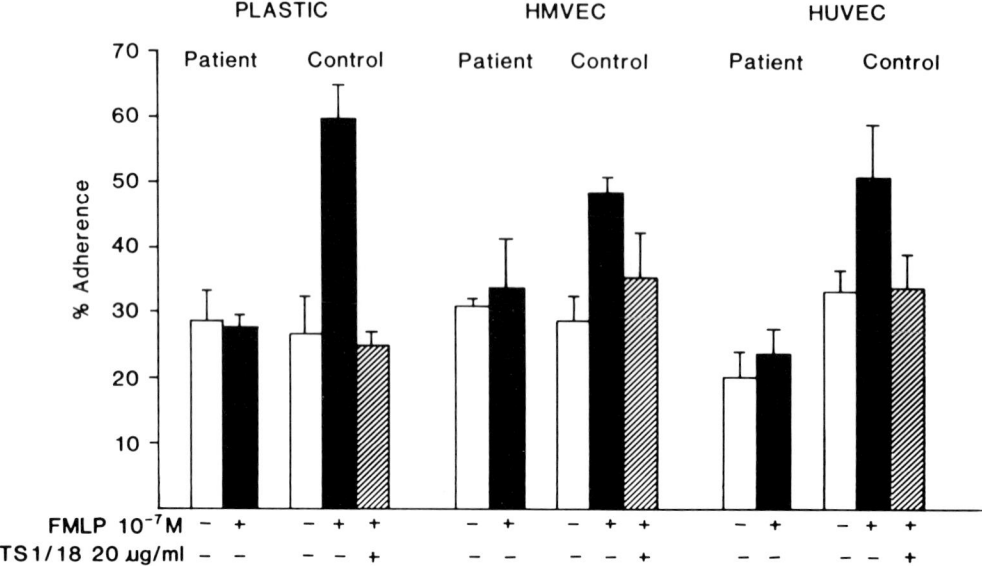

FIGURE 1.   Neutrophils isolated from a patient with leukocyte adhesion deficiency and stimulated with FMLP show diminished adherence to serum-coated plastic and to cultured human microvascular (HMVEC) and umbilical vein (HUVEC) endothelial cells compared to normal control neutrophils. The response of normal neutrophils to FMLP is inhibited by a monoclonal antibody (TS1/18) to the beta subunit of the Mac-1 glycoprotein family.

PMA[29] fail to stimulate normal neutrophils to increase adherence to cultured human endothelial cells in the presence of monoclonal antibodies to the common beta subunit of the glycoprotein family. Figure 1 illustrates the diminished adhesion response of neutrophils from a patient with leukocyte adhesion deficiency and the ability of a monoclonal antibody to the beta subunit to inhibit stimulated adherence of normal neutrophils. Despite their critical function in cell adhesion, the degree of active involvement of the glycoproteins in the actual process of adherence remains to be clarified.

In order to ensure appropriate localization of neutrophil interaction with the microvasculature adjacent to a site of inflammation in vivo, the process by which neutrophils are rendered adhesive by stimulation with chemotactic peptides and lipid mediators must be rapid and readily modulated. In addition, the failure of neutrophils to be stimulated to adhere in vivo should result in detectable pathology. In in vitro studies in which neutrophils were permitted to settle onto the surface of cultured endothelial monolayers prior to the addition of a chemotactic peptide or lipid mediator, the onset of enhanced adherence in response to the stimulus occurred quickly (within 30 sec), reaching maximum in less than 2 min.[13] The effect could be rapidly modulated, as demonstrated in Figure 2, since the capacity for enhanced adherence decreased upon removal of the stimulus but was rapidly regained upon readdition of the same stimulus.[13] Thus in vivo, presumably those neutrophils close to the endothelial surface would upregulate the Mac-1 glycoprotein family and adhere rapidly in response to local stimulation by a chemotactic peptide or lipid mediator generated at a nearby site of tissue injury. Neutrophils lacking the glycoprotein family, as occurs in leukocyte adhesion deficiency, would be expected not to adhere to endothelium in vivo. Indeed, such patients are characterized by their inability to mobilize neutrophils to inflammatory tissue sites, resulting in numerous chronic, nonpustular soft-tissue infections and early demise.[25] The ability of bone marrow transplantation to correct the defect and cure patients with this genetic disorder[31] strongly suggests that the enhanced adhesivity induced in the neutrophil by inflammatory mediators plays a primary role in neutrophil diapedesis and accumulation at tissue sites of inflammation.

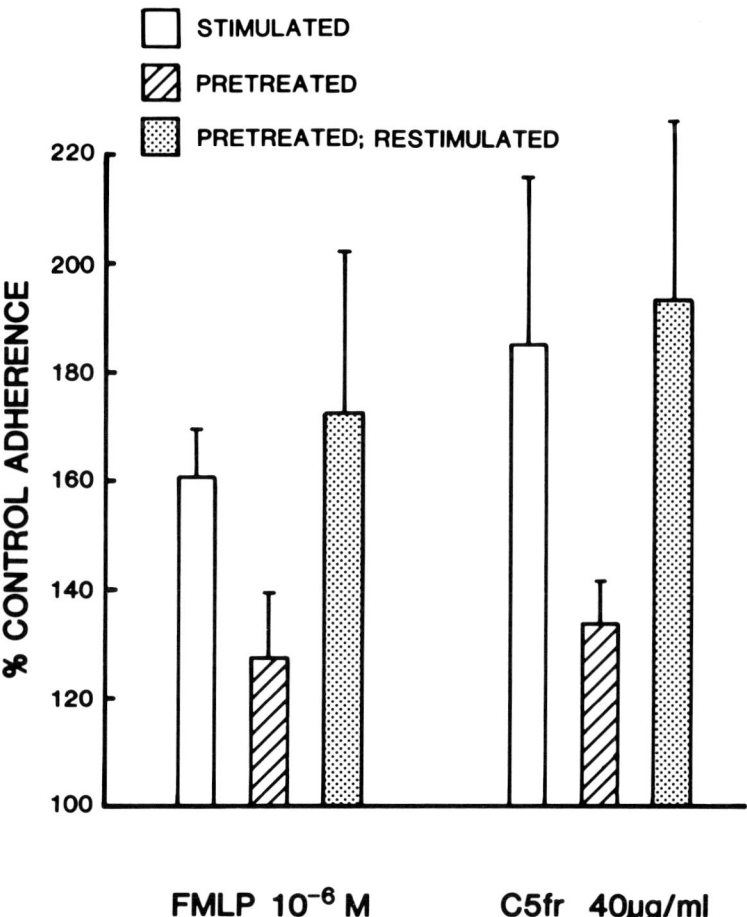

FIGURE 2.   Neutrophils stimulated by either FMLP or C5 fragments (C5fr) show increased adherence to human umbilical vein endothelial cells compared to control (open bars). Pretreatment followed by removal of the stimulus results in diminished stimulated adherence (striped bars), which is rapidly regained upon restimulation (dotted bars).

Recent experimental evidence suggests that endothelial cells may also actively contribute to the leukocyte-endothelial interaction, although with a delayed time course. Cultured human umbilical vein endothelial cells are induced by the human cytokines interleukin-1 (IL-1) and tumor necrosis factor (TNF) to increase their adhesivity for human neutrophils through a process which is time dependent (requiring 4 to 6 hr for maximal response), reversible, and appears to require *de novo* RNA and protein synthesis since it is inhibited by actinomycin D or cycloheximide.[32,33] Umbilical vein endothelial cells treated with thrombin or certain of the leukotrienes will also increase their adhesivity for neutrophils, possibly through increased synthesis of PAF.[34,35] It is as yet unclear whether these effects occur with human microvascular endothelial cells.

Thus, neutrophil-endothelial cell adhesion in vitro occurs spontaneously (perhaps related to margination), increases rapidly upon neutrophil stimulation by chemotactic peptides and lipid mediators (mediated by Mac-1 glycoprotein surface expression), and may be more slowly enhanced through active participation of the endothelium after stimulation by cytokines. In vivo physical forces may modulate the interaction, as discussed in the next section.

B. Hemodynamic and Physical Effects

1. Theoretical Considerations

While the biochemical mechanisms which affect neutrophil adherence to endothelial cells are beginning to be recognized, the physical principles which govern neutrophil behavior within the vasculature are still unclear. Current evidence suggests the following simplified scheme: in large vessels where neutrophils are not constrained geometrically, theoretical[36] and in vivo[37] data suggest that there exists a balance between shear stress (or velocity gradient) and neutrophil adhesiveness in determining adherence to endothelium. Although neutrophil viscoelasticity will play a role in determining the torque or force exerted on the neutrophil by the flowing blood, in a large vessel this will be a less important effect.[38]

In capillaries, on the other hand, particularly those of the lung where the marginating pool is localized, the effect of physical forces may be quite different. The *average* diameter of a pulmonary capillary is 7 to 8 μm,[39] while that of a neutrophil is 7 μm.[40] Considering the forces which affect the caliber of the pulmonary capillary, it appears likely that nearly every capillary path will include a region through which the neutrophil or capillary must deform to permit passage. The deforming neutrophil will be subject to several forces, including shear stress, modified by red cells that impact against the neutrophil.[41,42] The forces tending to retain neutrophils, however, include not only the adhesive interaction between neutrophils and endothelial cells, but also the viscoelastic properties of the neutrophil which resist deformation. The recent development of methods to study the physical properties of leukocytes offer insight into the process of deformation. Schmid-Schonbein and colleagues[43] have studied this process by measuring the displacement of neutrophils when sucked into micropipettes. The neutrophils so stressed are characterized by an initial rapid elastic displacement, followed by slow "creep" displacement due, in part, to cytoplasmic viscosity. Interestingly, although the *elastic* properties of neutrophils are similar to erythrocytes, the cellular viscosity of neutrophils is three orders of magnitude higher.

Thus, the time constant for deformation is much longer (on the order of minutes). Indeed, although mature neutrophils can actively migrate through 3-μm pores, they are avidly retained when passively perfused through sieves with a 5-μm pore diameter.[44] This is presumably because the time constant for deformation is too low to permit the alteration of shape necessary for passage. In addition, on the other side of the coin, is the deformability of the capillary, which is essentially unknown.

Thus, it seems apparent that different physical properties may be important in different areas of the vascular tree. Neutrophil surface adhesiveness is of primary importance in arteries and veins, while viscoelastic properties appear to be most important in determining the degree of neutrophil retention in capillaries, especially those of the lung.

2. Effects of Shear Stress in Vitro

Several groups have begun to study the effects of hydrodynamic parameters on neutrophil adherence. Doroszewski and colleagues[45] studied the effect of shear rate on adherence of leukemic leukocytes to glass and noted that adherence decreased as shear rate increased. Forrester and Lackie[46] studied adherence of neutrophils to various surface coatings in a flow chamber, and have calculated that the adhesive interaction between neutrophils and an albumin-coated surface is sufficient to resist a distractive force of 4×10^{-11} N. We have performed similar studies to examine the effect of shear stress on the interaction between blood neutrophils and pulmonary endothelial cells. Using a device similar to a cone-in-plate viscometer[47] we have subjected neutrophils and endothelial cells to shear stress during the process of adherence. As can be seen in Figure 3A, only low values of shear stress were required to markedly reduce adherence whether cells were stimulated or unstimulated. However, in Figure 3B it can be seen that allowing the cells to adhere to the monolayer for 6 min now permitted some of the stimulated cells to resist subsequent shear stress.

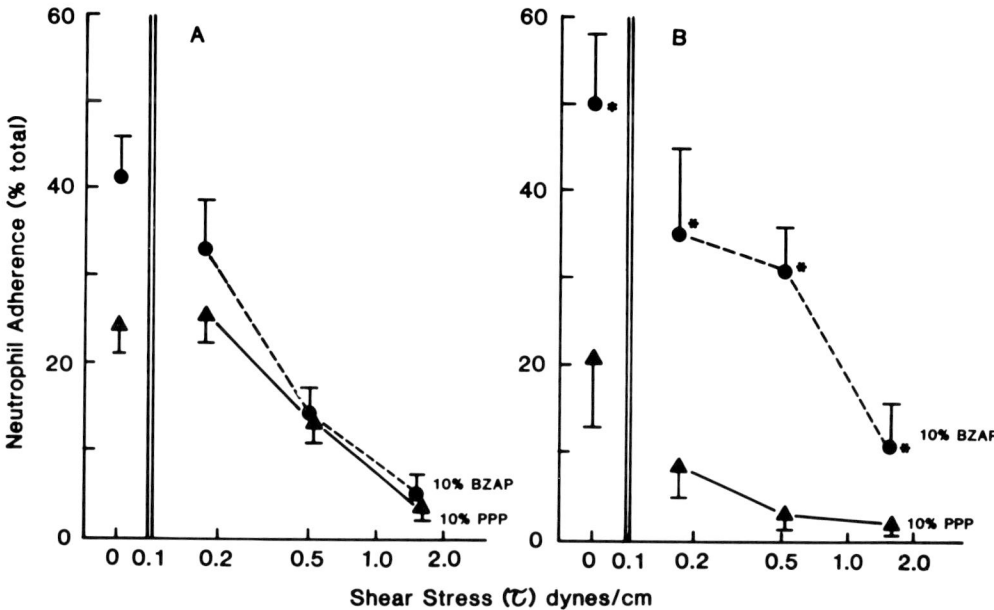

FIGURE 3.   Adherence of bovine neutrophils to bovine pulmonary artery endothelial cells in the presence of shear stress. (A) Shows adherence as a function of shear stress in dynes/cm² in the presence or absence of 10% bovine ZAP. Under these circumstances there is little effect of stimulus except under static (0 shear) conditions. In (B) the neutrophils are exposed to similar shear conditions after being exposed to ZAP under static conditions for 6 min. Now the ZAP is able to stimulate the neutrophils to resist shear. (From Tonnesen, M. G. and Worthen, G. S., *Pulmonary Endothelium in Health and Disease*, Ryan, U. S., Ed., Marcel Dekker, New York, in press. With permission.)

Thus, small degrees of shear stress effectively prevent neutrophil adherence to a flat endothelial monolayer in vitro. Consider how different the pulmonary microvasculature is in vivo, perfused under pulsatile conditions (rather than constant shear)[48] in the presence of red blood cells to alter neutrophil behavior.[41] Furthermore, neutrophils are constantly squeezed by extensive contact with the pulmonary capillary endothelium. The relative contributions of all these effects on neutrophil adherence remain to be determined. In light of these multiple factors, the requirement for a period of static adherence in order for neutrophil stimuli to increase the "strength" of adherence is of considerable interest. As we shall see in Section II.C, the pulmonary circulation provides such static contact for significant periods of time.

Thus neutrophil adherence to endothelium is a complex and as yet incompletely understood process apparently involving active participation by both the neutrophil and the endothelial cell. Several different mechanisms may be responsible for initiating and maintaining adhesive contact, depending upon the cell type and the nature of the stimulus. All aspects of the interaction may be modulated by local hemodynamic, physical, and geometric constraints.

C. Interactions In Vivo

1. The Marginating Pool

Under normal conditions, neutrophils may exist in one of two pools — a circulating pool and a marginating pool. Those neutrophils in the marginating pool, largely sequestered within the pulmonary vasculature,[49,50] are in dynamic equilibrium with circulating neutrophils and can be rapidly mobilized to join the circulating pool during periods of stress.[51-54]

Interest in the role of the lung in neutrophil margination began in the latter part of the last century.[55] It was noted that changes in circulating numbers of leukocytes occurred too rapidly to be explained by increased leukocyte production or destruction. It was postulated that they were a result of altered distribution of leukocytes within the vasculature.[56,57] Other investigators observed that when leukocytes were infused into the bloodstream of recipients, they rapidly disappeared from the circulating blood causing little or no change in circulating leukocyte counts.[58-60] It was subsequently shown, by methods which included infusion of radiolabeled or stained cells and comparison of leukocyte counts in blood entering or leaving the lung, that most of these cells were taken up in the lung.[49,59,61]

The vascular site at which neutrophils marginate and the mechanisms by which they do so are the subject of much current interest. In nonpulmonary vascular beds, intravital microscopists have observed that leukocytes roll slowly along the walls or margins of small vessels.[1,3,62] (Indeed, these observations gave rise to the term ''marginated pool'', which has become the term used to describe neutrophils which are localized within the vasculature but not circulating in the main bloodstream.) Outside the lung, this rolling phenomenon occurs principally in postcapillary venules and occasionally in small arterioles, while leukocytes spend little time in systemic capillaries. Recent quantitative work suggests that such rolling may be explicable by physical processes within the microvasculature. In brief, the process is as follows. Spherical shaped neutrophils, which are slightly larger than the cross-sectional area of capillaries, are slowly pushed through the capillaries with erythrocytes dammed up behind them. When they enter the postcapillary venule, the sudden widening allows a marked acceleration of erythrocytes and the resultant increase in hydrodynamic forces causes the slower-moving leukocytes to be pushed to the margins until they can later reenter the mainstream.[41,63]The rolling phenomenon is presumably governed by a balance between the intrinsic adhesive properties of the neutrophil and vascular endothelium, (see Section II.A) and the shear stress provided by the flowing blood (see Section II.B.2).[37]

Margination in this fashion almost certainly occurs in the lung as well. However, the lung vasculature differs radically from that of most organs perfused by the systemic circulation in that the pulmonary microcirculation has an extensive capillary bed perfused by blood under low pressure. This vascular bed provides a large potential area in which neutrophils could be retained. Although many observations have implicated the pulmonary capillaries as a site of pulmonary neutrophil margination, few studies have directly addressed the site of neutrophil margination in the lung. Morphological assessments were carried out by Perlo et al.[64] and Staub et al.[65] Perlo et al. performed a morphologic study using sections prepared from dog lungs rapidly frozen over liquid freon. They reported unexpectedly high numbers of leukocytes (although predominantly mononuclear cells) in alveolar capillaries when compared to erythrocytes, although arterioles and venules were not specifically examined. Staub et al. examined sheep lungs fixed *in situ*. They quantified the number and distribution of polymorphonuclear cells and found large numbers in alveolar wall capillaries as well as many along the walls of small arteries. More recently, Doerschuk et al.[66] used morphometric techniques to suggest that the predominant site of neutrophil retention within the rabbit lung was the alveolar capillaries. All of these studies have employed fixed tissues, however, which may alter the behavior of the cells within the lung, and can only provide ''static'' views.

A more dynamic picture of neutrophil margination in the lung can now be obtained through in vivo fluorescence videomicroscopy of the pulmonary microcirculation which avoids fixation. This is done by labeling purified canine neutrophils with fluorescein isothiocyanate (FITC) by methods which do not significantly alter their ability to circulate normally in the bloodstream of recipient animals. The transit of these FITC-neutrophils through the subpleural pulmonary microcirculation can then be observed and quantified through a transparent window inserted into the chest wall of anesthetized dogs. In ten experiments which we have performed using these techniques, some of the FITC-neutrophils passed directly through the microvasculature without stopping, while the majority were sequestered for variable periods of time. The FITC-neutrophils with prolonged transits were individually sequestered at

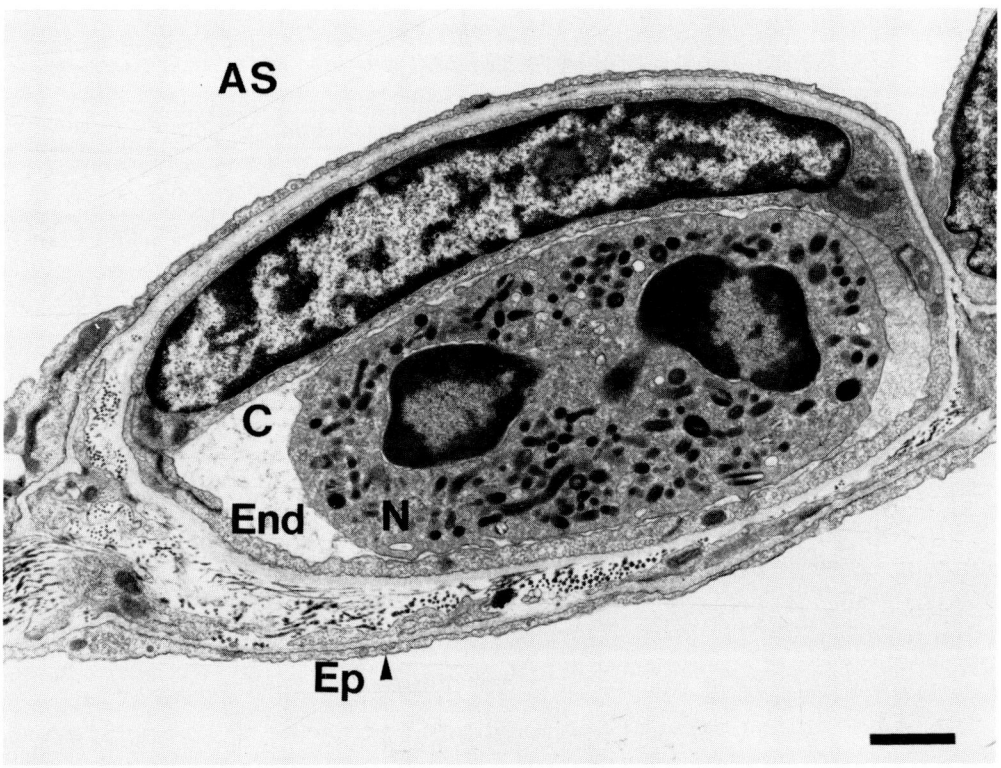

FIGURE 4. Electron-microscopic view of a neutrophil (N) lodged in an alveolar capillary (C) from a section of canine lung. Note areas of close contact between neutrophil and endothelium (End). Alveolar epithelium (Ep) is shown adjacent to the alveolar airspace (AS). (Magnification × 1080; bar = 1 μm).

discrete sites and were found exclusively within the capillaries rather than the arterioles or venules (see Figure 4). A total of 950 FITC-neutrophils were sequestered within the capillaries for a minimum of 2 sec with none significantly impeded in arterioles or venules. Although occasional cells were also noted to roll along the margins of venules (as described in nonpulmonary vascular beds), they were only minimally delayed in their transit. Each experiment was analyzed for a 20-min period. The time that each individual neutrophil required for transit through a microscopic field varied over a wide range from less than 2 sec to greater than 20 min, such that the time required for half of the cells to pass through the microscope field (median transit time) was approximately 26 sec (see Figure 5). This figure also shows that FITC-neutrophils which passed from the lung through the systemic vascular beds and recirculated through the microscopic field had the same range and distribution of transit times. Therefore, each time neutrophils pass through the lung, some move directly through the microcirculation without stopping while others are sequestered for variable periods of time, up to 20 min or longer.

These studies represent direct visual evidence of the dynamic exchange between circulating neutrophils and the marginated or sequestered pools within the lung. The wide range of neutrophil transit times is impressive when compared to the rate of capillary blood as assessed by the transit times of the plasma through the capillary bed. This was done by measuring the transit time of FITC-labeled dextran (mol wt 70,000) through the capillary bed by the method of Wagner et al.[67] The mean capillary plasma transit time was 1.4 ± 0.3 sec. Thus, although some neutrophils pass through the capillary network at the same speed as plasma (or red cells), the majority of neutrophils are impeded for much longer periods of time. The

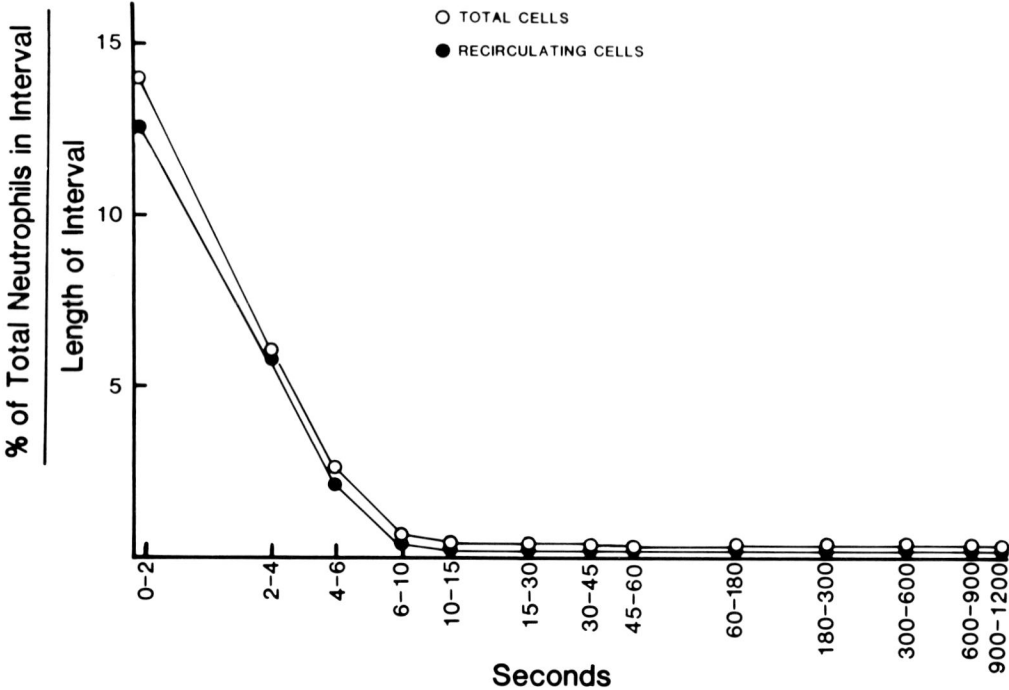

FIGURE 5.   Distribution of transit times of fluoresceinated canine neutrophils through the microvascular bed of dog lungs. The distribution is shown as the number of cells with transit times falling into each interval, and has a wide range from 2 to 1200 sec, with the total cells shown in open circles and recirculating cells which have already passed through the vascular bed once in closed circles. The distributions are identical.

discrepancy between blood flow and neutrophil flow may account for the formation of the marginating pool.

The factors which cause neutrophils to be sequestered within the capillaries are still unclear. Neutrophils could be impeded in pulmonary capillaries because they are larger than the anatomical dimensions of the capillaries (or of a different cross-sectional shape), because of intrinsic adhesiveness of the neutrophil or capillary endothelium, or because of the unique hemodynamic and flow characteristics of the pulmonary circulation. Evidence from in vitro studies have raised the possibility that intrinsic neutrophil adhesion (the spontaneous adhesion noted in in vitro studies) may in part account for the sequestration of neutrophils to form the marginating pool. However, we have seen that this adhesion is easily overcome by hemodynamic forces. An alternative notion is suggested by the in vitro studies showing rapid modulation of neutrophil adherence. If neutrophils are subjected to transient low-level stimulation of varying degrees within the lung, then this might be reflected in a distribution of transit times. The length of time each neutrophil spent sequestered would then be determined by the intensity and duration of the stimulus. To date, this conjecture has not been subject to experimental verification. A further suggestion to explain the same phenomenon arises from the in vivo microscopic observations reported here which can be used to argue that the major factors controlling neutrophil transit reside primarily within the microvasculature rather than the neutrophil itself. For example, if intrinsic neutrophil adhesiveness were the predominant factor, neutrophils which pass quickly through the lung because of low adhesiveness should do so repeatedly, while highly adhesive neutrophils should remain sequestered and not circulate. However, recirculating neutrophils in our study had the same range and distribution of transit times as first-pass cells, suggesting that it is the vascular pathways which impede neutrophil transit or that as mentioned above, rapid stimulation with variable effectiveness is present continually. These arguments are both consistent with the known geometrical and fluid-mechanical properties of lung and neutrophil which suggest that neutrophil deformation imposed by the characteristics of the microvascular bed plays a significant role in the sequestration of neutrophils within that bed, either to forcibly detain neutrophils, or to allow transient stimulation to be more effective in the face of hydrodynamic forces. These studies emphasize, however, that even unstimulated neutrophils may be, in the course of normal vascular events, forced into intimate contact with the vascular endothelium of the lung for significant periods of time. What then is the result when neutrophils are stimulated in the setting of the acute inflammatory response?

2. Neutrophil Interaction with Endothelium during Inflammation

The inflammatory process must, by definition, involve leukocyte interaction with vascular endothelium. Inflammatory reactions can be initiated by insults from the extravascular external environment, or from sites within the body with inflammatory mediators transported in the bloodstream. For example, in the lung, the presence of the pulmonary marginated neutrophil pool in the alveolar capillaries may be both beneficial and detrimental to the defense of the lung. These neutrophils, which are separated from the alveolar airspace only by the thin alveolar-capillary membrane, are in a position to respond rapidly to insults from the external environment. Similarly, the unique localization of neutrophils in the alveolar capillary to form the marginating pool may be accentuated by stimuli in the bloodstream, greatly enhancing neutrophil sequestration at this site. We will first focus on inflammation induced by the airway or extravascular delivery of chemotactic factors.

Neutrophils are thought to migrate along a gradient of chemotactic stimuli during acute extravascular inflammatory processes.[68] In order for this to occur, neutrophils circulating in the bloodstream must first come in contact with vascular endothelium, adhere to the endothelium, and then migrate through the vascular wall into the interstitial space in response to the stimulus. The acute inflammatory process has been observed in nonpulmonary tissues by several authors using intravital microscopic techniques in preparations such as rabbit ear chambers, hamster cheek pouches, and mesenteries (see, for example, References 5, 62, and 69). These investigators have described a characteristic group of events which include slowing of the local blood flow followed by the appearance of increased numbers of neutrophils in the marginal bloodstream, adherence to the endothelial wall, and migration into the inflammatory site. The site at which this neutrophil-endothelial interaction occurs outside the lung is predominantly the postcapillary venules (*vide supra*), the same site in which the marginated neutrophil pool is located in these tissues. The same group of events may also occur in the lung. However, since the site of neutrophil localization under physiological conditions in the lung appears to be the capillary (see Section II.C.1), we hypothesized that neutrophil-endothelial interaction would occur in the pulmonary capillary during acute inflammation.

A number of morphological studies have described changes occurring in the lung during acute alveolar inflammation mediated by inflammatory stimuli such as bacteria, immune complexes, C5a, and C5a des arg.[70-75] These studies have demonstrated accumulation of neutrophils within the alveolar walls and the alveolar airspace in association with endothelial and epithelial injury and increased vascular permeability. Larsen et al.[70] demonstrated in rabbits that in response to the intraalveolar administration of C5a des arg, neutrophils accumulated in the alveolar walls when examined at 30 min and within the alveolar airspaces by 2 hr. Shaw et al.[72] found that C5 fragments caused neutrophil accumulation in the alveolar walls and the alveolar airspaces of rabbits when examined at 1 hr. Transmission electron microscopy in their study showed neutrophils in the alveolar walls both within capillaries and the interstitial space, while both capillary endothelium and alveolar epithelium showed evidence of injury. However, these studies did not clearly delineate the time course of neutrophil localization after such stimulation.

Direct observation of neutrophil kinetics by in vivo microscopy of the pulmonary micro-vasculature has allowed us to demonstrate that FITC-labeled canine neutrophils are sequestered in the pulmonary capillaries extremely rapidly in response to the intraairways administration of C5 fragments. The transit times of FITC-labeled neutrophils infused as early as 5 min after the stimulus was given are dramatically prolonged. The prolongation of transit time was maximal at 15 to 30 min, and the cells generally remained individually sequestered for the entire 90-min observation period. In addition, there was a slowing in the velocity of capillary blood flow as measured by the transit time of FITC-labeled dextran from 1.35 ± 0.2 to 1.82 ± 0.3 sec and an increase in the number of neutrophils recoverable from bronchoalveolar lavage fluid, indicating that neutrophil migration had occurred.

The lung and other organs are susceptible to injury resulting from inflammatory mediators being delivered in the circulating bloodstream. It has been shown that neutrophils are extensively sequestered within the pulmonary vasculature in conditions associated with complement activation[76,77] or in response to the infusion of zymosan-activated plasma,[78] C5a,[79,80] or cobra venom factor to cause intravascular activation of the complement cascade.[81,82] Endotoxin infusion may also produce neutrophil sequestration.[83,84] Neutrophil sequestration may be associated with lung injury, as manifested by increased vascular permeability, and has been implicated in the pathogenesis of disorders such as adult respiratory distress syndrome. The localization of neutrophils appears to be primarily within the alveolar walls as shown in several studies. Till et al.,[81] using cobra venom factor to activate complement intravascularly in rats, found that 30-fold more neutrophils were sequestered in interstitial capillaries examined at 30 min when compared to controls. Meyrick and Brigham[78] found that an infusion of zymosan-activated plasma in sheep caused a threefold increase in peripheral lung granulocytes just 7.5 min after the infusion, which increased to sevenfold by 30 min. The same author also examined the effect of *Escherichia coli* endotoxin infusion on the lungs of sheep and found that there was a threefold increase in peripheral lung granulocytes at 15 min and a sixfold increase by 4 hr. In experiments designed to examine the mechanism of lung injury produced by the intravascular sequestration of granulocytes, Shaw and Henson[85] found that neutrophils pretreated in vitro with zymosan-activated serum were sequestered in the pulmonary capillaries and arterioles with light-microscopic evidence of lung injury.

Experiments in which the kinetics of FITC-labeled neutrophils were studied in the pulmonary microcirculation by in vivo microscopy show that neutrophils are rapidly sequestered in pulmonary capillaries in response to intravascular C5 fragments or endotoxin. FITC-labeled neutrophils infused within seconds of an infusion of C5 fragments were *immediately* sequestered in pulmonary capillaries (Figure 6). Only occasional cells were retained in the arterioles or venules. Neutrophils infused 15 or 30 min after the C5 fragments were also sequestered, and these cells generally remained for the full 90-min period over which these experiments were observed. Although these in vivo studies are not directly comparable to the in vitro ones previously reported, it is apparent that rapid modulation, within *seconds,* of neutrophil adherence is an important feature.

C5 fragments have been shown to increase neutrophil adherence in vitro by an effect on the neutrophil. Attempts to demonstrate this phenomenon in vivo are difficult, which led to studies using lipopolysaccharide (LPS) as an alternative stimulus. LPS is valuable for several reasons: it increases neutrophil adherence in vitro by a Mac-1-dependent mechanism,[86] and causes neutrophil sequestration in the lung in vivo,[51] but requires more time to act, thus making pretreatment studies feasible. FITC-labeled neutrophils exposed to *S. enteriditis* LPS, washed twice, and infused into recipient animals were immediately sequestered primarily within the pulmonary capillaries. Figure 7 shows the cumulative distribution of neutrophil transit times, indicating that many more neutrophils have long transit times after pretreatment. Therefore, in this system, LPS acted on the neutrophil, not the endothelium, to increase adherence.

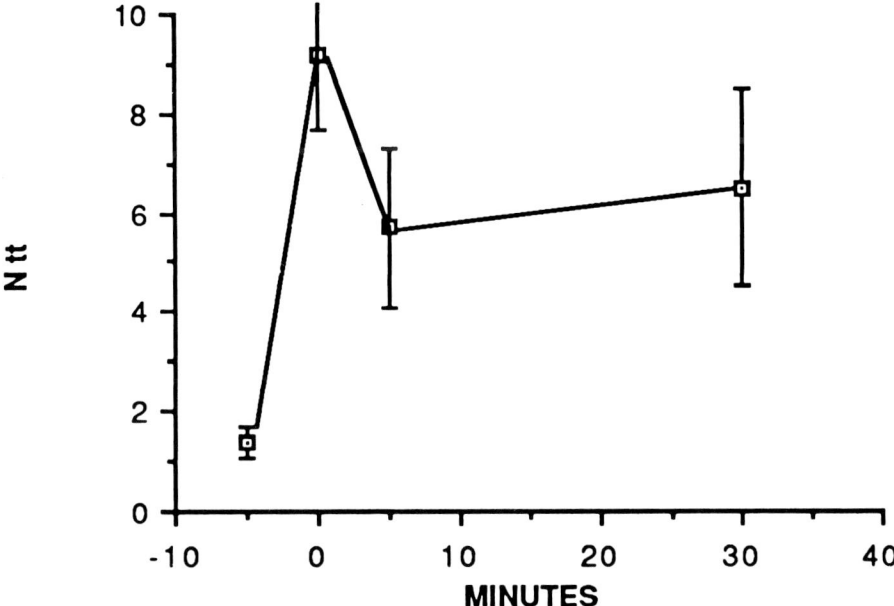

FIGURE 6. Time course of canine neutrophil transit times through the lung microcirculation after intravascular instillation of C5 fragments (C5f). The median neutrophil transit time is prolonged within seconds after instillation of C5f.

The sequestration of neutrophils in response to inflammatory stimuli is thus predominantly within the pulmonary capillaries; this localization may be facilitated by the large marginating pool of neutrophils which are stationary for varying periods of time, allowing greater potential for the inflammatory stimulus to enhance neutrophil-endothelial adherence (see Section II.B.2). Furthermore, the enhancement of neutrophil adherence within the lung induced by inflammatory mediators is rapidly modulated and appears to act primarily by an effect on the neutrophil rather than on the endothelium, at least in these short time-course experiments.

Finally, neutrophil localization within the lung may be modulated by local hemodynamic factors. It has been clearly shown in human subjects by Muir et al.[54] that exercise and catecholamine infusion accelerate the release of radiolabeled neutrophils from the lung in association with an increase in circulating neutrophil numbers. These agents may accelerate neutrophil transit through the lung by increasing pulmonary blood flow. The importance of blood flow on pulmonary neutrophil retention has been well shown in two studies by Martin et al.[87] and Thommasen et al.[88] which indicate that neutrophils are retained in the lung in direct relation to reductions in total blood flow and that an equivalent number of cells are released when blood flow is restored. These same authors have also shown that neutrophil retention increases in proportion to reductions in regional blood flow. This has important implications, as it indicates the existence of a gradient of neutrophil transit times through the lung. Because the pulmonary circulation is a low-pressure system, pulmonary blood flow is affected by gravitational forces. Regions of the lung which are the most dependent receive the majority of the blood flow, while the uppermost lung regions receive proportionally less. The data would indicate that neutrophils which enter these regions of low blood flow are retained for longer periods. This may be important in the pathogenesis of pulmonary disorders which may be neutrophil-dependent such as centrilobular emphysema, which has a predilection for upper lung zones.

Further insight into the effect of changes in pulmonary hemodynamics on neutrophil transit can be gained from in vivo microscopic observations of FITC-neutrophils in the canine microvasculature. Various stresses associated with demargination of neutrophils may cause

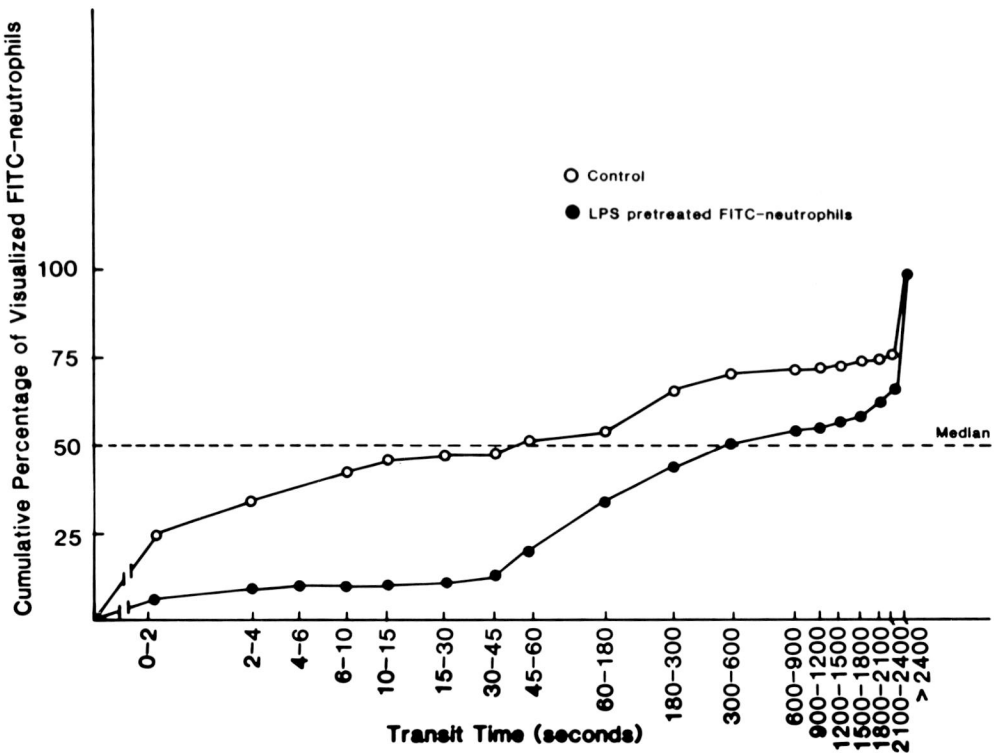

FIGURE 7.   Pretreatment of isolated neutrophils with LPS prior to reinfusion into recipient dogs. Treating canine neutrophils with 10 μg/mℓ *S. enteriditis* LPS at 37° for 5 min before washing and reinfusion results in an identical prolongation of neutrophil transit times through the capillaries as seen with intravascular infusion of LPS, suggesting that the effect of LPS is on the neutrophil. (From Tonnesen, M. G. and Worthen, G. S., *Pulmonary Endothelium in Health and Disease,* Ryan, U. S., Ed., Marcel Dekker, New York, in press. With permission.)

increases in either pressure, flow, or both within the pulmonary circulation. Figure 8 shows the effect on neutrophil transit time of the administration of epinephrine which increases both pulmonary artery pressure and cardiac output, or the inflation of an inferior vena cava balloon which decreases cardiac output. This cumulative function shows that increases in pulmonary artery pressure and cardiac output increase the cumulative percentage of cells with rapid transit times. These observations indicate that a shift has occurred in the dynamic equilibrium between the circulating and marginating neutrophils, resulting in fewer neutrophils being retained within the capillary bed. Similarly, a decrease in cardiac output resulting from inflation of an inferior vena cava (IVC) balloon increased the percentage of cells which were retained by the capillary network. The effect of these interventions on the velocity of blood flow through the capillary bed was also assessed by determining the transit time of FITC-labeled dextran. Increases in pulmonary artery pressure and cardiac output significantly decreased the transit time of plasma through the capillary bed. A significant correlation was present between the change in plasma capillary transit time and median neutrophil transit time, suggesting that increased pulmonary artery pressure and pulmonary blood flow act, at least in part, by increasing the rate of blood flow through the capillaries and mechanically dislodging neutrophils from the capillary bed.

Although it seems clear from these data that blood flow through the pulmonary microcirculation is an important determinant of neutrophil margination, it would seem likely that there are important contributions from other factors as well. These factors could include the intrinsic adhesiveness of the neutrophil or vascular endothelium (*vide supra*), or a unique

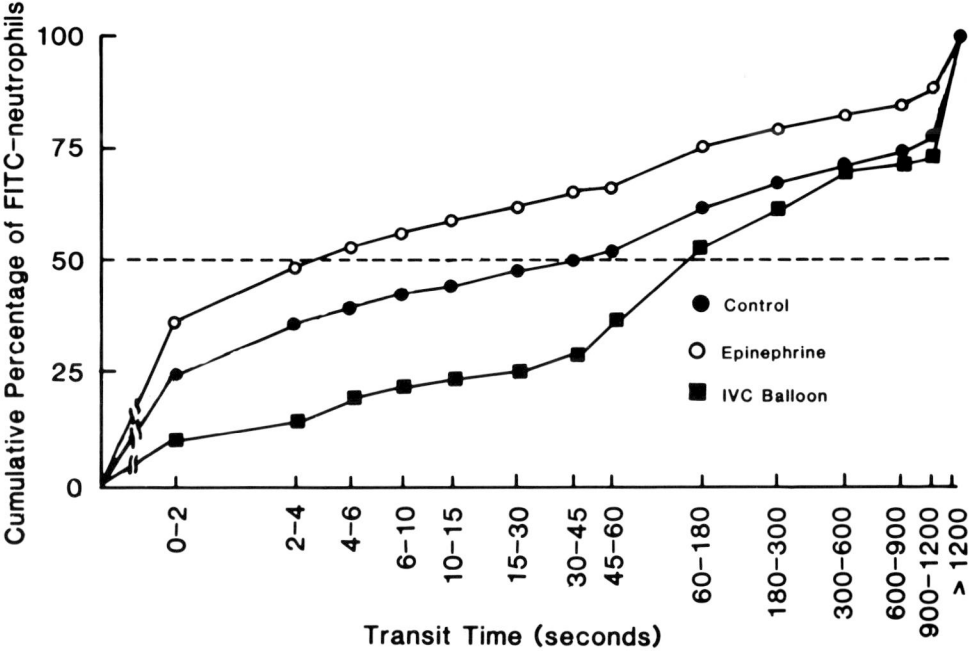

FIGURE 8. Cumulative distribution of canine neutrophil transit times after hemodynamic alterations to the recipient dog. Treating the recipient animal with epinephrine (open circles) to increase pulmonary artery pressure and cardiac output increases the percentage of cells with very rapid transit times, thus shifting the median transit time to low values, compared to cells in control animals (closed circles). In contrast, inflation of an inferior vena cava (IVC) balloon to decrease cardiac output markedly decreased the percentage of cells with rapid transit times and shifted the median transit time to higher values.(From Tonnesen, M. G. and Worthen, G. S., *Pulmonary Endothelium in Health and Disease*, Ryan, U. S., Ed., Marcel Dekker, New York, in press. With permission.)

interaction between the two. The complexity of the pulmonary capillary bed may also produce unique anatomical obstruction to neutrophil transit.

III. CONCLUSION

Neutrophil interaction with the vascular endothelium occurs physiologically to form the marginating pool and pathologically to initiate migration to tissue sites of inflammation. In vitro studies have demonstrated that nonstimulated neutrophils express a low level of spontaneous adhesivity for endothelial cells which can be enhanced by stimulation with chemotactic factors and lipid mediators. Stimulated adherence appears to be mediated by a primary effect on the neutrophil, with the induction of increased expression of adhesive glycoproteins on the surface of the neutrophil. The quick onset and capacity for rapid modulation of the stimulated neutrophil adhesivity contribute to the appropriate localization of the neutrophil-endothelial interaction within the microvasculature. Cytokines may render the endothelial cell more adhesive for neutrophils, also aiding in localization. Hemodynamic and physical effects modulate the interaction between neutrophils and endothelial cells: shear stress acts to diminish adhesion while low blood flow and geometric constraints serve to enhance adhesion. It can be postulated that all these factors interact in vivo to determine the site and degree of neutrophil sequestration and emigration in various organs. Since in vivo effects presumably result from an interplay of these and perhaps additional factors, it is clear that much work remains to be done to further our understanding of the complex and fascinating interaction between circulating leukocytes and the vascular endothelium.

REFERENCES

1. **Grant, L.,** The sticking and emigration of white blood cells in inflammation, in *The Inflammatory Process,* Vol. II, Zweifach, B. W., Grant, L., and McCluskey, R. T., Eds., Academic Press, New York, 1973, 205.
2. **Wilkinson, P. C.,** *Chemotaxis and Inflammation,* 2nd ed., Churchill Livingstone, New York, 1982, 90 and 183.
3. **Clark, E. R. and Clark, E. L.,** Observations on changes in blood vascular endothelium in the living animal, *Am. J. Anat.,* 57, 385, 1935.
4. **Metchnikoff, E.,** *Lectures on the comparative pathology of inflammation,* Kegan, Paul, Trench, Trubner, London, 1893.
5. **Allison, F., Jr., Smith, M. R., and Wood, W. B.,** Studies on the pathogenesis of acute inflammation. I. The inflammatory reaction to thermal injury as observed in the rabbit ear chamber, *J. Exp. Med.,* 102, 655, 1955.
6. **Ryan, U. S., Mortara, M., and Whitaker, C.,** Methods for microcarrier culture of bovine pulmonary artery endothelial cells avoiding the use of enzymes, *Tissue Cell,* 12, 619, 1980.
7. **Gimbrone, M. A., Jr.,** Culture of vascular endothelium, *Prog. Hemostasis Thromb.,* 3, 1, 1976.
8. **Kern, P. A., Knedler, A., and Eckel, R. H.,** Isolation and culture of microvascular endothelium from human adipose tissue, *J. Clin. Invest.,* 71, 1822, 1983.
9. **Lackie, J. M. and DeBono, D.,** Interactions of neutrophil granulocytes and endothelium in vitro, *Microvasc. Res.,* 13, 107, 1977.
10. **MacGregor, R. R., Macarak, E. J., and Kefalides, N. A.,** Comparative adherence of granulocytes to endothelial monolayers and nylon fiber, *J. Clin. Invest.,* 61, 697, 1978.
11. **Hoover, R. L., Briggs, R. T., and Karnovsky, M. J.,** The adhesive interaction between polymorphonuclear leukocytes and endothelial cells *in vitro, Cell,* 14, 423, 1978.
12. **Beesley, J. E., Pearson, J. D., Carleton, J. S., Hutchings, A., and Gordon, J. L.,** Interaction of leukocytes with vascular cells in culture, *J. Cell Sci.,* 33, 85, 1978.
13. **Tonnesen, M. G., Smedly, L. A., and Henson, P. M.,** Neutrophil-endothelial cell interactions: modulation of neutrophil adhesiveness induced by complement fragments C5a and C5a des arg and formyl-methionyl-leucyl-phenylalanine *in vitro, J. Clin. Invest.,* 74, 1581, 1984.
14. **Beesley, J. E., Pearson, J. D., Hutchings, A., Carleton, J. S., and Gordon, J. L.,** Granulocyte migration through endothelium in culture, *J. Cell Sci.,* 38, 237, 1979.
15. **Hoover, R. L., Folger, R., Haering, W. A., Ware, B. R., and Karnovsky, M. J.,** Adhesion of leukocytes to endothelium: roles of divalent cations, surface charge, chemotactic agents and substrate, *J. Cell Sci.,* 45, 73, 1980.
16. **Gimbrone, M. A., Jr., Brock, A. F., and Schafer, A. L.,** Leukotriene B$_4$ stimulates polymorphonuclear leukocyte adhesion to cultured vascular endothelial cells, *J. Clin. Invest.,* 74, 1552, 1984.
17. **Ingraham, L. M., Coates, T. D., Allen, J. M., Higgins, C. P., Baehner, R. L., and Boxer, L. A.,** Metabolic, membrane and functional responses of human polymorphonuclear leukocytes to platelet-activating factor, *Blood,* 59, 1259, 1982.
18. **Tonnesen, M. G., Anderson, D. C., Springer, T. A., Knedler, A., Avdi, N., and Henson, P. M.,** Mac-1 glycoprotein family mediates adherence of neutrophils to endothelial cells stimulated by leukotriene B4 and platelet activating factor, *Fed. Proc.,* 45, 379, 1986.
19. **Marchesi, V. T. and Florey, H. W.,** Electron micrographic observations on the emigration of leukocytes, *Q. J. Exp. Physiol.,* 45, 343, 1960.
20. **Florey, H. W. and Grant, L. H.,** Leukocyte migration from small blood vessels stimulated with ultraviolet light: an electron-microscope study, *J. Pathol. Bacteriol.,* 82, 13, 1961.
21. **Anderson, D. C., Schmalstieg, F. C., Arnaout, M. A., Kohl, S., Tosi, M. F., Dana, N., Buffone, G. J., Hughes, B. J., Brinkley, B. R., Dickey, W. D., Abramson, J. S., Springer, T., Boxer, L. A., Hollers, J. M., and Smith, C. W.,** Abnormalities of polymorphonuclear leukocyte function associated with a heritable deficiency of high molecular weight surface glycoproteins (GP138): common relationship to diminished cell adherence, *J. Clin. Invest.,* 74, 536, 1984.
22. **Bernstein, I. D. and Self, S.,** The joint report of the myeloid section of the Second International Workshop on Human Leukocyte Differentiation Antigens, in *Leukocyte Typing II: Report of the Second International Workshop on Human Leukocyte Differentiation Antigens,* Reinherz, E. L., Haynes, B. S., Nadler, L. M., and Bernstein, I. D., Eds., Springer-Verlag, Berlin, 1985, 1.
23. **Sanchez-Madrid, F., Nagy, J. A., Robbins, E., Simon, P., and Springer, T. A.,** A human leukocyte differentiation antigen family with distinct alpha subunits and a common beta subunit: the lymphocyte function-associated antigen (LFA-1), the C3bi complement receptor (OKM1/Mac-1) and the p 150,95 molecule, *J. Exp. Med.,* 158, 1785, 1983.

24. **Springer, T. A., Thompson, W. S., Miller, L. J., Schmalstieg, F. C., and Anderson, D. C.,** Inherited deficiency of the Mac-1, LFA-1, p150,95 glycoprotein family and its molecular basis, *J. Exp. Med.,* 160, 1901, 1984.

25. **Anderson, D. C. and Springer, T. A.,** Leukocyte adhesion deficiency: an inherited defect in the Mac-1, LFA-1, and p150,95 glycoproteins, *Ann. Rev. Med.,* 38, 175, 1987.

26. **Todd, R. F., Arnaout, M. A., Rosin, R. E., Crowley, C. A., Peters, W. A., and Babior, B. M.,** Subcellular localization of the large subunit of Mol (Molα; formerly gp 110), a surface glycoprotein associated with neutrophil adhesion, *J. Clin. Invest.,* 74, 1280, 1984.

27. **Anderson, D. C., Schmalstieg, F. C., Finegold, M. J., Hughes, B. J., Rothlein, R., Miller, L. J., Kohl, S., Tosi, M. F., Jacobs, R. L., Waldrop, T. C., Goldman, A. S., Shearer, W. T., and Springer, T. A.,** The severe and moderate phenotypes of heritable Mac-1, LFA-1 deficiency: their quantitative definition and relation to leukocyte dysfunction and clinical features, *J. Infect. Dis.,* 152, 668, 1985.

28. **Arnaout, M. A., Spits, H., Terhorst, C., Pitt, J., and Todd, R. F.,** Deficiency of a leukocyte surface glycoprotein (LFA-1) in two patients with Mol deficiency: effects of cell activation on Mol/LFA-1 surface expression in normal and deficient leukocytes, *J. Clin. Invest.,* 74, 1291, 1984.

29. **Harlan, J. M., Killen, P. D., Senecal, F. M., Schwartz, B. R., Yee, E. K., Taylor, R. F., Beatty, P. G., Price, T. H., and Ochs, H. D.,** The role of neutrophil membrane glycoprotein GP-150 in neutrophil adherence to endothelium *in vitro, Blood,* 66, 167, 1985.

30. **Tonnesen, M. G., Anderson, D. C., Springer, T. A., Knedler, A., Avdi, N., and Henson, P. M.,** Mac-1 glycoprotein family mediates adherence of neutrophils to endothelial cells stimulated by chemotactic peptides, *Clin. Res.,* 34, 419, 1986.

31. **Fischer, A., Descamps-Latscha, B., Gerota, I., Scheinmetzler, C., Virelizier, J. L., Trung, P. H., Lisowska-Grospierre, B., Perez, N., Durandy, A., and Griscelli, C.,** Bone-marrow transplantation for inborn error of phagocytic cells associated with defective adherence, chemotaxis and oxidative response during opsonised particle phagocytosis, *Lancet,* 2, 473, 1983.

32. **Bevilacqua, M. P., Pober, J. S., Wheeler, M. E., Cotran, R. S., and Gimbrone, M. A., Jr.,** Interleukin-1 acts on cultured human vascular endothelium to increase the adhesion of polymorphonuclear leukocytes, monocytes and related leukocyte cell lines, *J. Clin. Invest.,* 76, 2003, 1985.

33. **Gamble, J. R., Harlan, J. M., Klebanoff, S. J., and Vadas, M. A.,** Stimulation of the adherence of neutrophils to umbilical vein endothelium by human recombinant tumor necrosis factor, *Proc. Natl. Acad. Sci. U.S.A.,* 82, 8667, 1985.

34. **Zimmerman, G. A., McIntyre, T. M., and Prescott, S. M.,** Thrombin stimulates the adherence of neutrophils to human endothelial cells in vitro, *J. Clin. Invest.,* 76, 2235, 1985.

35. **McIntyre, T. M., Zimmerman, G. A., and Prescott, S. M.,** Leukotrienes C_4 and D_4 stimulate human endothelial cells to synthesize platelet-activating factor and bind neutrophils, *Proc. Natl. Acad. Sci. U.S.A.,* 83, 2204, 1986.

36. **Schmid-Schonbein, G. W., Fung, Y. C., and Zweifach, B. W.,** Vascular endothelium-leukocyte interaction. Sticking shear force in venules, *Circ. Res.,* 36, 173, 1975.

37. **Atherton, A. and Born, G. V. R.,** Relationship between velocity of rolling granulocytes and that of the blood flow in venules, *J. Physiol.,* 233, 157, 1973.

38. **Gaehtgens, P., Pries, A. R., and Nobis, U.,** Flow behavior of white cells in capillaries, in *White Cell Mechanics: Basic Science and Clinical Aspects,* Alan R. Liss, New York, 1984, 147.

39. **Weibel, E. R.,** *Morphometry of the Human Lung,* Academic Press, New York, 1963.

40. **Wintrobe, M. M.,** *Clinical Hematology,* Lea & Febiger, Philadelphia, 1981, 192.

41. **Schmid-Schonbein, G. S. W., Usami, S., Skalak, R., and Chein, S.,** Interaction of leukocytes and erythrocytes in capillary and post-capillary vessels, *Microvasc. Res.,* 19, 45, 1980.

42. **Gaehtgens, P., Duhrssen, C., and Albrecht, K. H.,** Motion, deformation and interaction of blood cells and plasma during flow through narrow capillaries, *Blood Cells,* 6, 799, 1980.

43. **Schmid-Schonbein, G. W., Sung, K. L. P., Tozeren, H., Skalak, R., and Chien, S.,** Passive mechanical properties of human leukocytes, *Biophys. J.,* 36, 243, 1981.

44. **Chien, S., Schmalzer, E. H., Lee, M. M. L., Impulluso, T., and Skalak, R.,** Role of white blood cells in filtration of blood cell suspensions, *Biorheology,* 20, 11, 1983.

45. **Doroszewski, J. J., Skierski, M., and Przadka, L.,** Interaction of neoplastic cells with glass surface under flow conditions, *Exp. Cell Res.,* 104, 335, 1977.

46. **Forrester, J. V. and Lackie, J. M.,** Adhesion of neutrophil leukocytes under conditions of flow, *J. Cell Sci.,* 70, 93, 1984.

47. **VanGrondelle, A., Worthen, G. S., Ellis, D., Mathias, M. M., Murphy, R. C., Strife, R. J., Reeves, J. T., and Voelkel, N. F.,** Increased prostacyclin production in endothelial cell during shear stress and in rat lungs at high flow, *J. Appl. Physiol.,* 57, 388, 1984.

48. **Karatzais, N. B. and Lee, G.,** Propagation of blood flow pulse in the normal human pulmonary arterial system: analysis of the pulsatile capillary flow, *Circ. Res.,* 25, 11, 1969.

49. **Ambrus, C. M., Ambrus, J. L., Johnson, G. C., Packman, E. W., Chernick, W. S., Back, A. N., and Harrisson, J. W. E.,** Role of the lungs in the regulation of the white blood cell level, *Am. J. Physiol.,* 178, 33, 1954.

50. **Bierman, H. R., Kelly, K. H., and Cordes, F. L.,** The sequestration and visceral circulation of leukocytes in man, *Ann. N.Y. Acad. Sci.,* 59, 850, 1955.

51. **Athens, J. W., Raab, S. O., Haab, O. P., Mauer, A. M., Ashenbrucker, H., Cartwright, G. E., and Wintrobe, M. M.,** Leukokinetic studies. IV. The total blood, circulating and marginating granulocyte turnover rate in normal subjects, *J. Clin. Invest.,* 40, 989, 1961.

52. **Bierman, S. A., Kelly, K. H., Cordes, F. L., Byron, R. L., Polhemus, J. A., and Rappart, S.,** The release of leukocytes and platelets from the pulmonary circulation by epinephrine, *Blood,* 7, 683, 1952.

53. **Joyce, R. A., Boggs, D. R., Hasiba, U., and Strodes, C. H.,** Marginal neutrophil pool size in normal subjects and neutropenic patients as measured by epinephrine infusion, *J. Lab. Clin. Med.,* 88, 614, 1984.

54. **Muir, A. L., Cruz, M., Martin, B. A., Thommasen, H., Belzberg, A., and Hogg, J. C.,** Leukocyte kinetics in the human lung: role of exercise and catecholamines, *J. Appl. Physiol.,* 57, 711, 1984.

55. **Goldscheider, S. and Jacob, P.,** Veber die variationen der leukocytose, *Z. Klin. Med.,* 25, 373, 1894.

56. **Vejlens, G.,** The distribution of leukocytes in the vascular system, *Acta Pathol. Microbiol. Scand. Suppl.,* 33, 11, 1938.

57. **Lawrence, J. S., Ervin, D. M., and Wetrich, R. M.,** Life cycle of white blood cells, *Am. J. Physiol.,* 144, 284, 1945.

58. **Erf, L. A.,** The disappearance of intravenously injected leukocytes in pulmonary marginated pool of neutrophils and comparison of the distribution of labeled and unlabeled neutrophils in the rabbit pulmonary vasculature, *Am. Rev. Resp. Dis.,* 133 (Abstr.), A272, 1986.

59. **Bierman, H. R., Byron, R. L., Jr., Kelly, K. H., and Petriki, N.,** The lung removal mechanism for leukocytes, *Blood,* 6, 770, 1951.

60. **Lannan, J. T., Bierman, H. R., and Byron, R. L.,** Transfusion of leukemic leukocytes in man, *Blood,* 5, 1099, 1950.

61. **Weisberger, A. S., Heinle, R. W., Storaasli, J. P., and Hannah, R.,** Transfusion of leukocytes labeled with radioactive phosphorus, *J. Clin. Invest.,* 29, 336, 1950.

62. **Connheim, J.,** *Lectures in General Pathology,* 2nd ed., New Syndenham Society, London, 1889.

63. **Chien, S.,** Role of blood cells in microcirculatory regulation, *Microvas. Res.,* 29, 129, 1985.

64. **Perlo, S., Jalowayski, A. A., Durand, C. M., and West, J. B.,** Distribution of red and white blood cells in alveolar walls, *J. Appl. Physiol.,* 38, 117, 1975.

65. **Staub, N. E., Schultz, E. L., and Albertine, K. H.,** Leukocytes and pulmonary microvascular injury, *Ann. N.Y. Acad. Sci.,* 38, 332, 1982.

66. **Doerschuk, C. M., Martin, B. A., Mackenzie, A., Walker, D. A., and Hogg, J. C.,** Localization of the pulmonary marginated pool of neutrophils and comparison of the distribution of labeled and unlabeled neutrophils in the rabbit pulmonary vasculature, *Am. Rev. Resp. Dis.,* 133 (Abstr.), A272, 1986.

67. **Wagner, W. W., Jr., Lathan, L. P., Gillespie, M. N., Guenther, J. P., and Capen, R. L.,** Direct measurement of pulmonary capillary transit times, *Science* 218, 379, 1982.

68. **Zigmond, S. H.,** Chemotaxis by polymorphonuclear leukocytes, *J. Cell Biol.,* 77, 269, 1978.

69. **Bjork, J.,** *Microvascular Reactions in Acute Inflammation Intravital Microscopy Study in the Hamster,* Ph.D. thesis, Uppsala University, Uppsala, Sweden, 1984.

70. **Larsen, G. L., McCarthy, K., Webster, R. O., Henson, J., and Henson, P. M.,** A differential effect of C5a and C5a des arg in the production of pulmonary inflammation, *Am. J. Pathol.,* 100, 179, 1980.

71. **Scherzer, H. and Ward, P. A.,** Lung and dermal vascular injury produced by preformed immune complexes, *Am. Rev. Resp. Dis.,* 117, 551, 1978.

72. **Shaw, J. O., Henson, P. M., Henson, J. E., and Webster, R. O.,** Lung inflammation induced by complement derived chemotactic fragments in the alveolus, *Lab. Invest.,* 42, 547, 1980.

73. **Henson, P. M., McCarthy, K., Larsen, G. L., Webster, R. O., Giclas, P. C., Dreisin, R. B., King, T. E., and Shaw, J. O.,** Complement fragments, alveolar macrophages, and alveolitis, *Am. J. Pathol.,* 97, 93, 1979.

74. **Desai, U., Kreutzer, D. L., Showell, H., Arroyave, C. V., and Ward, P. A.,** Acute inflammatory pulmonary reactions induced by chemotactic factors, *Am. J. Pathol.,* 96, 71, 1979.

75. **Lipscomb, M. F., Onofrio, J. M., Nash, J., Pierce, A. K., and Toews, G. B.,** A morphologic study of the role of phagocytes in the clearance of Staphylococcus aureus from the lung, *J. Reticuloendo. Soc.,* 33, 429, 1983.

76. **Craddock, P. R., Fehr, J., Dalmasso, A. P., Brigham, K. L., and Jacob, H. S.,** Hemodialysis leukopenia. Pulmonary vascular leukostasis resulting from complement activation by dialyzer cellophane membranes, *J. Clin. Invest.,* 59, 879, 1977.

77. **Craddock, P. R., Fehr, J., Brigham, K. L., Kronenberg, R. S., and Jacob, H. S.,** Complement and leukocyte mediated pulmonary dysfunction in hemodialysis, *N. Engl. J. Med.,* 296, 769, 1977.
78. **Meyrick, B. and Brigham, K. L.,** Effect of a single infusion of zymosan-activated plasma on the pulmonary microcirculation of sheep: structure-function relationships, *Am. J. Pathol.,* 114, 32, 1984.
79. **Snyderman, R., Phillips, J. K., and Mergenhagen, S. E.,** Biological activity of complement *in vivo:* role of C5 in the accumulation of polymorphonuclear leukocytes in inflammatory exudation, *J. Exp. Med.,* 134, 1131, 1971.
80. **Henson, P. M., Larsen, G. L., Webster, R. O., Mitchell, B. C., Goins, A. J., and Henson, J. E.,** Pulmonary microvascular alterations and injury induced by complement fragments: synergistic effect of complement activation, neutrophil sequestration and prostaglandins, *Ann. N.Y. Acad. Sci.,* 384, 287, 1982.
81. **Till, G., Johnson, K., Kunkel, R., and Ward, P.,** Intravascular activation of complement and acute lung injury: dependency on neutrophils and toxic oxygen metabolites, *J. Clin. Invest.,* 69, 1126, 1982.
82. **Larsen, G. L., Webster, R. O., Worthen, G. S., Gumbay, R. S., and Henson, P. M.,** The additive effect of intravascular complement activation and brief episodes of hypoxia in producing increased permeability in the rabbit lung, *J. Clin. Invest.,* 75, 902, 1985.
83. **Meyrick, B. and Brigham, K. L.,** Acute effects of *E. coli* endotoxin on the pulmonary microcirculation of anesthetized sheep: structure-function relationships, *Lab. Invest.,* 48, 458, 1983.
84. **Snapper, J., Bernard, G., Hinson, J., Hutchison, A., Loyd, J., Ogletree, M., and Brigham, K.,** Endotoxemia induced leukopenia in sheep — correlation with vascular permeability and hypoxemia but not with pulmonary hypertension, *Am. Rev. Resp. Dis.,* 127, 306, 1983.
85. **Shaw, J. O. and Henson, P. M.,** Pulmonary intravascular sequestration of activated neutrophils, *Am. J. Pathol.,* 108, 17, 1982.
86. **Tonnesen, M. G., Anderson, D. C., Springer, T. A., Knedler, A., Avdi, N., and Henson, P. M.,** Endotoxin directly enhances neutrophil adherence to endothelial cells by a Mac-1 glycoprotein-dependent mechanism, *Clin. Res.,* 35, 722A, 1987.
87. **Martin, B. A., Wright, J. L., Thommasen, H., and Hogg, J. C.,** Effect of pulmonary blood flow on the exchange between the circulating and marginating pool of polymorphonuclear leukocytes in dog lungs, *J. Clin. Invest.,* 69, 1277, 1982.
88. **Thommasen, H. V., Martin, B. A., Wiggs, B. R., Quiroga, M., Baile, E. M., and Hogg, J. C.,** Effect of pulmonary blood flow on leukocyte uptake and release by dog lung, *J. Appl. Physiol.,* 56, 966, 1984.

Chapter 24

MECHANISMS OF LEUKOCYTE ADHERENCE TO ENDOTHELIUM

Barbara R. Schwartz and John M. Harlan

TABLE OF CONTENTS

I. INTRODUCTION

Leukocyte traffic across endothelium is a central event in host response to tissue injury and defense against foreign antigens.[1] Leukocyte emigration to tissue occurs not only during inflammatory and immune reactions, but also during the physiologic circulation of lymphocytes and probably during the seeding of tissues with resident macrophages. An early event in emigration is leukocyte adherence to the vessel wall.[2] Many investigators have examined the adhesive interaction of leukocytes and endothelium using in vitro and in vivo models. Increasing evidence indicates that both leukocytes and endothelial cells are capable of playing an active role in their interaction. Leukocyte-dependent augmentation of the adhesive interaction is usually marked by morphological or biochemical evidence of leukocyte "activation". Endothelial cells may also be "activated", with or without morphologic alterations, to express increased adhesiveness for leukocytes. We will briefly summarize the contributions of both leukocytes and endothelial cells in their adhesive interactions.

II. LEUKOCYTE MECHANISMS

Two distinct families of leukocyte surface molecules, the CD11/CD18 complex and the Lymphocyte Homing Receptor glycoprotein, participate in the adhesive interaction with endothelium.

A. The CD11/CD18 Complex

Our understanding of leukocyte adhesion to endothelium has been facilitated by multiple disparate and almost simultaneous observations. Several laboratories developed monoclonal antibodies (MoAbs) that immunoprecipitate a family of leukocyte membrane glycoproteins.[3] This family of glycoproteins has been designated CD11/CD18 complex by the Third International Workshop on Human Leukocyte Differentiation Antigens (Figure 1).[4] The complex contains three distinct subunits: LFA-1, Mac-1 and p150,95.[5] Each subunit consists of a common light-chain (β) polypeptide (M_r 95 kdaltons) (CD18) that is noncovalently associated with a distinct heavy-chain (α) polypeptide (CD11a-c). The lymphocyte function-associated (LFA-1) subunit (CD11a/CD18) described by Springer and colleagues[6] contains a heavy-chain polypeptide of M_r 177 kdaltons (α_L), and is found on T and B lymphocytes, NK cells, monocytes, and neutrophils. The Mac-1[7] (Mo1)[8] (CD11b/CD18) subunit has an α polypeptide of M_r 165 kdaltons (α_M) and is found on NK cells, monocytes, and neutrophils. It also functions as the C3bi receptor (CR3).[9] The p150,95 subunit (CD11c/CD18) contains a heavy-chain polypeptide of M_r 150 kdaltons (α_X) and is found on monocytes and neutrophils.[10,11]

At the same time, a number of patients were described with a syndrome marked by delayed separation of the umbilical cord, recurrent bacterial infections, and persistent leukocytosis.[12-26] Biopsies from infected tissues do not contain neutrophils or mononuclear phagocytes.[17,25] Their phagocytes also fail to migrate to skin windows or skin chambers.[17,22,23,27] Phagocytes from these patients show greatly depressed responses in tests of spreading, aggregation, chemotaxis, phagocytosis, adherence and antibody-dependent cell-mediated cytotoxicity. The adherence deficit is clearly due to defective leukocytes, since transfused allogeneic leukocytes migrate normally to sites of infection or skin windows.[17]

In contrast to acute inflammatory reactions, immune responses in these patients are apparently intact. Even though T-lymphocyte-mediated cytotoxicity, NK cell-mediated cytotoxicity, interferon generation, and mitogen-induced lymphocyte transformation are depressed in patient cells in vitro, severely affected patients do not suffer from diseases characteristic of impaired cell-mediated immunity.[25,26] They have normal delayed-type hypersensitivity skin test responses,[17,25] and biopsies of tissues from sites of chronic inflammation contain plasma cells and lymphocytes.[26,85]

Leukocyte Adhesion Complex*

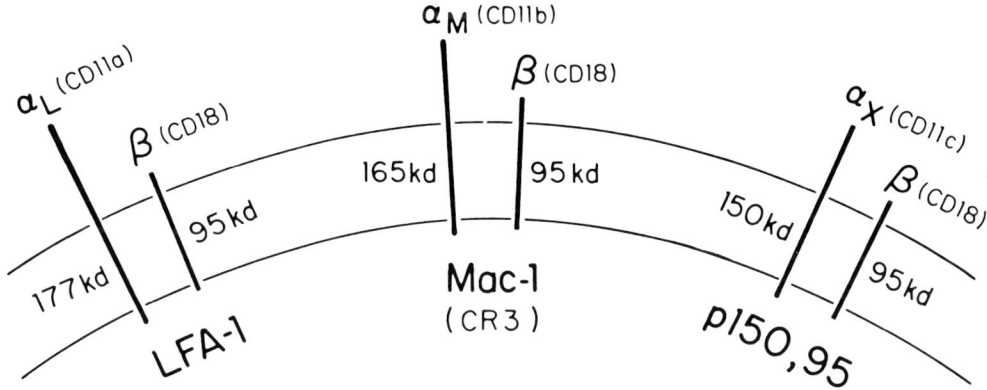

*3rd International Workshop on Human
Leukocyte Differentiation Antigens

FIGURE 1.   CD11/CD18 complex.

Several laboratories have demonstrated that leukocytes from these patients are deficient in membrane expression of one or more of the glycoproteins of the CD11/CD18 complex.[28-31] Moreover, treatment of normal leukocytes with some of the MoAbs that recognize epitopes on the CD11/CD18 complex induces defects in adherence-dependent functions similar to those observed in patient cells,[27-33] in particular, decreased leukocyte adhesion to endothelium.[28,34,35] Furthermore, treatment of experimental animals with an anti-CD18 MoAb markedly decreases phagocyte migration to areas of induced inflammation in vivo.[36,37] These studies have yielded important insights into the molecular basis of leukocyte adherence to endothelium.

1. Neutrophils

Peripheral blood neutrophils express all three subunits (LFA-1, Mac-1, and p150,95) of the CD11/CD18 complex. This complex appears to be involved primarily in stimulated neutrophil adherence. The basal, unstimulated adherence of CD11/CD18-deficient neutrophils or normal cells treated with anti-CD18 MoAb to cultured endothelium is quantitatively low and similar to that observed with normal neutrophils.[34] Normal neutrophil adherence to endothelial monolayers increases 5- to 30-fold following stimulation by a variety of soluble agonists. In marked contrast to normal neutrophils, CD11/CD18-deficient or anti-CD18-treated neutrophils fail to increase adherence when stimulated.[34] Other neutrophil functions that are dependent on adhesive responses of stimulated cells, including C3bi binding, chemotaxis, phagocytosis of particles, antibody-dependent cytotoxicity, and aggregation, are impaired in deficient cells or normal neutrophils treated with anti-CD18 MoAb.[22-33] Since CD11/CD18-deficient neutrophils or normal neutrophils treated with anti-CD18 MoAb retain the ability to depolarize, generate superoxide anion, release granule constituents and undergo shape change in response to soluble stimuli, it appears that the CD11/CD18 complex is not required for stimulant binding and neutrophil activation, but is important only for adherence-dependent responses.

Some MoAbs bind epitopes on the α_M and β polypeptides of the CD11/CD18 complex and inhibit adherence-dependent functions, suggesting that the site(s) relevant to adherence functions are contained in the Mac-1 subunit or perhaps in the quarternary structural created by their noncovalent association.[38,39] When treated with stimulants such as phorbol ester, bacterial chemotactic peptide, or calcium ionophore, neutrophils exhibit increased binding of MoAb directed to α_M, α_X, and β polypeptides.[30,34,38,40,41] Neutrophil stimulation by soluble agonists induces translocation of the Mac-1 subunit from intracellular granules to the cell surface.[40] This "upregulation" of antigen may explain in part the increase in neutrophil adhesiveness following stimulation.[40,41] Alternatively, allosteric or other qualitative changes in the heterodimer may be necessary for increased adhesiveness, and the mobilization of antigen from intracellular stores may be an epiphenomenon. Indeed, unstimulated and non-adherent neutrophils isolated from peripheral blood by Ficoll-Hypaque centrifugation express relatively large amounts of the CD11/CD18 complex on their surface. The increase in membrane expression observed with phorbol ester stimulation (usually 2- to 10-fold) does not parallel the marked increase in adherence (5- to 30-fold).[34] Moreover, neutrophil-derived cytoplasts that are markedly depleted of intracellular granules fail to increase Mac-1 (Mo1) antigen expression after stimulation,[42,43] yet they still adhere and aggregate by a CD11/CD18-dependent mechanism.[86]

2. Monocytes

Less is known about the adherence of peripheral blood monocytes, in part because of the difficulty in obtaining sufficient numbers of cells in pure preparations for adherence assays. Preparative methods involving adherence to plastic render interpretation of subsequent adherence assays difficult. In addition, adherence cannot be used to isolate monocytes from CD11/CD18-deficient patients because their cells adhere poorly to artificial substrates.[17] Although the monocyte-like U937 cell line has been investigated,[44] the expression of membrane glycoproteins and adherence behavior of these cells does not resemble those observed in peripheral blood monocytes.[35] It seems likely that monocytes purified by counter-current centrifugal elutriation or Percoll gradient centrifugation without an adherence step more closely resemble the phenotype and "activation state" of circulating monocytes.

Like neutrophils, monocytes express all three subunits of the CD11/CD18 complex. Baseline monocyte adherence to endothelial cells is greater than that observed with neutrophils when cells from the same donor are tested simultaneously.[35,45] When monocytes are stimulated, adherence to endothelium is augmented, albeit to a lesser degree than observed with neutrophils.[35] Monocyte adherence to human umbilical vein endothelial cells is significantly greater than that observed to bovine aortic endothelial cells.[35] The basal adherence of unstimulated peripheral blood monocytes to cultured endothelium is significantly, but not completely, reduced by anti-CD18 MoAb.[35] Augmentation of monocyte adherence by phorbol ester, however, is completely abrogated by anti-CD18 MoAbs.[35] It thus appears that unstimulated monocyte adherence is mediated by both CD11/CD18-dependent and -independent mechanisms, while stimulated adherence is only CD11/CD18 dependent.

3. Lymphocytes

Normal peripheral blood B- and T-lymphocytes express the LFA-1, but not the Mac-1 and p150-95 subunits. Antibodies to the CD11/CD18 inhibit T-lymphocyte mediated cytotoxicity by preventing T-lymphocyte adherence to target cells.[46] Anti-CD18 MoAb only partially inhibit basal T-lymphocyte binding to cultured endothelium,[47,48] but completely inhibit stimulated T-lymphocyte binding.[48] Thus, as with monocytes, basal T-lymphocyte adherence occurs by both CD11/CD18-dependent and -independent mechanisms, whereas stimulated adherence is largely CD11/CD18 dependent. Because MoAb directed to the α_L (CD11a) and β (CD18) polypeptides inhibit stimulated lymphocyte binding to endothelium in vitro, LFA-1 is the subunit involved in stimulated lymphocyte adherence to endothelium in vivo.

As outlined previously, the CD11/CD18 complex appears to be the predominant mechanism for stimulated neutrophil and monocyte adherence to endothelium. Patients whose leukocytes lack the CD11/CD18 complex do have lymphocytes in extravascular tissues and intact delayed hypersensitivity responses, suggesting that other (non CD11/CD18) mechanisms are primarily involved in lymphocyte interaction with endothelium.

4. Relationship of CD11/CD18 to Other Cytoadhesion Molecules

A family of glycoproteins has recently been described that functions as receptors for matrix adhesion proteins containing the tripeptide recognition sequence arg-gly-asp-(RGD).[49] These "integrins" include the fibronectin receptor, the vitronectin receptor, and platelet glycoprotein IIb/IIIa (Gp IIb/IIIa).[50] These cytoadhesion molecules are all heterodimers with molecular weights similar to LFA-1, Mac-1, and p150,95. There is some homology between the N-terminal amino acid sequence of α_L and α_M polypeptides of the leukocyte receptor and that of platelet GpIIb.[51] The beta subunit (CD18) of the leukocyte receptor has been cloned and shows significant homology to the light chain of integrin.[52] CD11/CD18 complex thus appears to be a member of the integrin receptor family. However, polyclonal and monoclonal antibodies to GpIIb or GpIIIa, and synthetic RGD-containing oligopeptides that complete for binding of adhesive proteins to RGD-receptor heterodimers, do not inhibit stimulated neutrophil adherence to cultured endothelium.[87]

B. Lymphocyte Homing Receptor

Lymphocytes recirculate through specific lymphoid tissues permitting constant surveillance for foreign antigens. Lymphocyte recirculation involves both lymphocyte- and endothelial-dependent mechanisms. Lymphocyte emigration in lymphoid tissue occurs at specialized endothelium of postcapillary venules designated high endothelium (HEV) because it is cuboidal rather than thin and flattened as in other vessels.[53] During maturation, subpopulations of lymphocytes develop membrane receptors that recognize determinants on HEV and thereby facilitate recirculation.[54] B lymphocytes develop receptors that recognize HEV of mucosal lymphoid tissue (e.g., Peyer's patch), whereas T-lymphocytes bind preferentially to HEV of lymph nodes. Butcher and colleagues recently described a third lymphocyte-HEV recognition system in inflamed synovium that is distinct from those involved in lymphocyte recirculation through mucosal or peripheral lymph node lymphoid tissue.[55]

One lymphocyte "homing receptor" has been well characterized by Gallatin et al.[56] MEL-14 is a rat MoAb that blocks binding of mouse lymphocytes to peripheral lymph node HEV without affecting binding to mucosal HEV in vitro, and blocks homing of treated lymphocytes to peripheral lymph nodes in vivo. In general, only mouse lymphocytes that express the MEL-14 antigen bind to lymph node endothelium in vitro or home to lymph nodes in vivo. Antigen-activated lymphocytes concurrently lose surface MEL-14 glycoprotein and their ability to home to lymph nodes.[54] MEL-14 also inhibits neutrophil binding to lymph node endothelium ex vivo,[57] although the significance of this is unknown, since neutrophils do not normally emigrate via this route. The lymphocyte homing receptor precipitated by MEL-14 is a single-chain ubiquitinated 90-kdaltons glycoprotein.[58,59] Another MoAb designated Hermes-1 binds an analogous glycoprotein on human lymphocytes and blocks adherence to lymphoid tissue HEV.[57]

III. ENDOTHELIAL CELL MECHANISMS

Increased leukocyte adherence to endothelium occurs during inflammatory reactions (e.g., tissue injury) and during immune reactions (e.g., allograft rejection and delayed type hypersensitivity). Increasing evidence indicates that the endothelial cell contributes to these adhesive interactions.

Table 1
ENDOTHELIAL-DEPENDENT
MECHANISMS INDUCED IN VITRO

	Mechanisms	Ref.
Neutrophils and Monocytes		
Rapid response	PMA	63
	LTB_4	64
	Thrombin	65
	LTC_4/LTD_4	66
Delayed response	IL-1	63,69,70
	TNF	69,71
	LPS	63,69
Lymphocytes		
	IL-1	80
	TNF	88
	LPS	88
	IFN-γ	81,82

A. Phagocyte Adherence

Investigations using intravital microscopy suggest that the endothelium of microvessels in vivo is normally "leuko-resistant". Leukocytes appear to roll along the vessel wall, occasionally hesitating, but rarely if ever "sticking" firmly.[60] With tissue damage or inflammation, however, the situation is rapidly and dramatically altered.[2] Over a period of minutes, neutrophils adhere to the vessel wall adjacent to the site of injury, emigrate between the endothelial cells and migrate through subendothelial matrix to the site of injury. It has long been a matter of controversy whether this acute adhesive interaction is mediated by changes in the endothelium or the leukocyte. Cohnheim (1889) emphasized a "molecular change of the vessel wall",[61] whereas Metchnikoff (1893) stressed the influence of soluble factors on the leukocyte.[62] It now appears likely that both leukocyte and endothelial cells are modified during the acute inflammatory reaction. The importance of leukocyte glycoproteins in phagocyte adherence to endothelium has been convincingly established by the "experiment of nature", CD11/CD18 deficiency. Increasing evidence from in vitro studies indicates that "molecular alterations" in the endothelial cell surface may also augment leukocyte adherence.

Cultured human endothelial cells (HEC) respond to a variety of stimuli by changes in surface membrane components that result in increased adhesiveness for leukocytes. There appear to be both rapid and delayed responses (Table 1). Rapid responses produce an increase in leukocyte binding within minutes of stimulation and are maximal in less than 1 hr. This pattern is independent of RNA and protein synthesis and has been observed after pretreatment of HEC with PMA,[63] LTB_4,[64] thrombin,[65] and LTC_4 or LTD_4.[66] PMA[34] and LTB_4[67] also increase neutrophil adherence by direct stimulation of the neutrophil, i.e., when added with the neutrophil without pretreatment of HEC.[1] Alpha-thrombin and LTC_4 or LTD_4 are of particular interest since they do not act directly on the neutrophil, but increase adherence only with pretreatment of HEC.[65,66] The nature of the adherence-promoting surface change induced by these agents remains unknown. Some studies suggested that endothelial surface expression of platelet-activating factor (PAF) is the means by which α-thrombin,[65] LTC_4, and LTD_4[66] augment endothelial cell adhesiveness since PAF expression and increased adhesiveness occur simultaneously. PMA and LTB_4, however, do not induce PAF expression in HEC,[66,68] but do increase adhesiveness.

The delayed form of surface proadhesive activity induces maximal leukocyte binding at 4 to 6 hr, and gradually declines over the ensuing 24 hr. It requires synthesis of both RNA and protein by HEC. This pattern of response is observed after HEC treatment with lipopolysaccharide (LPS),[63,69] interleukin-1 (IL-1),[63,69,70] and tumor necrosis factor (TNF).[69,71] IL-1 does not act directly on the leukocyte to augment adherence.[70] In contrast, both LPS and TNF can also augment neutrophil adherence to HEC by a direct effect on the neutrophil.[71]

The endothelial cell surface proadhesive activity induced by LPS, IL-1, or TNF acts in part via the CD11/CD18 complex since MoAb 60.3 (directed to a function-associated epitope on the β-chain polypeptide) incompletely inhibits the increase in neutrophil adherence.[69] Reduced binding is also observed in assays using IL-1-, LPS- or TNF-pretreated HEC and CD11/CD18-deficient neutrophils.[69] A MoAb to the α_M heterodimer inhibits the increased adhesion observed with PMA-treated neutrophils, but has no effect on the increased leukocyte binding observed after HEC pretreatment with IL-1, TNF, or LPS.[69] This finding suggests that the leukocyte and endothelial cell contributions to the adhesive interaction involve different domains within the CD11/CD18 complex. It is worth emphasizing, however, that the adherence of neutrophils to Il-1-, LPS-, or TNF-pretreated HEC is still significantly greater than their adherence to untreated HEC, even if neutrophils lack the CDw18 complex or have been treated with anti-CD18 MoAbs.

The endothelial cell surface molecule(s) induced by IL-1, TNF, or LPS has not been identified. The requirement for RNA and protein synthesis implies that the surface activity is a protein, but endothelial cell synthesis of PGI_2[72] and PAF[73] is also prevented by RNA and protein synthesis inhibitors. In addition, it is not yet known whether the adhesion molecules induced in rapid responses are the same as those induced in slower responses.

MoAb H4/18 recognizes a surface antigen induced on HEC by IL-1 and TNF.[74] Its expression after stimulation follows a similar time course as the induction of the proadhesive activity, and also requires RNA synthesis and protein synthesis. MoAb H4/18 partially inhibits HL-60 binding to IL-1-pretreated HEC, but it remains uncertain whether the antigen recognized by H4/18 and the proadhesive activity reside on the same molecule.

DiCorleto and de la Motte showed that U937 cells and peripheral blood monocytes (but not neutrophils) adhered preferentially to subconfluent or wounded endothelial cell monolayers.[44] Increased surface adhesivity for monocytes in migrating or proliferating endothelial cells may represent another endothelial-dependent mechanism of particular importance to wound healing.

B. Lymphocyte Adherence

While phagocytes emigrate through postcapillary venules with flattened endothelial morphology similar to that of the rest of the vascular tree, some lymphocytes emigrate through venules with cuboidal endothelium. Lymphocytes bind to the HEV of lymphoid tissue via leukocyte surface glycoproteins (e.g., MEL-14 antigen) that recognize tissue specific determinants on HEV. The surface molecules on HEV recognized by lymphocyte homing receptors have not yet been identified. A MoAb (MECA 325) that recognizes HEV specifically has recently been reported, although the MoAb does not inhibit lymphocyte binding to HEV.[57]

HEV of lymph nodes become thin and flattened if afferent vessels are ligated.[75,76] The HEV phenotype is observed in vivo in delayed hypersensitivity reactions and at sites of chronic inflammation.[77-79] These observations suggest that the HEV phenotype is not constitutive, but is instead induced at sites of chronic inflammation by cytokines from macrophages or lymphocytes. Pretreatment of cultured HEC with macrophage-derived IL-1[80] or TNF[88] increases lymphocyte adherence. Of note, enhanced lymphocyte binding to IL-1-[48] or TNF-pretreated[88] HEC is not inhibited by anti-CD18 MoAb. The nature of this CD18- and endothelial cell-dependent mechanism of lymphocyte adherence is not known, but it may account for the normal lymphocyte traffic and intact cell-mediated immunity oberved in patients whose leukocytes lack CD11/CD18.

T-lymphocyte-derived immune interferon (IFN-gramma) also augments lymphocyte binding to endothelium in vitro. Treatment of cultured HEC with IFN-gamma increases the adherence of T-lymphocytes[81] but not neutrophils.[63] Increased binding may also result from interaction of the lymphocyte CD3 determinant and the endothelial cell Class II MHC antigens (induced by IFN-gamma).[82] IFN-gamma may also induce the HEV phenotype, since treatment of HEC with IFN-gamma (but not IL-1 or IL-2) causes an increased binding of the HEV-specific MoAb MECA 325.[57]

Springer and colleagues[83] have recently identified another membrane glycoprotein involved in adhesive interactions. This intercellular adhesion molecule-1 (ICAM-1) is expressed on hematopoietic and nonhematopoietic cells including endothelial cells. ICAM-1 is a single-chain polypeptide with M_r heterogeneity in different cell types. A MoAb to ICAM-1 (RR1/1) inhibits lymphoblast binding to dermal fibroblasts by an effect on the fibroblast rather than on the lymphoblast.[84] ICAM-1 expression on endothelial cells is increased by IL-1 and TNF. A possible role for ICAM-1 in lymphocyte diapedesis is suggested by the strong expression of ICAM-1 on vascular endothelial cells in T-lymphocyte areas, e.g., tonsils, lymph nodes, and delayed hypersensitivity reaction in skin.[84]

IV. CONCLUSION

Our understanding of the molecular biology to leukocyte/endothelial adherence has recently undergone considerable expansion. Recent studies have identified a lymphocyte homing receptor involved in lymphocyte recirculation, and a phagocyte membrane complex that mediates phagocyte adherence and emigration at sites of inflammation. Cell culture studies have demonstrated that the adhesiveness of the endothelial cell surface for leukocytes can be increased by inflammatory mediators. Further insights into the mechanisms of leukocyte adherence to endothelium may provide new approaches to the therapy of inflammatory and immune disorders.

ACKNOWLEDGMENTS

Supported by grants HL 18645 from USPHS and a grant from R. J. Reynolds Industries, Inc.

We thank Dr. Wayne Jack Wallis for his helpful comments and Lou Limtiaco for her skillful word processing.

REFERENCES

1. **Harlan, J. M.,** Leukocyte-endothelial interactions, *Blood,* 65, 513, 1985.
2. **Grant, L.,** The sticking and emigration of white blood cells in inflammation, in *The Inflammatory Process,* Vol. II, Zweifach, B. W., Grant, L., and McCluskey, R. T., Eds., Academic Press, Orland, Fla., 1973, 205.
3. **Todd, R. F., III and Arnaout, M. A.,** Monoclonal antibodies that identify Mo1 and LFA-1, two human leukocyte membrane glycoproteins: a review, in *Leukocyte Typing II,* Reinherz, E. L., Haynes, B. S., Nadler, L. M., and Bernstein, I. D., Eds., Springer-Verlag, Berlin, 1985, 95.
4. **Shaw, S.,** Characterization of human leukocyte differentiation antigens *Immunol. Today* 8, 1, 1987.
5. **Sanchez-Madrid, F., Nagy, J. A., Robbins, E., Simmon, P., and Springer, T. A.,** A human leukocyte differentiation antigen family with distinct α-subunits and a common β-subunit: the lymphocyte function-associated antigen (LFA-1), the C3bi complement receptor (OKM1/Mac-1), and p150,95 molecule, *J. Exp. Med.,* 158, 1785, 1983.
6. **Davignon, D., Martz, E., Reynolds, T., Kurzinger, K., and Springer, T. A.,** Lymphocyte function-associated antigen 1 (LFA-1): a surface antigen distinct from Lyt-2,3 that participates in T lymphocyte-mediated killing, *Proc. Natl. Acad. Sci. U.S.A.,* 78, 4535, 1981.

7. **Ho, M. K. and Springer, T. A.,** Mac-1 antigen: quantitative expression in macrophage populations and tissues, and immunofluorescent localization in spleen, *J. Immunol.,* 128, 2281, 1982.

8. **Todd, R. F., Van Agthovern, A., Schlossman, S. F., and Terhorst, C.,** Structural analysis of differentiation antigens Mo1 and Mo2 on human monocytes, *Hybridoma,* 1, 329, 1982.

9. **Wright, S. D., Rao, P. E., Van Voorhis, W. C., Craigmyle, L. S., Lida, K., Talle, M. A., Westberg, E. F., Goldstein, G., and Silverstein, S. C.,** Identification of the C3bi receptor of human monocytes and macrophages by using monoclonal antibodies, *Proc. Natl. Acad. Sci. U.S.A.,* 80, 5699, 1983.

10. **Lanier, L. L., Arnaout, M. A., Schwarting, R., Warner, N. L., and Ross, G. D.,** p150, 95, third member of the LFA-1/CR$_3$ polypeptide family identified by anti-Leu M5 monoclonal antibody, *Eur. J. Immunol.,* 15, 713, 1985.

11. **Springer, T. A., Miller, L. J., and Anderson, D.,** p150,95, the third member of the Mac-1, LFA-1 human leukocyte adhesion glycoprotein family, *J. Immunol.,* 136, 240, 1986.

12. **Boxer, L. A., Hedley-Whyte, E. T., and Stossel, T. P.,** Neutrophil actin dysfunction and abnormal neutrophil behavior, *N. Engl. J. Med.,* 291, 1093, 1974.

13. **Hayward, A. R., Leonard, J., Wood, C. B. S., Harvey, B. A. M., Greenwood, M. C., and Soothill, J. F.,** Delayed separation of the umbilical cord, widespread infections, and defective neutrophil mobility, *Lancet,* 1, 1099, 1979.

14. **Crowley, C. A., Curnutte, J. T., Rosin, R. E., Andre-Schwartz, J., Gallin, J. I., Klempner, M., Snyderman, R., Southwick, F. S., Stossel, T. P., and Babior, B.,** An inherited abnormality of neutrophil adhesion: its genetic transmission and its association with a missing protein, *N. Engl. J. Med.,* 302, 1163, 1980.

15. **Abramson, J. S., Mills, E. L., Sawyer, M. K., Regelmann, W. R., Nelson, J. D., and Quie, P. G.,** Recurrent infections and delayed separation of the umbilical cord in an infant with abnormal phagocytic cell locomotion and oxidative response during particle phagocytosis, *J. Pediatr.,* 101, 932, 1981.

16. **Bissenden, J. G., Haeney, M. R., Tarlow, M. J., and Thompson, R. A.,** Delayed separation of the umbilical cord, severe widespread infections, and immunodeficiency, *Arch. Dis. Child.,* 56, 397, 1981.

17. **Bowen, T. J., Ochs, H. D., Altman, L. C., Price, T. H., Van Epps, D. E., Brautigan, D. L., Rosen, R. E., Perkins, W. D., Babior, B. M., Klebanoff, S. J., and Wedgwood, R. J.,** Severe recurrent bacterial infections associated with defective adherence and chemotaxis in two patients with neutrophils deficient in a cell-associated glycoprotein, *J. Pediatr.,* 101, 932, 1982.

18. **Arnaout, M. A., Pitt, J., Cohen, H. J., Melamed, J., Rosen, F. S., and Colten, H.,** Deficiency of a granulocyte-membrane glycoprotein (gp150) in a boy with recurrent bacterial infections, *N. Engl. J. Med.,* 306, 693, 1982.

19. **Davies, E. G., Isaacs, D., and Levinsky, R. J.,** Defective immune interferon production and natural killer activity associated with poor neutrophil mobility and delayed umbilical cord separation, *Clin. Exp. Immunol.,* 50, 454, 1982.

20. **Fischer, A., Descamps-Latscha, B., Gerota, I., Scheinmetzler, C., Virelizier, J. L., Trung, P. H., Lisowska-Grospierre, B., Perez, N., Durandy, A., and Griscelli, C.,** Bone-marrow transplantations for inborn error of phagocytic cells associated with defective adherence, chemotaxis, and oxidative response during opsonised particle phagocytosis, *Lancet,* 2, 473, 1983.

21. **Kobayashi, K., Fujita, K., Okino, F., and Kajii, T.,** An abnormality of neutrophil adhesion: autosomal recessive inheritance associated with missing neutrophil glycoproteins, *Pediatrics,* 73, 606, 1984.

22. **Weisman, S. J., Berkow, R. L., Plautz, G., Torres, M., McGuire, W. A., Coates, T. D., Haak, R. A., Floyd, A., Jersild, R., and Baehner, R. L.,** Glycoprotein-180 deficiency: genetics and abnormal neutrophil activation, *Blood,* 65, 696, 1985.

23. **Buescher, E. S., Gaither, T., Nath, J., and Gallin, J. I.,** Abnormal adherence-related functions of neutrophils, monocytes, and Epstein-Barr virus-transformed B cells in a patient with C3bi receptor deficiency, *Blood,* 65, 1382, 1985.

24. **Ross, G. D., Thompson, R. A., Walport, M. J., Springer, T. A., Watson, J. V., Ward, R. H. R., Lida, J., Newman, S. L., Harrison, R. A., and Lachman, P. J.,** Characterization of patients with an increased susceptibility to bacterial infections and a genetic deficiency of leukocyte membrane complement receptor type 3 and the related membrane antigen LFA-1, *Blood,* 66, 882, 1985.

25. **Anderson, D. C., Schmalsteig, F. C., Finegold, M. J., Hughes, B. J., Rothlein, R., Miller, L. J., Kohl, S., Tosi, M. F., Jacobs, R. L., Waldrop, T. C., Goldman, A. S., Shearer, W. T., and Springer, T. A.,** The severe and moderate phenotypes of heritable Mac-1, LFA-1 deficiency: their quantitative definition and relation to leukocyte dysfunction and clinical features, *J. Infect. Dis.,* 152, 668, 1985.

26. **Fischer, A., Seger, R., Durandy, A., Grospierre, B., Virelizier, J. L., Le Deist, P., Griscelli, C., Fischer, E., Kazatchkine, M., Bohler, M-C., Descamps-Latscha, B., Trung, P. H., Springer, T. A., Olive, D., and Mawas, C.,** Deficiency of the adhesive protein complex lymphocyte function antigen 1, complement receptor type 3, glycoprotein p150,95 in a girl with recurrent bacterial infections, *J. Clin. Invest.,* 76, 2385, 1985.

27. **Anderson, D. C., Schmalstieg, F. C., Arnaout, M. A., Kohl, S., Tosi, M. F., Dana, N., Buffone, G. J., Hughes, B. J., Brinkley, B. R., Dickey, W. D., Abramson, J. S., Springer, T., Boxer, L. A., Hollers, J. M., and Smith, C. W.,** Abnormalities of polymorphonuclear leukocyte function associated with a heritable deficiency of high molecular weight surface glycoproteins (GP138): common relationship to diminished cell adherence, *J. Clin. Invest.,* 74, 536, 1984.

28. **Beatty, P. G., Harlan, J. M., Rosen, H., Hansen, J. A., Ochs, H. D., Price, T. H., Taylor, R. F., and Klebanoff, S. J.,** Absence of monoclonal-antibody-defined protein complex in boy with normal leukocyte function, *Lancet,* 1, 535, 1984.

29. **Dana, N., Todd, R. F., III, Pitt, J., Springer, T. A., and Arnaout, M.,** Deficiency of a surface membrane glycoprotein (Mo1) in man, *J. Clin. Invest.,* 73, 153, 1984.

30. **Arnaout, M. A., Spits, H., Terhorst, C., Pitt, J., and Todd, R. F., III,** Deficiency of a leukocyte surface glycoprotein (LFA-1) in two patients with Mo1 deficiency. Effects of cell activation on Mo1/LFA-1 surface expression in normal and deficient leukocytes, *J. Clin. Invest.,* 74, 1291, 1984.

31. **Springer, T. A., Thompson, W. S., Miller, L. J., Schmalstieg, F. C., and Anderson, D. C.,** Inherited deficiency of the Mac-1, LFA-1, p150,95 glycoprotein family and its molecular basis, *J. Exp. Med.,* 160, 1901, 1984.

32. **Arnaout, M. A., Todd, R. F., III, Dana, N., Melamed, J., Schlossman, S. F., and Colten, H. R.,** Inhibition of phagocytosis of complement C3- or immunoglobulin G-coated particles and of C3bi binding by monoclonal antibodies to a monocyte-granulocyte membrane glycoprotein (Mo1), *J. Clin. Invest.,* 72, 171, 1983.

33. **Schwartz, B. R., Ochs, H. D., Beatty, P. G., and Harlan, J. M.,** A monoclonal antibody-defined membrane antigen complex is required for neutrophil-neutrophil aggregation, *Blood,* 65, 1553, 1985.

34. **Harlan, J. M., Killen, P. D., Senecal, F. M., Schwartz, B. R., Yee, E. K., Taylor, R. F., Beatty, P. G., Price, T. H., and Ochs, H. D.,** The role of neutrophil membrane glycoprotein GP-150 in neutrophil adherence to endothelium in vitro, *Blood,* 66, 167, 1985.

35. **Wallis, W. J., Beatty, P. G., Ochs, H. D., and Harlan, J. M.,** Human monocyte adherence to cultured vascular endothelium: monoclonal antibody-defined mechanisms, *J. Immunol.,* 135, 2323, 1985.

36. **Price, T., Beatty, P., and Corpuz, S.,** In vivo inhibition of neutrophil function using monoclonal antibody to the CDw18 complex, *Clin. Res.,* 34, 467A, 1986.

37. **Lindbom, L., Lundberg, C., Lundberg, K., Harlan, J., and Arfors, K-E.,** Prevention of experimentally induced neutrophil accumulation in vivo, *Int. J. Microcirc.,* 5, 232A, 1986.

38. **Wallis, W. J., Hickstein, D. D., Schwartz, B. R., June, C. H., Ochs, H. D., Beatty, P. G., Klebanoff, S. J., and Harlan, J. M.,** Monoclonal antibody-defined functional epitopes on the adhesion-promoting glycoprotein complex (CDw18) of human neutrophils, *Blood,* 67, 1007, 1986.

39. **Anderson, D. C., Miller, L. J., Schmalstieg, F. C., Rothlein, R., and Springer, T. A.,** Contributions of the Mac-1 glycoprotein family to adherence-dependent granulocyte functions: structure-function assessments employing subunit-specific monoclonal antibodies, *J. Immunol.,* 137, 15, 1986.

40. **Todd, R. F., III, Arnaout, M. A., Rosin, R. E., Crowley, C. A., Peters, W. A., and Babior, B. M.,** Subcellular localization of the large subunit of Mo1 (Mo1$_\alpha$, formerly gp 110), a surface glycoprotein associated with neutrophil adhesion, *J. Clin. Invest.,* 74, 1280, 1984.

41. **Arnaout, M. A., Hakim, R. M., Todd, R. F., III, Dana, N., and Colten, H. R.,** Increased expression of an adhesion-promoting surface glycoprotein in the granulocytopenia of hemodialysis, *N. Engl. J. Med.,* 312, 457, 1985.

42. **O'Shea, J. J., Brown, E. J., Seligmann, B. E., Metcalf, J. A., Frank, M. M., and Gallin, J. I.,** Evidence for distinct intracellular pools of receptors for C3b and C3bi in human neutrophils, *J. Immunol.,* 134, 2580, 1985.

43. **Petrequin, P. R., Todd, R. F., III, Smolen, J. E., and Boxer, L. A.,** Expression of specific granule markers on the cell surface of neutrophil cytoplasts, *Blood,* 67, 1119, 1986.

44. **DiCorleto, P. E. and de la Motte, C. A.,** Characterization of the adhesion of the human monocytic cell line U937 to cultured endothelial cells, *J. Clin. Invest.,* 75, 1153, 1985.

45. **Pawlowski, N. A., Abraham, E. L., Pontier, S., Scott, W. A., and Cohn, Z. A.,** Human monocyte-endothelial cell interaction *in vitro, Proc. Natl. Acad. Sci. U.S.A.,* 82, 8208, 1985.

46. **Krensky, A. M., Robbins, E., Springer, T. A., and Burakoff, S. J.,** LFA-1, LFA-2 and LFA-3 antigens are involved in CTL-target conjugation, *J. Immunol.,* 132, 2180, 1984.

47. **Mentzer, S. J., Burakoff, S. J., and Faller, D. V.,** Adhesion of T lymphocytes to human endothelial cells is regulated by the LFA-1 membrane molecule, *J. Cell. Physiol.,* 126, 285, 1986.

48. **Haskard, D., Cavender, D., Beatty, P., Springer, T., and Ziff, M.,** T lymphocyte adhesion to endothelial cells: mechanisms demonstrated by anti-LFA-1 monoclonal antibodies, *J. Immunol.,* 137, 2901, 1986.

49. **Ruoslahti, E. and Pierschbacher, M. D.,** Arg-gly-asp: a versatile cell recognition signal, *Cell,* 44, 517, 1986.

50. **Hynes, R. O.,** Integrins: A family of cell surface receptors, *Cell,* 48, 549, 1987.

51. **Charo, I. F. et al.,** Platelet glycoproteins IIb and IIIa: evidence for a family of immunologically and structurally related glycoproteins in mammalian cells, *Proc. Natl. Acad. Sci. U.S.A.,* 83, 8351, 1986.

52. **Kishimoto, T. K. et al.,** Cloning of the subunit of the leukocyte adhesion proteins: homology to an extracellular matrix receptor defines a novel supergene family, *Cell,* 48, 681, 1987.

53. **Gowans, J. L. and Knight, E. J.,** The route of re-circulation of lymphocytes in the rat, *Proc. R. Soc. London Ser.* B, 159, 257, 1964.

54. **Gallatin, M., St. John, T. P., Siegelman, M., Reichert, R., Butcher, E. C., and Weissman, I. L.,** Lymphocyte homing receptors, *Cell,* 44, 673, 1986.

55. **Jalkanen, S., Steere, A. C., Fox, R. I., and Butcher, E. C.,** A distinct endothelial cell recognition system that controls lymphocyte traffic into inflamed synovium, *Science,* 233, 556, 1986.

56. **Gallatin, W. M., Weissman, I. L., and Butcher, E. C.,** A cell-surface molecule involved in organ-specific homing of lymphocytes, *Nature (London),* 304, 30, 1983.

57. **Butcher, E. C., Lewinsohn, D., Duijvestijn, A., Bargatze, R., Wu, N., and Jalkanen, S.,** Interactions between endothelial cells and leukocytes, *J. Cell. Biochem.,* 30, 121, 1986.

58. **Siegelman, M., Bond, M. W., Gallatin, W. M., St. John, T., Smith, H. T., Fried, V. A., and Weissman, I. L.,** Cell surface molecule associated with lymphocyte homing is a ubiquitinated branched-chain glycoprotein, *Science,* 231, 823, 1986.

59. **St. John, T., Gallatin, W. M., Siegelman, M., Smith, H. T., Fried, V. A., and Weissman, I. L.,** Expression and cloning of a lymphocyte homing receptor cDNA: ubiquitin is the reactive species, *Science,* 231, 845, 1986.

60. **Mayrovitz, H., Wiedeman, M., and Tuma, R.,** Factors influencing leukocyte adherence in microvessels, *Thromb. Haemostasis,* 38, 823, 1977.

61. **Cohnheim, J.,** *Lectures in General Pathology,* Vol. 1, 2nd ed., New Sydenham Society, London, 1889.

62. **Metchnikoff, E.,** *Lecture on Comparative Pathology of Inflammation,* Starling, F. A., Trans., Kegan Paul, London, 1893.

63. **Schleimer, R. P. and Rutledge, B. K.,** Cultured human vascular endothelial cells acquire adhesiveness for neutrophils after stimulation with interleukin 1, endotoxin and tumor-promoting phorbol diesters, *J. Immunol.,* 136, 649, 1986.

64. **Hoover, R. L., Karnovsky, M. J., Austen, K. F., Corey, E. J., and Lewis, R. A.,** Leukotriene B_4 action on endothelium mediates augmented neutrophil/endothelial adhesion, *Proc. Natl. Acad. Sci. U.S.A.,* 81, 2191, 1984.

65. **Zimmerman, G. A., McIntyre, T. M., and Prescott, S. M.,** Thrombin stimulates the adherence of neutrophils to human endothelial cells in vitro, *J. Clin. Invest.,* 76, 2235, 1985.

66. **McIntyre, T. M., Zimmerman, G. A., and Prescott, S. M.,** Leukotrienes C_4 and D_4 stimulate human endothelial cells to synthesize platelet-activating factor and bind neutrophils, *Proc. Natl. Acad. Sci. U.S.A.,* 83, 2204, 1986.

67. **Gimbrone, M. A., Jr., Brock, A. F., and Schafer, A. I.,** Leukotriene B_4 stimulates polymorphonuclear leukocyte adhesion to cultured vascular endothelial cells, *J. Clin. Invest.,* 74, 1552, 1984.

68. **Zimmerman, G. A., McIntyre, T. M., and Prescott, S. M.,** Production of platelet-activating factor by human vascular endothelial cells: evidence for a requirement for specific agonists and modulation by prostacyclin, *Circulation,* 72, 718, 1985.

69. **Pohlman, T. H., Stanness, K. A., Beatty, P. G., Ochs, H. D., and Harlan, J. M.,** An endothelial cell surface factor(s) induced in vitro by lipopolysaccharide, interleukin-1, and tumor necrosis factor-α increases neutrophil adherence by a CDw18-dependent mechanism, *J. Immunol.,* 136, 4548, 1986.

70. **Bevilacqua, M. P., Pober, J. S., Wheeler, M. E., Cotran, R. S., and Gimbrone, M. A., Jr.,** Interleukin-1 acts on cultured human vascular endothelium to increase the adhesion of polymorphonuclear leukocytes, monocytes, and related leukocyte cell lines, *J. Clin. Invest.,* 76, 2003, 1985.

71. **Gamble, J. R., Harlan, J. M., Klebanoff, S. J., and Vadas, M. A.,** Stimulation of the adherence of neutrophils to umbilical vein endothelium by human recombinant tumor necrosis factor, *Proc. Natl. Acad. Sci. U.S.A.,* 82, 8667, 1985.

72. **Rossi, V., Brevario, F., Ghezzi, P., Dejana, E., and Mantovani, A.,** Prostacyclin synthesis induced in vascular cells by interleukin-1, *Science,* 229, 174, 1985.

73. **Bussolino, F., Breviario, F., Tetta, C., Aglietta, M., Mantovani, A., and Dejana, E.,** Interleukin-1 stimulates platelet activating factor (1-O-alkyl-2-acetyl-*sn*-glycero-3-phosphocholine) production in cultured human endothelial cells, *J. Clin. Invest.,* 77, 2027, 1986.

74. **Pober, J. S., Bevilacqua, M. P., Mendrick, D. L., Lapierre, L. A., Fiers, W., and Gimbrone, M. A., Jr.,** Two distinct monokines, interleukin 1 and tumor necrosis factor, each independently induced biosynthesis and transient expression of the same antigen on the surface of cultured human vascular endothelial cells, *J. Immunol.,* 136, 1680, 1986.

75. **Hendriks, H. R. and Eestermans, I. L.,** Disappearance and reappearance of high endothelial venules and immigrating lymphocytes in lymph nodes deprived of afferent lymphatic vessels: a possible regulatory role of macrophages in lymphocyte migration, *Eur. J. Immunol.,* 13, 663, 1983.

76. **Hendricks, H. R., Eestermans, I. L., and Hoefsmit, E. C. M.,** Depletion of macrophages and disappearance of postcapillary high endothelial venules in lymph nodes deprived of afferent lymphatic vessels, *Cell Tissue Res.,* 211, 375, 1980.

77. **Nightingale, G. and Hurley, J. V.,** Relationship between lymphocyte emigration and vascular endothelium in chronic inflammation, *Pathology,* 10, 27, 1978.

78. **Freemont, A. J. and Ford, W. L.,** Functional and morphological changes in postcapillary venules in relation to lymphocytic infiltration site BCG-induced granulomata in rat skin, *J. Pathol.,* 147, 1, 1985.

79. **Iguchi, T. and Ziff, M.,** Electron microscopic study of rheumatoid synovial vasculature. Intimate relationship between tall endothelium and lymphoid aggregation, *J. Clin. Invest.,* 77, 355, 1986.

80. **Cavender, D. E., Haskard, D. O., Joseph, B., and Ziff, M.,** Interleukin 1 increases the binding of human B and T lymphocytes to endothelial cell monolayers, *J. Immunol.,* 136, 203, 1986.

81. **Yu, C.-L., Haskard, D. O., Cavender, D., Johnson, A. R., and Ziff, M.,** Human gamma interferon increases the binding of T lymphocytes to endothelial cells, *Clin. Exp. Immunol.,* 62, 554, 1985.

82. **Masuyama, J.-I., Minato, N., and Kano, S.,** Mechanisms of lymphocyte adhesion to human vascular endothelial cells in culture. T lymphocyte adhesion to endothelial cells through endothelial HLA-DR antigens induced by gamma interferon, *J. Clin. Invest.,* 77, 1596, 1986.

83. **Rothlein, R., Dustin, M. L., Marlin, S. D., and Springer, T. A.,** A human intercellular adhesion molecule (ICAM-1) distinct from LFA-1, *J. Immunol.,* 137, 1270, 1986.

84. **Dustin, M. L., Rothlein, R., Bhan, A. K., Dinarello, C. A., and Springer, T. A.,** Induction of IL-1 and interferon-γ: tissue distribution, biochemistry, and function of a natural adherence molecule (ICAM-1), *J. Immunol.,* 137, 245, 1986.

85. **Ochs, H.,** personal communication, University of Washington, Seattle.

86. **Diener, A. and Harlan, J.,** unpublished observation.

87. **Schwartz, B., Ginsberg, M., Thiagarajan, P., and Harlan, J.,** unpublished observations.

88. **Schwartz, B. and Harlan, J.,** unpublished observations.

89. **Schwartz, B. and Harlan, J.,** unpublished observations.

Immunologic Factors

Chapter 25

COMPLEMENT AND IMMUNOGLOBULIN LIGAND-RECEPTORS ON PULMONARY ENDOTHELIAL CELLS

Duane R. Schultz and Una S. Ryan

TABLE OF CONTENTS

I. INTRODUCTION

A natural question for investigations of ligand-receptors on lung vascular endothelium is their physiologic relevance in normal conditions and in pulmonary injury. Endothelial cells are a barrier between the circulation and the vessel wall. In various immunopathologic processes where the lung is a target for the deposition of circulating immune complexes, the role of endothelial cell receptors is poorly understood. In this chapter we will describe and assess the information available on the interaction of two proteins in the complement (C) system and immunoglobulin with receptors on endothelium and other cells. Although we and others have described receptors on endothelial cells for subcomponent C1q of macromolecular C1 and C3b of the C system as well as the Fc region of IgG,[1-9] the chemistry, topology, and biomedical functions of these receptor molecules have not been fully elucidated.[10,11] In particular, the question remains open as to whether these receptors are obligate for damage in acute pulmonary injury and in conditions leading to extensive microvascular injury following the deposition of immune complexes with or without bound C components in the lung.

At present, much of the biochemical and immunological data on ligand-receptors for C components and Fc-IgG have come from studies of a variety of somatic cells from many different species, and it is not known whether all of these data apply to endothelium. Nevertheless, we will draw on the studies of receptors on other cells when information pertaining directly to endothelial cells is lacking or incomplete.

II. THE FIRST COMPONENT OF COMPLEMENT (C1)

C1, the first component of the classical C pathway, is a macromolecular complex,[12] and consists of one molecule of subcomponent C1q, a $C1r_2$ dimer, a $C1s_2$ dimer, and calcium ions.[13-16] Native pentamolecular C1 exists in plasma in a inactive precursor form.[17] Although both $C1r_2$ and $C1s_2$ are glycoproteins, they are antigenically and functionally distinct proenzymes which exist as single chains of molecular weight 85 kdaltons. Equimolar concentrations of the C1r and C1s form a long flexible asymmetric structure called the $C1r_2$-$C1s_2$ tetramer,[13] and subcomponent C1q contains the attachment sites for the weakly binding tetramer (Figure 1).

The subcomponent C1q, an 11S, 460-kdalton mol wt cationic glycoprotein[18,19] has no known enzymic activity, and is the subunit within the C1 macromolecule which recognizes and combines reversibly with the immunoglobulin portion of antibody-antigen complexes. The binding of C1 occurs on sites of the Fc portion of IgG (C_H2 domain) or on the $(Fc)_5$ disk of IgM (C_H4 domains), and this is the initial event in the activation of the classical C pathway.[14,20,21] From electron microscopy studies, we know that C1q is an unusual structure composed of six peripheral globular heads, each of which is joined by a flexible connecting strand to a fibril-like central core.[22-24] Each globular head contains the carboxyterminal ends of three similar but distinct polypeptide chains termed A, B, and C of approximately 20 to 22 kdaltons that occur six times in the molecule. The structure consists of an A-B heterodimer noncovalently linked with a C polypeptide chain. The A and B polypeptide chains are linked by a single disulphide bond near the central core, and the C polypeptide chain is disulphide linked with the C polypeptide chain of an adjacent subunit. Thus, one molecule of C1q contains six copies (three disulphide bonded pairs of A,B,C chain units) of each individual chain.[25,26] The strands and central core are characterized by an amino acid sequence and carbohydrate composition characteristic of collagen, and this makes up approximately 40% of the molecule. Thus, the collagenous portion has an unusually high glycine composition, contains hydroxylysine, hydroxyproline, and galactose-glucose disaccharides linked to hydroxylysine, and is susceptible to collagenase.[27] The six globular heads are

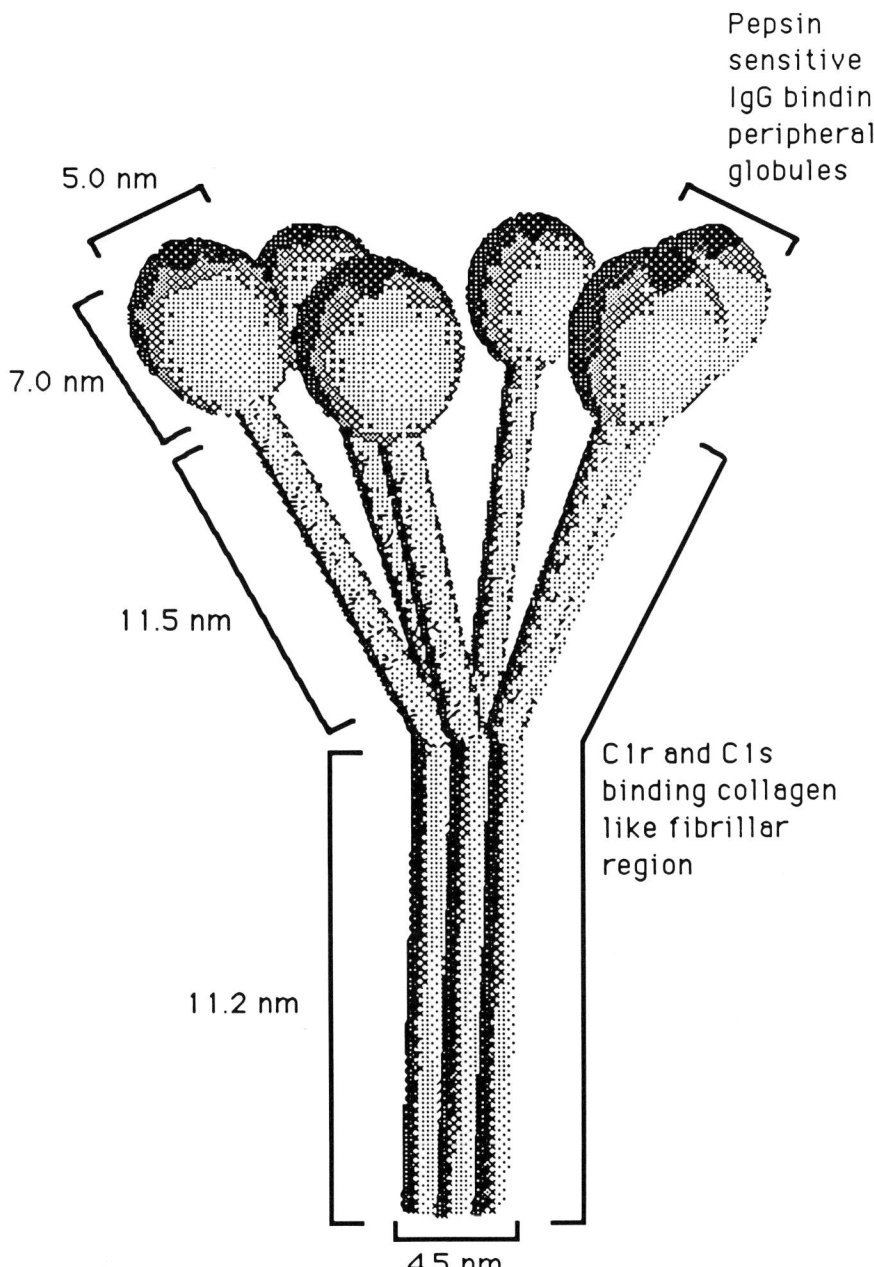

Pepsin
sensitive
IgG binding
peripheral
globules

5.0 nm

7.0 nm

11.5 nm

11.2 nm

4.5 nm

C1r and C1s
binding collagen
like fibrillar
region

FIGURE 1. A pictorial representation of the structure of Clq. The six globular heads contain the binding sites for immunoglobulin. The collagen-like portion is located in the stem-like tail region and bears the C1r and C1s binding sites.

the portion of the molecule which bind to the antibody in immune complexes, and it appears that each of the six heads possesses a binding site for immunoglobulins.[28-30] The collagen-like amino terminal connecting strands of Clq form a helical configuration, and here is inserted the tetramer $C1r_2$-Ca^{++}-$C1s_2$;[25,31] each end of the tetramer is thought to be curved around two opposite Clq strands, allowing the distal catalytic domains of the two C1s molecules to be in contact with the central domains of $C1r_2$.[32-34] Binding of C1 via the globular heads of Clq to Ig usually leads to activation of the molecule, although there are cases where binding occurs without activation.[35] The Clr_2 and Cls_2 proenzymes after activation are converted to serine proteases. The active serine protease site results from a cleavage of the molecules into heavy (57 kdaltons) and light (28 kdaltons) chains held together by a disulphide bond.[16] The active sites are present on the light chain.

Native Cl is activated by many substances in the absence of antibody, including certain bacterial capsular and viral components, lipopolysaccharides, many polyanionic substances and phlogistic agents, polysaccharides, and small synthetic chemically defined saccharides linked to carriers.[36-40] Native Cl is a very labile molecule which is readily dissociated into its subunits simply by dilution of serum.[41] With the use of immune aggregates[42] or nucleic acids,[43] the binding of subcomponent Clq appears to be mostly ionic.

Binding of Clq to immune complexes results in conformational changes within the Clq molecule. The change in structure was shown by binding studies with a polysaccharide[44] and with monoclonal antibodies.[45] The binding of Clq is quite weak with monomeric ligands but much stronger with associated or polymerized ligands.[46,47] Before native Cl binds to an activating substance, spontaneous activation is inhibited through a noncovalent interaction with a 100-kdalton glycoprotein, the Cl inhibitor. Without the inhibitor, spontaneous activation of Cl occurs with a half-life of approximately 4 min at 37°C.[48,49]

Although a great deal of information exists on the molecular and biochemical mechanisms of Cl activation, it is still the subject of considerable research and theoretical discussions.[24,32-34,50,51] The currently proposed mechanism of how sufficient quantities of subcomponent Clq are liberated from the Clr_2-Cls_2 tetramer so that it is available in vivo is biomedically important when addressing the Clq receptor. In addition, the region of free Clq (i.e., globular heads or collagenous tail) which binds to receptor sites on endothelial cells and a variety of somatic cells is discussed below, together with other aspects of the Clq receptor.

III. Clq AND THE Clq RECEPTOR: BACKGROUND

A. Liberation of Subcomponent Clq

The first problem that must be addressed is the concentration of uncombined Clq not associated with macromolecular Cl that is available to bind to receptors in biological fluids. Only negligible amounts of free Clq are ordinarily present in normal plasma because almost all of the molecules are found as part of the Cl macromolecular complex.[52] Thus, if a Clq receptor binds specifically to the collagenous tail portion of Clq and not the globular heads (see below), the tail is apparently not available for binding to the receptor when the Clr_2-Cls_2 tetramer is linked to this region.[9,53,54] However, a mechanism has been described which results in the liberation of free Clq.[48,55,56] Following the activation of Cl, the Cl inhibitor rapidly disassembles the Clr_2 and Cls_2 subunits from the macromolecular complex so that free Clq is now available to interact with other molecules via either the globular or collagenous end. It was demonstrated that complexes of Clr_2 and Cls_2 and the Cl inhibitor together with free Clq signify activation of macromolecular Cl.[48,55,56] Both Clr_2 and Cls_2 are functionally inactivated after binding to the Cl inhibitor. If macromolecular Cl is bound to circulating or stationary immune complexes or other activating substances, after Clr_2 and Cls_2 are released into the fluid phase, the subcomponent Clq remains associated with the activator. It was calculated by Ziccardi that a continuous equilibrium between free Clq and Clq associated with Clr_2-Cls_2 could result in the availability of 25 to 30% of the Clq as free subcomponent, provided no other serum factors affect the reaction.[57]

There also is evidence in some diseases such as primary biliary cirrhosis,[58] rheumatoid arthritis,[59-61] and other disorders,[62] that free Clq not in complex with Clr_2-Cls_2 together with Cl inhibitor-Clr, -Cls complexes are present in sera and other biological fluids, suggesting Cl activation. For example, free and functionally active Clq was found in the synovial fluid of patients with rheumatoid arthritis together with low concentrations of functionally active C1 and low C4 activities.[61] These studies suggested that C1q would be free to interact with the C1q receptor on cells such as polymorphonuclear leukocytes in the environment of a joint to stimulate biologic functions of these cells that may or may not benefit the patient. Accordingly, subcomponent C1q has been shown to stimulate an oxidative response in PMN leukocytes,[63] and there is evidence that relates the production of toxic oxygen products by activated PMNs to tissue injury.[64]

B. Methods of Assay

Three of the most common methods to show that subcomponent C1q binds to the surface of different cells are rosette assays, fluorescent methods, and radiochemical assays. An example of the first method is the rosette assay described by Gabay et al.[54] It is based on the observation by Hughes-Jones[65] that purified, radioiodinated C1q binds specifically in an antibody-independent reaction to glutaraldehyde-treated erythrocytes. The binding sites involved in the reaction between C1q and the treated red cells appeared to involve the globular heads, the same portion of the molecule that binds to sites of the Fc region of IgG. The rosette assay developed by Gabay et al.[54] with glutaraldehyde-treated bovine red cells and human C1q (EC1q) exposed the collagenous portion of C1q, and was a useful assay for testing cell receptors that bind C1q via the aminoterminal collagen-like region.

A second method for detecting C1q uses immunofluorescence. A technique was published by Linder,[5,66] who first reacted sections of various tissues with diluted normal human serum or isolated C1q, and binding was detected with fluorescein isothiocyanate-labeled (FITC) anti-C1q and microscopy. The method was used to show that C1q, C4, and C3 bind to the endothelium of vessel walls in an apparent antibody-independent reaction. Loos et al.[67] used FITC-labeled anti-C1q and lissamine rhodamine B 200 (RB-200)-labeled anti-C1s to locate C1q and C1s in rectum biopsy material and in monolayers of peritoneal macrophages in culture. This double staining method in contrasting colors was useful to show that the subcomponents of macromolecular C1 are synthesized independently of each other.

A third method to show that C1q binds to receptors on cell surfaces is radioiodination. Most studies of C1q receptors have relied on this method to show that ^{125}I-C1q binds to a variety of cells. Since there are many different methods for radiolabeling proteins, the selection of a method that maintains the binding capacity and the functional activity and causes no aggregation of C1q is an important consideration. Heusser et al.[68] compared the chloramine T[69] and lactoperoxidase[70] methods of radioiodination. With the lactoperoxidase method, C1q was iodinated to 0.2 atoms I per molecule with no loss of hemolytic activity. With the chloramine T method, 50% of the functional activity was lost with 0.02 atoms I per molecule. However, neither method impaired the ability of the labeled C1q to bind to immune complexes.

Tenner et al.[71] compared five different methods for radiolabeling C1q, including two that label tyrosine, two that label the epsilon amino group of lysine and terminal amino groups, and a method that labels the terminal sialic acid residue of carbohydrate groups. Some of the data was useful for determining whether C1q "heads" or "tails" bind to a specific receptor. For example, the lactoperoxidase-glucose oxidase method[70] selectively labeled the globular head regions of the C polypeptide chain of C1q. Heusser et al.[68] earlier reported essentially the same results. By ^{3}H borohydride reduction, the A polypeptide chain was predominantly labeled with ^{3}H in the globular region.[71] We have determined that when purified human C1q was labeled with ^{125}I by the use of Iodo-beads,[72] most of the radiolabel was associated with the globular subunits of C1q and not the collagenous portion.[9] Thus, by enzymic digestion of the collagenous portion of C1q with collagenase and separating the digest from the globular heads by chromatography, the radiolabeled heads were shown not to be the portion of the molecule which bound to a C1q receptor on lung endothelial cells.[9]

C. Significance of Ionic Strength

A conceptual problem with discussions of the biologic importance of the Clq receptor on cells is that very little or no binding of subcomponent Clq occurred in media at normal ionic strength (u = 0.15), and hypotonic media had to be used for experiments. It is well known that macromolecular Cl activity is optimal at ionic strengths less than normal,[73] and Cl and Clq activities which depend on binding via the globular heads are usually assayed in hemolytic assays at an ionic strength of 0.065 and pH 7.4. In the original report of Sobel and Bokisch[74] describing the Clq receptor on peripheral human lymphocytes and human lymphoblastoid cells, binding was optimal between μ = 0.06 to 0.08 and pH 6 to 7.5. In the report by Tenner and Cooper[53] describing monomeric Clq binding via the collagenous region by human mononuclear leukocytes, most of their experiments were carried out at μ = 0.09; almost no binding occurred at μ = 0.15 and pH 7.2. However, with multi-site and/or cross-linking of the Clq, it bound readily to cellular receptors at μ = 0.15.[54,75] When Clq is bound to immune complexes, it is probably presented to cells such as endothelium as an array or aggregate of forms, and this multi-site attachment would greatly increase its avidity to receptors. Although human ^{125}I-Clq bound slightly better to bovine endothelial cells in culture at low ionic strength compared to normal, Zhang et al.[9] reported that adequate binding occurred at μ = 0.15.

D. Biologic Activities

Relatively little information is available on the biological activity of the Clq receptor, and only a few reports have addressed this subject. A hypothesis by Gupta et al.[76] based on their experiments and earlier work by Sundqvist et al. with Clq[77] stated that receptors found on populations of B and T lymphocytes may play a role in physiologic functions of these cells depending on the binding of soluble immune complexes via C ligand-receptors to the cell surface. This stemmed from the fact that binding of aggregated IgG to lymphocytes or Raji and Molt 4 lymphoblastoid cell lines was enhanced by preincubation of the cells with Clq.[76,77]

A report by Ghebrehiwet and Müller-Eberhard[78] described Clq acting in an antibody-independent cytotoxicity reaction. Chromium-labeled chicken erythrocytes as target cells bearing bound Clq were lysed by lymphocytes or lymphoblastoid cells in a 20-hr cytolytic assay. Although the nature of the ligand was unknown, the cells appeared to be lysed by a Clq-mediated cellular reaction.

In another study of a biological function, Tenner and Cooper[63] stimulated a respiratory burst (oxidative metabolism) in human polymorphonuclear leukocytes with latex beads coated with purified Clq. The burst was measured by chemiluminescence and NADPH-oxidase-dependent hexose monophosphate shunt activity. Activated products of oxygen in stimulated leukocytes are accompanied by emission of energy in the form of light (chemiluminescence). In the case of polymorphonuclear leukocytes, the generation of chemical light occurs when singlet oxygen ($^1O_2$) decays to ground state triplet O_2. The chemiluminescent response of the PMNs to aggregated IgG was enhanced two- to tenfold by purified Clq, and the response to the Clq-coated beads was partially inhibited by free monomeric Clq. The possibility was discussed that Clq has a role in the PMN-mediated destruction and clearance of factors which activate the classic C pathway.

E. The Clq Receptor

Although Clq is the recognition unit of the classical C pathway, it also serves as a ligand for a number of cellular receptors. The first report to show that Clq binds to cell membranes is ascribed to Dickler and Kunkel in 1972,[79] who located the protein by staining lymphocytes with fluoresceinated anti-Clq. In 1975, Sobel and Bokisch[74] were the first to describe a receptor for human Clq on peripheral human lymphocytes and lymphoblastoid cells by immunofluorescent methods as well as with radioiodinated Clq. The latter was inhibited from binding to its receptor by unlabeled Clq, but not inhibited by unrelated proteins. The receptor was saturable, although they had to use hypotonic medium. After treatment of cells with trypsin, binding was reduced by approximately 50%.

A number of studies were published describing the action of Cl or Clq with platelets. These include a report by Cazenave et al.[80] showing that native Clq inhibited both collagen-induced platelet aggregation and the release of radiolabeled serotonin from prelabeled platelets; heated Clq (56°C, 30 min) was inactive. Similar experiments by Suba and Csako[81] showed that Clq added to human platelets did not induce aggregation, but it inhibited the aggregation of the platelets by collagen. They suggested that platelets had a specific surface receptor that bound Clq or collagen in a competitive reaction. From the experiments of Wautier et al.[82] it was suggested that Clq was attached to platelet-bound subcomponent Cls by its collagenous region, leaving the globular heads to react with aggregated IgG. In a later report, Wautier et al.[83] attempted to locate the region of Clq which inhibited the collagen-induced adhesion and aggregation of platelets. Evidence was obtained that the active area was located on the A chain. The latter is more highly charged than the B and C chains of Clq, containing twice the number of basic residues, and the A chain is also the most highly glycosylated of the three chains.[84]

Compared to other types of cells, relatively few papers have been published describing Clq as a ligand for receptors on endothelial cells. In 1979 Shadforth et al.[4] and later Andrews et al.[6] investigated endothelial cells in tissue sections of human umbilical vein and artery, and cultured, viable umbilical vein endothelial cells. They showed that a Clq surface receptor was located on these cells. Not only did Clq bind to endothelial cells in umbilical vein and artery, but also the endothelial basement membrane. At physiologic ionic strength, Raji cells did not form rosettes with erythrocytes coated with antibody and Clq (EAClq). However, the EAClq bound to endothelial cells at physiologic ionic strength, but the binding was increased at lower ionic strength. They found that directly conjugating Clq with fluorochrome significantly reduced its binding to the receptor. With human endothelial cells in culture, immunofluorescence was used to visualize surface patching and capping of Clq at 37°C which diminished after incubating the cells at 4°C.

As noted earlier, the binding of Clq to endothelium was also investigated by Linder and co-workers[3,5] in human tissue frozen sections and cultured cells treated with the nonionic detergent NP40 so that the cells were permeable to serum proteins. Cytoskeletal intermediate filaments of the cells were the binding sites of either Clq in diluted human serum or purified Clq as well as C4 and C3 in an antibody-independent mechanism of deposition. Immunoglobulins and fibrin did not bind to the filaments. No C-component binding occurred if the cells were not treated with detergent. The cytoskeletal sites which bound Clq on these cells were destroyed following treatment with trypsin.

In 1980, Tenner and Cooper[53] reported quantitative binding studies with radioiodinated human Clq and human peripheral blood mononuclear cells. Approximately 26% of the cell population bound the Clq. The reaction was specific, saturable at ionic strength 0.09, reversible, and of moderately high affinity. Binding of Clq to the receptor occurred via the collagenous region of the molecule, and to show specificity, type I collagen inhibited the binding reaction. The Clq receptor was distinct from the C3b, Fc-IgG and fibronectin receptors on these cells. In a later study,[75] monocytes, B lymphocytes and null cells bound Clq, but T lymphocytes did not have that property. Polymorphonuclear leukocytes also had a specific receptor for Clq which varied from donor to donor of the cells (mean, $3.7 \pm 2.0 \times 10^5$ binding sites per PMN, determined from Scatchard plots). To further investigate the Clq receptor, Bobak et al.[85] tested its surface expression by treating human PMNs and monocytes with either the chemoattractant N-formyl-methionylleucylphenylalanine or the phorbol ester, phorbol dibutyrate. These substances alter the expression of some receptors

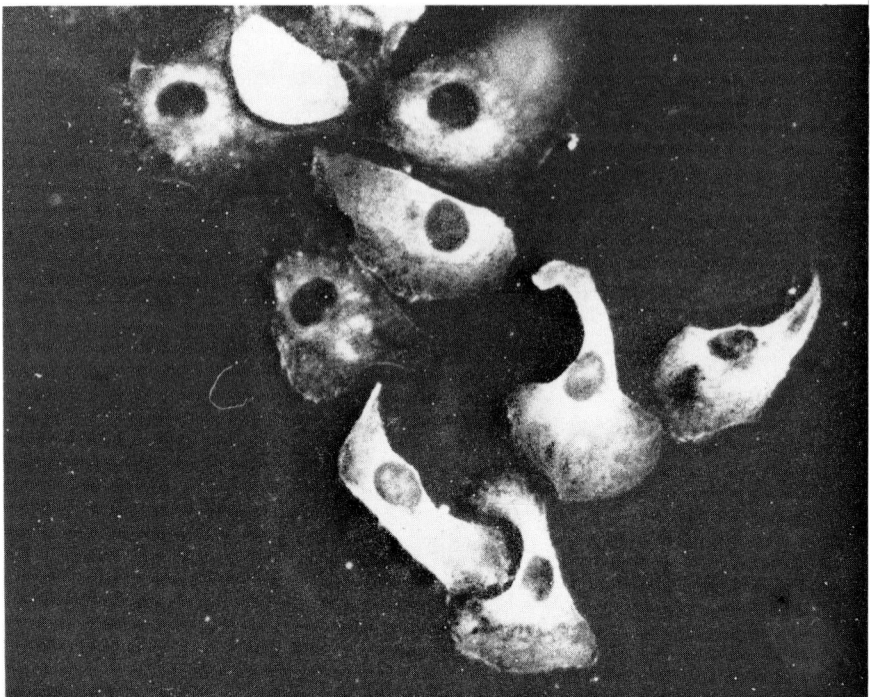

FIGURE 2.   The immunofluorescence localization of Clq binding sites of endothelial cells is illustrated. Bovine pulmonary artery endothelial cells on glass coverslips were incubated with 250 $\mu\ell$ of human Clq (in 5 mM phosphate-buffered saline, pH 7.3 diluted to OD_{280} = 0.117) for 60 min at 37°C. They were washed by dipping three times in PBS, pH 7.4, then incubated with 250 $\mu\ell$ of antihuman Clq/FITC diluted 1:20 with PBS at 37°C for 60 min. The coverslips were washed again three times, the backs were wiped clean, and they were mounted on a slide using nine parts glycerol to one part PBS, pH 8.5. Bright fluorescence indicates binding sites for Clq. Control cells, incubated in PBS without Clq followed by anti-Clq/FITC, did not fluoresce. (Magnification × 450). (From Ryan, U. S. and Ryan, J. W., Clin. Lab. Med., 3, 577, 1983. With permission.)

on the surface of cells. For example, Fearon and Collins[86] showed that the chemoattractant increased the expression of the receptor for C3b on PMNs, and similar results were reported by Yancy et al. with monocytes.[87] In contrast, there was no significant increase in the receptor for Clq on PMNs or monocytes after preincubation with the chemoattractant or the phorbol ester.[85] From these data it was suggested[85] that factors controlling the expression of receptors for C3b must differ from those regulating the receptor for Clq on PMNs and monocytes.

Cultured human gingival fibroblasts also have receptors for Clq, and the binding occurs via the collagenous portion of the molecule.[88] With ^{125}I-Clq, the binding was specific, saturable, and reversible. The number of binding sites per cell from the Scatchard analyses varied from 2.6 to 17.7 × 10^6, a figure higher than that reported with many other types of cells,[9,53,74,85] but in the same range as the human macrophage cell line U937.[89] The fibroblast receptor was not susceptible to digestion by trypsin, a result also obtained by Arvieux et al.[89] with the U937 cells. However, pronase digestion partially reduced the number of Clq binding sites of the U937 cells.[89] The Clq receptors on lymphocytes[74] and lymphoblastoid cell lines[78] were sensitive to trypsin digestion. Binding of ^{125}I-Clq to fibroblasts occurred optimally at low ionic strength (μ = 0.05), and only 18 to 20% were bound at μ = 0.15.[88] A similar dependence on ionic strength was reported with the U937 cells.[89]

Preliminary studies using antibodies to Clq conjugated to fluorescein indicated the presence of Clq binding sites on endothelial cells (Figure 2).[89a] In a more definitive study, Zhang et

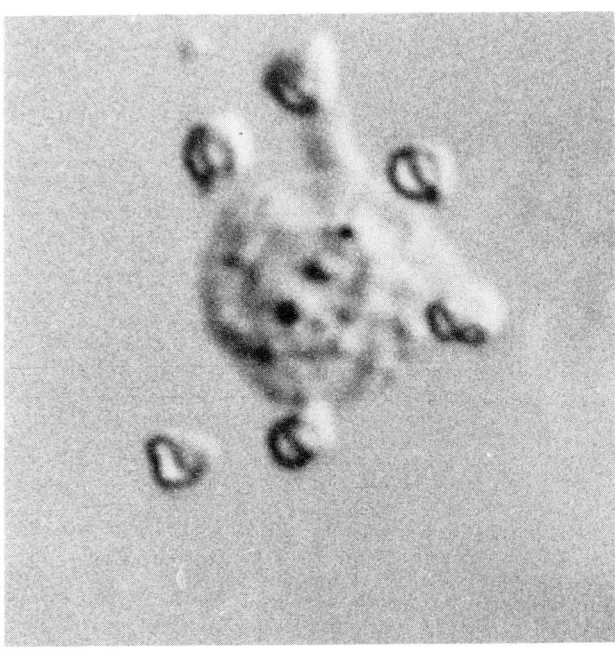

FIGURE 3. Rosettes, indicating the presence of Fc receptors, are formed between endothelial cells incubated first with Clq and then reacted with EA7S (sheep erythrocytes sensitized with rabbit IgG; Cordis Laboratories Inc., Miami, Fla.). Control cells, not incubated with Clq did not form rosettes with EA7S. Neither control endothelial cells nor those treated with Clq showed any binding with E (nonsensitized sheep erythrocytes). (Nomarski optics: magnification × 1600.) (From Zhang, S. C., Schultz, D. R., and Ryan, U. S., Tissue Cell, 18, 13, 1986. With permission.)

al.[9] reported that normal undamaged bovine pulmonary endothelial cells have a receptor for Clq. The binding of ^{125}I-Clq was dose dependent, reversible, and saturable. The binding which occurred via the collagenous portion of the molecule was shown by two different protocols: first, purified ^{125}I-Clq heads did not bind to the cells; and second, cell-bound native ^{125}I-Clq was removed after treatment of the cells with collagenase at 37°C. Macromolecular Cl did not bind to the cells. In addition, rosettes were formed between endothelial cells incubated first with native Clq followed by sheep erythrocytes sensitized with specific IgG antibody (EA). This indicated that the Clq tails were bound to receptors and the globular heads were free to bind the Fc-IgG on the erythrocytes. Without pretreatment of the cells with Clq, no rosettes were formed since undamaged cells have no Fc receptors[7,8] (Figure 3). Although the binding of Clq to its receptor was optimal at $\mu = 0.066$, sufficient binding occurred to carry out experiments at physiologic ionic strength.

Finally, the binding of Clq is not always restricted to an accessible receptor on cell membranes or on soluble aggregates of proteins. For example, Isenman et al.[90] showed that the binding of Clq is not restricted to the C_H2 domain of IgG; the C_H3 domain also contains Clq binding regions that are buried in native IgG. Due to tertiary folding within unaltered IgG, the Clq-binding regions are not expressed except in the C_H2 domain of native IgG.

F. Purification and Characterization of the Clq Receptor

A Clq receptor was partially purified from human polymorphonuclear leukocyte membranes by Tenner and Cooper.[91] The method was to bind ^{125}I-membrane material in the presence of NP40 detergent to Clq Sepharose at low ionic strength and neutral pH. From their results, the receptor appeared to be different from the serum Clq inhibitor originally reported by Conradie et al.[92] and later characterized by Silvestri and co-workers as a chondroitin 4-sulfate proteoglycan.[93] Binding of the membrane material to Clq-latex beads was reduced following treatment of the material with trypsin.

Ghebrehiwet et al.[94] applied detergent-solubilized membrane proteins from Raji cells to a Clq Sepharose 4B column, and after extensive washing, eluted two peaks that were capable of binding and precipitating free Clq. Initial evidence was obtained that the Raji cell receptor was similar to the Clq inhibitor.[93] However, after chemical analyses and also because of a lack of digestion with chondroitinase, the membrane receptor appeared to be different from the serum Clq inhibitor. In addition, Silvestri et al.[93] found that the Clq inhibitor interacted with Clq at physiological ionic strength and neutral pH, whereas the binding of Clq to the Raji cell receptor occurred in hypotonic medium and neutral pH.

In a later publication, Ghebrehiwet characterized a IgM-lambda monoclonal antibody that was produced in mice immunized with viable lymphoblastoid Raji cells.[95] The IgM reacted specifically with not only the Clq receptor on Raji cells, but also with U937 cell membranes and other lymphoblastoid cells, but not with sheep erythrocytes. The uptake of the antibody by the 85-kdalton receptor was saturable, and after Scatchard analysis, the affinity constant was 2.9×10^{-10} M with an estimated 7.8×10^{3} antibody molecules bound per cell. Since most studies with the Clq receptor report values in the range of 10^{5} to 10^{6} sites per cell (e.g., 9, 53, 88), the disparity was explained by the steric hindrance caused by the size and valance of the monoclonal IgM.

G. Receptors for C3b and Fc-IgG on Endothelial Cells

In addition to the Clq receptor, there are other receptors on endothelial cells that undoubtedly have important roles in normal events and in inflammatory reactions. Some of these receptors are expressed as part of the normal cellular membrane, but others appear to be available only after an alteration or injury to the cell.

There seems to be a consensus among the few papers published that C3b receptors are not present on endothelial cells unless the cell is injured or altered by a microbial infection. For example, Andrews et al.[6] were not able to show that human endothelial cells in culture or in frozen sections of umbilical vessels had receptors for two products of C3 activation, C3b and C3d. Linder and co-workers[3,5] demonstrated that C3 from diluted human serum bound to vascular endothelium independent of antibodies, but the cells were first made permeable by treatment with the nonionic detergent NP40 before treatment with serum. The C3 was detected on cytoskeletal intermediate filaments by fluorochrome-labeled antibodies to C3.

A receptor on endothelial cells has been demonstrated for Fc-IgG.[1,2] Receptors for aggregated immunoglobulin and immune complexes were reported by Johnson et al.[1] on vascular endothelial cells of fetal stem vessels in human placentas. However, as with the Clq receptor, a physiological role for immunoglobulin receptors on nonimmune cells can only be conjectured at this time.

In initial studies with healthy bovine pulmonary endothelial cells in monolayer culture, we were not able to show receptors for either the activated third component of complement (C3b) or for Fc-IgG.[7,8] With certain alterations to these cells, however, we induced the expression of both the receptors for C3b and for Fc-IgG. Thus, for the "hidden" C3b and Fc-IgG receptors, endothelial cells from control cultures were compared with cells that had been infected with influenza virus or cytomegalovirus, and the cells incubated with a lysate of bovine cells rich in PMNs. The rationale for the latter approach was that in many inflammatory conditions initiated by immune complexes, the predominant cell type associated with tissue damage is the PMN. The cytoplasmic granules of PMNs contain large amounts of potent proteases and other factors that are responsible for much of the tissue damage found in inflammatory diseases of the lung.[96] For the Fc-IgG receptor, a rosette assay was used with antibody-coated sheep erythrocytes (EA), and for the C3b receptor, EA with activated C components C1,C4,C2,C3 (EAC1,4b2a3b). In contrast to unexposed control cells, the endothelial cells exposed to the white cell lysate formed large rosettes with both EA and EAC1,4b,2a,3b, indicating the presence of receptors for both Fc-IgG and C3b. The cells infected with cytomegalovirus were positive for both receptors, but to a lesser degree than the lysate-exposed cells. Although the influenza infection produced the Fc-IgG receptor, the assay for the C3b receptor was negative. These results indicated that latent or otherwise unexpressed receptors for both C3b and Fc-IgG were expressed when the cells were either exposed to the cell lysate or infected with virus. By light microscopy, no rupture of the endothelial cell membranes was detected after treatment of the cells.

IV. CONCLUSIONS

From the above studies, we speculated that circulating immune complexes with or without bound C components may initiate or contribute to the pulmonary inflammatory response after lung endothelial cells are altered, exposing the C3b and Fc-IgG receptors. The ability of lysates of PMN white blood cells and two viruses capable of infecting endothelial cells to unmask these receptors is likely to be of pathophysiologic significance. Firstly, white blood cells activated by an anaphylatoxin such as C5a (a small peptide released from the activated fifth component of complement)[97] or by immune complexes (as may occur in IgG aggregate anaphylaxis, chronic serum sickness, lupus erythematosis, eosinophilic granuloma of the lung, and idiopathic pulmonary fibrosis)[98] are known to marginate and attach to endothelial cells in vivo.[99] Secondly, endothelial cells infected with certain viruses acquire the capacity to bind PMNs, and the PMNs thus bound become adherent for other PMNs.[100] White blood cell aggregates, for example, those formed in anaphylaxis,[101] might be expected to unmask receptors on endothelial cells. The stagnant or reduced blood flow caused by the mechanical blockage by white cell aggregates and emboli would provide favorable conditions for binding immune complexes with or without bound C components. In turn, endothelial cell-bound immune complexes would provide further stimuli for complement activation and complement-linked immune lysis. Presumably, this series of reactions would occur most prominently at the level of the microcirculation where mechanical blockage of blood flow would be most pronounced and where, in fact, the inflammatory response to, for example, anaphylaxis, is most pronounced.[101] This concept became even more plausible when we described a receptor on *normal*, unaltered endothelial cells for free subcomponent Clq of the C system[9] (see above). Binding of circulating immune complexes to lung endothelial cells via the Clq ligand-receptor and activation of the complement and other inflammatory pathways could result in exposing the C3b and Fc-IgG receptors. This would enhance the binding to the endothelial cells of other immune complexes and eventually result in the influx of PMNs. With these series of events, inflammation and tissue damage would occur at the level of the microcirculation. Thus, it is quite feasible that the overall pathophysiologic effect on the lung caused by certain diseases characterized by circulating immune complexes would have as its origin the endothelial cell-bound immune complexes.

ACKNOWLEDGMENTS

This work was supported by grants HL 21568 and HL 33064 from the National Heart Lung and Blood Institute, and 814 from the Council for Tobacco Research.

REFERENCES

1. **Johnson, P. M., Trenchev, P., and Fault, W. P.,** Immunological studies of human placentae. Binding of complexed immunoglobulin by stromal endothelial cells, *Clin. Exp. Immunol.,* 22, 133, 1975.
2. **Matre, R.,** Similarities of Fc receptors on trophoblast and placental endothelial cells, *Scand. J. Immunol.,* 6, 953, 1977.
3. **Linder, E., Lehto, V.-P., and Stenman, S.,** Activation of complement by cytoskeletal intermediate filaments, *Nature (London),* 278, 176, 1979.
4. **Shadforth, M. F., Cunningham, P. H., and Andrews, B. S.,** The demonstration of complement Clq and Fc-IgG receptors on the surface of human endothelial cells, *Fed. Proc.,* 38, 1075, 1979.
5. **Linder, E.,** Binding of Clq and complement activation by vascular endothelium, *J. Immunol.,* 126, 648, 1981.
6. **Andrews, B. S., Shadforth, M., Cunningham, P., and Davis, J. S.,** Demonstration of a Clq receptor on the surface of human endothelial cells, *J. Immunol.,* 127, 1075, 1981.
7. **Ryan, U. S., Schultz, D. R., Del Vecchio, P. J., and Ryan, J. W.,** Endothelial cells of bovine pulmonary artery lack receptors for C3b and for the Fc portion of IgG, *Science,* 208, 748, 1980.
8. **Ryan, U. S., Schultz, D. R., and Ryan, J. W.,** Fc and C3b receptors on pulmonary endothelial cells: induction by injury, *Science,* 214, 557, 1981.
9. **Zhang, S. C., Schultz, D. R., and Ryan, U. S.,** Receptor-mediated binding of Clq on pulmonary endothelial cells, *Tissue Cell.,* 18, 13, 1986.
10. **Fearon, D. T. and Wang, W. W.,** Complement ligand-receptor interactions that mediate biological responses, *Annu. Rev. Immunol.,* 1, 243, 1983.
11. **Schreiber, R. D.,** The chemistry and biology of complement receptors, in *Complement,* Muller-Eberhard, H. J. and Miescher, P. A., Eds., Springer-Verlag, Berlin, 1985, 115.
12. **Lepow, I. H., Naff, G. B., Todd, E. W., Pensky, J., and Hinz, C. F.,** Chromatographic resolution of the first component of human complement into three activities, *J. Exp. Med.,* 117, 983, 1963.
13. **Tschopp, J., Villiger, W., Fuchs, H., Kilchherr, E., and Engel, J.,** Assembly of subcomponents Clr and Cls of first component of complement: electron microscope and ultracentrifugation studies, *Proc. Natl. Acad. Sci. U.S.A.,* 77, 7014, 1980.
14. **Reid, K. B. M. and Porter, R. R.,** The proteolytic activation systems of complement, *Annu. Rev. Biochem.,* 50, 433, 1981.
15. **Loos, M.,** The classical complement pathway: mechanism of activation of the first component by antigen-antibody complexes, *Prog. Allergy,* 30, 135, 1982.
16. **Cooper, N. R.,** The classical complement pathway: activation and regulation of the first complement component, *Adv. Immunol.,* 37, 151, 1985.
17. **Borsos, T. and Rapp, H. J.,** Chromatographic separation of the first component of complement and its assay on a molecular basis, *J. Immunol.,* 91, 851, 1963.
18. **Calcott, M. A. and Muller-Eberhard, H. J.,** Clq protein of human complement, *Biochemistry,* 11, 3443, 1972.
19. **Reid, K. B. M.,** Proteins involved in the activation and control of the two pathways of human complement, *Biochem. Soc. Trans.,* 11, 1, 1983.
20. **Hurst, M. M., Volanakis, J. E., Stroud, R. M., and Bennett, J. C.,** Cl fixation and classical complement pathway activation by a fragment of the Cμ4 domain of IgM, *J. Exp. Med.,* 142, 1322, 1975.
21. **Feinstein, A., Richardson, N., and Taussig, M. J.,** Immunoglobulin flexibility in complement activation, *Immunol. Today,* 7, 169, 1986.
22. **Shelton, E., Yonemasu, K., and Stroud, R. M.,** Ultrastructure of the human complement component Clq, *Proc. Natl. Acad. Sci. U.S.A.,* 69, 65, 1972.
23. **Knobel, H. R., Villiger, W., and Isliker, H.,** Chemical analysis and electron microscopy studies of human Clq prepared by different methods, *Eur. J. Immunol.,* 5, 78, 1972.
24. **Borsos, T., Loos, M., Chapuis, R. M., Medicus, R., and Isliker, H.,** A novel way of relating the structure of Clq to the hemolytic activity of the first component of complement, *Mol. Immunol.,* 17, 1415, 1980.
25. **Reid, K. B. M., Sims, R. B., and Faiers, A. P.,** Inhibition of the reconstitution of the haemolytic activity of the first component of human complement by a pepsin-derived fragment of subcomponent Clq, *Biochem. J.,* 161, 239, 1977.
26. **Reid, K. B. M., Gagnon, J., and Frampton, J.,** Completion of the amino acid sequences of the A and B chains of subcomponent Clq of the first component of human complement, *Biochem. J.,* 203, 559, 1982.
27. **Reid, K. B. M.,** A collagen-like amino acid sequence in a polypeptide chain of human Clq (a sub component of the first component of complement), *Biochem. J.,* 141, 189, 1974.
28. **Knobel, H. R., Heusser, C., Rodrick, M. L., and Isliker, H.,** Enzymatic digestion of the first component of human complement (Clq), *J. Immunol.,* 112, 2094, 1974.

29. **Hughes-Jones, N. C., and Gardner, B.,** Reaction between the isolated globular sub-units of the complement component Clq and IgG-complexes, *Mol. Immunol.,* 16, 697, 1979.

30. **Paques, E. P., Huber, R., Priess, H., and Wright, J. K.,** Isolation of the globular region of the subcomponent q of the Cl component of complement, *Hoppe-Seyler's Z. Physiol. Chem.,* 360, 177, 1979.

31. **Siegel, R. C., Schumaker, V. N., and Poon, P. H.,** Stoichiometry and sedimentation properties of the complex formed beween Clq and Clr₂ Cls₂ subcomponents of the first component of complement, *J. Immunol.,* 127, 2447, 1981.

32. **Colomb, M. G., Arlaud, G.-J., and Villiers, C. L.,** Structure and activation of Cl: current concepts, *Complement,* 1, 69, 1984.

33. **Schumaker, V. N., Hanson, D. C., Kilchherr, E., Phillips, M. L., and Poon, P. H.,** A molecular mechanism for the activation of the first component of complement by immune complexes, *Mol. Immunol.,* 23, 557, 1986.

34. **Kilchherr, E., Schumaker, V. N., Phillips, M. L., and Curtiss, L. K.,** Activation of the first component of human complement, Cl, by monoclonal antibodies directed against different domains of subcomponent Clq, *J. Immunol.,* 137, 255, 1986.

35. **Allan, R. and Isliker, H.,** Studies on the complement binding site of rabbit IgG. II. The reaction of rabbit IgG and its fragments with Clq, *Immunochemistry,* 11, 243, 1974.

36. **Loos, M., Bitter-Suermann, D., and Dierich, M. P.,** Interaction of the first (C1), the second (C2) and the fourth (C4) component of complement with different preparations of bacterial lipopolysaccharides and with lipid A, *J. Immunol.,* 112, 935, 1974.

37. **Cooper, N. R., Jensen, F. C., Welsh, R. M., and Oldstone, M. B. A.,** Lysis of RNA tumor viruses by human serum: direct antibody independent triggering of the classical pathway, *J. Exp. Med.,* 144, 970, 1978.

38. **Loos, M., Wellek, B.,Thesen, R., and Opferkuch, W.,** Antibody-independent interaction of the first component of complement with Gram-negative bacteria, *Infect. Immunol.,* 22, 5, 1978.

39. **Giclas, P. G., Ginsberg, M. H., and Cooper, N. R.,** Immunoglobulin G independent activation of the classical pathway by monosodium urate crystals, *J. Clin. Invest.,* 63, 759, 1979.

40. **Schultz, D. R. and Arnold, P. I.,** The first component of human complement: on the mechanisms of activation by some carbohydrates, *J. Immunol.,* 126, 1994, 1981.

41. **Loos, M. and Hill, H. U.,** Different behaviour of native and purified Clq, a subcomponent of the first component of complenet (C1), *Monogr. Allergy,* 12, 46, 1977.

42. **Burton, D. R., Boyd, J., Brampton, A. D., Easterbrook-Smith, S. B., Emmanuel, E. J., Novotny, J., Rademacher, T. W., Van Schravendiik, M. R., Sternberg, M. J. E., and Dwek, R. A.,** The Clq receptor site on immunoglobulin G, *Nature (London),* 288, 338, 1980.

43. **Van Schravendiik, M. R., and Dwek, R. A.,** Interaction of Clq with DNA, *Mol. Immunol.,* 19, 1179, 1982.

44. **Schultz, D. R., Loos, M., Bub, F., and Arnold, P. I.,** Differentiation of hemolytically active fluid-phase and cell-bound human Clq by an ant venon-derived polysaccharide, *J. Immunol.,* 124, 1251, 1980.

45. **Golan, M. D., Burger, R., and Loos, M.,** Conformational changes in Clq after binding to immune complexes: detection of neoantigens with monoclonal antibodies, *J. Immunol.,* 129, 445, 1982.

46. **Schumaker, V. N., Calcott, M. A., Spiegelberg, M. L., and Müller-Eberhard, H. J.,** Ultracentrifuge studies of the binding of IgG of different subclasses to the Clq subunit of the first component of complement, *Biochemistry,* 16, 5175, 1976.

47. **Tschopp, J., Schulthess, T., Engel, J., and Jaton, J. C.,** Antigen-independent activation of the first component of complement C1 by chemically cross-linked rabbit IgG-oligomers, *FEBS Lett.,* 112, 152, 1980.

48. **Ziccardi, R. J. and Cooper, N. R.,** Active disassembly of the first complement component C1, by C1 inactivator, *J. Immunol.,* 123, 788, 1979.

49. **Ziccardi, R. J.,** A new role for C1 inhibitor in homeostasis: control of activation of the first component of human complement, *J. Immunol.,* 128, 2505, 1982.

50. **Heinz, H.-P., Burger, R., Golan, M. D., and Loos, M.,** Activation of the first component of complement C1, by a monoclonal antibody recognizing the C-chain of Clq, *J. Immunol.,* 132, 804, 1984.

51. **Heinz, H.-P., Dlugonska, H., Rude, E., and Loos, M.,** Monoclonal antimouse macrophage antibodies recognize the globular portions of Clq, a subcomponent of the first component of complement, *J. Immunol.,* 133, 400, 1984.

52. **Ziccardi, R. J. and Cooper, N. R.,** Direct demonstration and quantitation of the first complement component in human serum, *Science,* 199, 1080, 1978.

53. **Tenner, A. J. and Cooper, N. R.,** Analysis of receptor-mediated Clq binding to human peripheral blood mononuclear cells, *J. Immunol.,* 125, 1658, 1980.

54. **Gabay, Y., Perlmann, H., Perlmann, P., and Sobel, A. T.,** A rosette assay for the determination of Clq receptor-bearing cells, *Eur. J. Immunol.,* 9, 797, 1979.

55. **Laurell, A.-B., Johnson, U., Martensson, U., and Sjoholm, A. G.,** Formation of complexes composed of Clr, Cls, and C1 inactivator in human serum on activation of C1, *Acta Pathol. Microbiol. Scand. Sect. C,* 86, 299, 1978.

56. **Sim, R. B., Arlaud, G. J., and Colomb, M. G.,** C1 inhibitor-dependent dissociation of human complement component C1 bound to immune complexes, *Biochem. J.,* 179, 449, 1979.

57. **Ziccardi, R. J.,** The role of immune complexes in the activation of the first component of human complement, *J. Immunol.,* 132, 283, 1984.

58. **Lindgren, S., Laurell, A.-B., and Eriksson, S.,** Complement components and activation in primary biliary cirrhosis, *Hepatology,* 4, 9, 1984.

59. **Innan, R. D., and Harpel, P. C.,** C1 inactivator-Cls complexes in inflammatory joint disease, *Clin. Exp. Immunol.,* 53, 521, 1983.

60. **Ochi, T., Yonemasu, K., and Ono, K.,** Immunochemical quantitation of complement components Clq and C3 in sera and synovial fluids of patients with bone and joint diseases, *Ann. Rheum. Dis.,* 39, 235, 1980.

61. **Sjoholm, A. G., Berglund, K., Johnson, U., Laurell, A.-B., and Sturfelt, G.,** C1 activation, with Clq in excess of functional C1 in synovial fluid from patients with rheumatoid arthritis, *Int. Arch. Allergy Appl. Immunol.,* 79, 113, 1986.

62. **Laurell, A.-B., Martensson, U., and Sjoholm, A. G.,** Studies on C1 subcomponents in chronic urticaria and angioedema, *Int. Arch. Allergy Appl. Immunol.,* 54, 434, 1977.

63. **Tenner, A. J. and Cooper, N. R.,** Stimulation of a human polymorphonuclear leukocyte oxidative response by the Clq subunit of the first complement component, *J. Immunol.,* 128, 2547, 1982.

64. **Johnson, K. J. and Ward, P. A.,** Biology of disease. Newer concepts in the pathogenesis of immune complex-induced tissue injury, *Lab. Invest.,* 47, 218, 1982.

65. **Hughes-Jones, N. C.,** Functional affinity constants of the reaction between [125]-I-labelled Clq and Clq binders and their use in the measurement of plasma Clq concentrations, *Immunology,* 32, 191, 1977.

66. **Linder, E.,** A simple immunofluorescence assay for Clq binding, *J. Immunol. Meth.,* 32, 239, 1980.

67. **Loos, M., Storz, R., Muller, W., Lemmel, E.-M.,** Immunofluorescence studies on the subcomponents of the first component of complement (C1): detection of Clq and Cls in different cells of biopsy material and on human as well as on guinea pig peritoneal macrophages, *Immunobiology,* 158, 213, 1981.

68. **Heusser, C., Boesman, M., Nordin, J. H., and Isliker, H.,** Effect of chemical and enzymatic radioiodinization on in vitro human Clq activities, *J. Immunol.,* 110, 820, 1973.

69. **McConahey, P. J. and Dixon, F. J.,** A method of trace iodination of proteins for immunologic studies, *Int. Arch. Allergy Appl. Immunol.,* 29, 185, 1966.

70. Bio Rad Laboratories, Richmond, Calif.

71. **Tenner, A. J., Lesavre, P. H., and Cooper, N. R.,** Purification and radiolabeling of human Clq, *J. Immunol.,* 127, 648, 1981.

72. **Markwell, M. A. K.,** A new solid-state reagent to iodinate proteins. I. Conditions for the efficient labeling of antiserum, *Anal. Biochem.,* 125, 427, 1982.

73. **Rapp, H. J. and Boroso, T.,** Effects of low ionic strength on immune hemolysis, *J. Immunol.,* 91, 826, 1963.

74. **Sobel, A. T. and Bokisch, V. A.,** Receptor for Clq on peripheral human lymphocytes and human lymphoblastoid cells, *Fed. Proc.,* 34 (Abstr.), 965, 1975.

75. **Tenner, A. J. and Cooper, N. R.,** Identification of types of cells in human peripheral blood which bind Clq, *J. Immunol.,* 126, 1174, 1981.

76. **Gupta, R. C., McDuffie, F. C., Tappeiner, G., and Jordan, R. E.,** Binding of soluble immune complexes to Raji lymphocytes. Role of receptors for complement components Clq and C3-C3b, *Immunology,* 34, 751, 1978.

77. **Sundqvist, K. G., Suchag, S. E., and Thorstensson, R. T.,** Dynamic aspects of the interaction between antibodies and complement at the cell surface, *Scand. J. Immunol.,* 3, 237, 1974.

78. **Ghebrehiwet, B. and Müller-Eberhard, J. H.,** Lysis of Clq coated chicken erythrocytes by human lymphoblastoid cell lines, *J. Immunol.,* 120, 27, 1978.

79. **Dickler, H. B. and Kunkel, H. G.,** Interaction of aggregated-globulin with B lymphocytes,, *J. Exp. Med.,* 136, 191, 1972.

80. **Cazenave, J.-P., Assimeh, S. N., Painter, R. H., Packham, M. A., and Mustard, J. F.,** Clq inhibition of the interaction of collagen with human platelets, *J. Immunol.,* 116, 162, 1976.

81. **Suba, E. A. and Csako, G.,** Clq (C1) receptor on human platelets: inhibition of collagen-induced platelet aggregation by Cla (C1) molecules, *J. Immunol.,* 117, 304, 1976.

82. **Wautier, J. L., Souchon, H., Cohen-Solal, L., Peltier, A. P., and Caen, J. P.,** C1 and human platelets. III. Role of C1 subcomponents in platelet aggregation induced by aggregated IgG, *Immunology,* 31, 595, 1976.

83. **Wautier, J. L., Reid, K. B. M., Legrand, Y., and Caen, J. P.,** Region of the Clq molecule involved in the interaction between platelets and subcomponent Clq of the first component of complement, *Mol. Immunol.,* 17, 1399, 1980.

84. **Reid, K. B. M.,** Complete amino acid sequences of the three collagen-like regions present in subcomponent Clq of the first component of human complement, *Biochem. J.,* 179, 367, 1979.

85. **Bobak, D. A., Frank, M. M., and Tenner, A. J.,** Characterization of Clq receptor expression on human phagocytic cells: effects of PDBu and FMLP, *J. Immunol.,* 136, 4604, 1986.

86. **Fearon, D. T. and Collins, L. A.,** Increased expression of C3b receptors on polymorphonuclear leukocytes induced by chemotactic factors and by purification procedures, *J. Immunol.,* 130, 370, 1983.

87. **Yancey, K. B., O'Shea, J. J., Chased, T., Brown, E., Takahashi, T., Frank, M. M., and Lawley, T. J.,** Human C5a modulates monocyte Fc and C3 receptor expression, *J. Immunol.,* 135, 465, 1985.

88. **Bordin, S., Kolb, W. P., and Page, R. C.,** Clq receptors on cultured human gingival fibroblasts: analysis of binding properties, *J. Immunol.,* 130, 1871, 1983.

89. **Arvieux, J., Reboul, A., Bensa, J.-C., and Colomb, M. G.,** Characterization of the Clq receptor on a human macrophage cell line, U937, *Biochem. J.,* 218, 547, 1984.

89a. **Ryan, U. S. and Ryan, J. W.,** Endothelial cells and inflammation, *Clin. Lab. Med.,* 3, 577, 1983.

90. **Isenman, D. E., Ellerson, J. R., Painter, R. H., and Dorrington, K. J.,** Correlation between the exposure of aromatic chromophores at the surface of the Fc domains of immunoglobulin G and their ability to bind complement, *Biochemistry,* 16, 233, 1977.

91. **Tenner, A. J. and Cooper, N. R.,** Purification and partial characterization of a human polymorphonuclear leukocyte Clq receptor, *Fed. Proc.,* 41, 965, 1982.

92. **Conradie, J. D., Volanakis, J. E., and Stroud, R. M.,** Evidence for a serum inhibitor of Clq, *Immunochemistry,* 12, 967, 1975.

93. **Silvestri, L., Baker, J. R., Roden, L., and Stroud, R. M.,** The Clq inhibitor in serum is a chondroitin 4-sulfate proteoglycan, *J. Biol. Chem.,* 256, 7383, 1981.

94. **Ghebrehiwet, B., Silvestri, L., and McDevitt, C. D.,** Identification of the Raji cell membrane-derived Clq inhibitor as a receptor for human Clq. Purification and immunochemical characterization, *J. Exp. Med.,* 160, 1375, 1984.

95. **Ghebrehiwet, B.,** Production and characterization of a murine monoclonal IgM antibody to human Clq receptor (ClqR), *J. Immunol.,* 137, 618, 1986.

96. **Baggiolini, M., Bretz, U., Dewald, B., and Feigenson, M. E.,** The polymorphonuclear leukocyte, *Agents Actions,* 8, 3, 1978.

97. **Hugli, T. E. and Müller-Eberhard, H. J.,** Anaphylatoxins C3a and C5a, *Adv. Immunol.,* 26, 1, 1978.

98. **Ward, P. A.,** Immune complex injury of the lung, *Am. J. Pathol.,* 97, 85, 1979.

99. **Henson, P. M., McCarthy, K., Larsen, G. L., Webster, R. O., Giclas, P. C., Dreisin, R. B., King, T. E., and Shaw, J. O.,** Complement fragments, alveolar macrophages, and alveolitis, *Am. J. Pathol.,* 97, 93, 1979.

100. **MacGregor, R. R., Friedman, H. M., Macarak, E. J., Kefalides, N. A.,** Virus infection of endothelial cells increases granulocyte adherence, *J. Clin. Invest.,* 65, 1469, 1980.

101. **Goodman, M. L., Way, B. A., and Irwin, J. W.,** The inflammatory response to anaphylaxis and intravenous antibody or antigen, *Virchows. Arch. A,* 383, 271, 1979.

Chapter 26

EFFECT OF THE ANAPHYLATOXINS ON VASCULAR TISSUE IN VIVO AND IN VITRO

Claes Lundberg, Marco Gardinali, Francois Marceau, and Tony E. Hugli

TABLE OF CONTENTS

I. INTRODUCTION

Intravascular complement activation induces hemodynamic and hematologic changes in both man[1-4] and experimental animals.[5-7] These effects have been shown to require the third component of complement, C3, but occur in the absence of the sixth component of complement, C6.[6] The anaphylatoxins (i.e., C3a, C4a, and C5a) are cleaved from the complement components C3, C4, and C5, respectively, when the classical pathway of the complement system is activated. However, C4a is not produced when the alternative (properdin) pathway is activated.

Recent experiments have shown that purified anaphylatoxins induce hemodynamic and hematologic effects very similar to those observed after intravascular complement activation. These biologically active polypeptides exhibit proinflammatory effects and have an ability to release mediators such as histamine, prostaglandins (PGs), and leukotrienes (LTs). In fact, many of the in vivo effects of anaphylatoxins can be successfully inhibited by antagonists or inhibitors of the above-mentioned mediators. Other anaphylatoxin-induced mediators (e.g., PAF acether) are presumably also associated with in vivo actions of the anaphylatoxins. Therefore, it appears that many anaphylatoxin effects are indirect, and that affected cells do not necessarily possess receptors for the anaphylatoxins.

C5a has a broader spectrum of activities than C3a and C4a, in that C5a activates leukocytes as well as being spasmogenic. A particularly important role for C5a is the ability to sequester leukocytes to a site of injury. This chemotactic activity is retained by human C5a even after removal of the C-terminal arginine by plasma carboxypeptidase N, which virtually eliminates spasmogenic effects of human anaphylatoxins. C5a is the most active of the anaphylatoxins and C4a is the least potent. In this context, only the effects of C5a, and to some extent those of C3a, will be reviewed here.

C3a and C5a interact with different cellular receptors as judged by an absence of cross-desensitization,[8] whereas C4a seems to utilize the C3a receptor.[9] Cell surface receptors for C5a have been identified on the polymorphonuclear leukocyte (PMN) at a density of approximately 50,000 to 80,000 per cell;[10-13] this number is comparable to that of the formyl-methionyl-leucyl-phenylalanine (fMLF) receptor density on PMNs.[11]

Reports on vascular effects of the anaphylatoxins in experimental animals are somewhat confusing and contradictory. It is important to keep in mind that significant species differences exist in terms of anaphylatoxin responsiveness. Guinea pigs, for example, show a dramatic systemic response to anaphylatoxins, whereas rats show only a minor reaction. Furthermore, large variations in relative activity of anaphylatoxins exist depending on the animal source of these polypeptides. $C5a_{desArg}$ derived from the rat, for example, possesses approximately 1000-fold greater spasmogenic activity than does human $C5a_{desArg}$ on the guinea-pig isolated ileum preparation. The reader is referred to Hugli[14] for an extensive reivew of anaphylatoxin chemistry and biology.

II. IN VIVO HEMODYNAMIC RESPONSES TO THE ANAPHYLATOXINS

A. Systemic Hemodynamics

The hemodynamic effects of anaphylatoxins, as well as intravascular complement activation (ICA) have been examined. C5a or $C5a_{desArg}$ have been reported to cause a transient arterial hypotension in guinea pigs,[15] cats,[16] dogs,[17] and rabbits.[18] Furthermore, hypotension has been observed after ICA in the rabbit,[6-7] and infusion of zymosan-activated plasma (ZAP) into pigs[19] also induced a hypotensive response. In guinea pigs, cats, and pigs, the hypotension is followed by a rebound hypertensive phase. The hypertension in this biphasic response is tachyphylactic and mediated by catechloamines, presumably released by histamine.[15,20] A histamine/catecholamine-dependent mechanism for hypertension was concluded from results of experiments using alpha-adrenergic and histamine H_1-antagonists.

The hypotensive response, however, is not tachyphylactic and can be reproduced by repeatedly injecting $C5a_{desArg}$ for as many as ten times.[15] We have recently demonstrated that the hypotensive response induced in rabbits by purified human or porcine C5a is paralleled by increased plasma levels of prostaglandin (PG) I_2 (prostacyclin; measured as 6-keto-$PGF_{1\alpha}$), thromboxane (Tx) A_2 (measured as TxB_2), and PGE_2.[18] Furthermore, the cyclooxygenase inhibitor indomethacin completely inhibits the C5a-induced hypotension in rabbits as well as prevents the increase in plasma prostanoid levels (Figure 1). Histamine appeared to be involved in this reaction via the H_2-receptor since the H_2-receptor antagonist cimetidine reduced both the hypotensive response and prostanoid release. The histamine H_1-receptor antagonist pyrilamine did not prevent the C5a effect. In agreement with these results, Bodammer and Vogt[15] and Ulevitch and Cochrane[6] reported that histamine H_1-receptor antagonists fail to influence the $C5a_{desArg}$- or ICA-induced hypotension in either guinea pigs or rabbits, whereas the histamine H_2-receptor antagonist burimamide significantly reduced hypotension after ICA.[6] Furthermore, involvement of prostanoids in C5a-induced hypotension has been suggested by Bult et al.[7] These authors reported that the hemodynamic changes that follow ICA parallel the increase in plasma levels of prostacyclin and TxA_2. They suggested a causal relationship between prostacyclin formation and arterial hypotension. Pavek et al.[17] suggested that cyclooxygenase products were involved in the C5a-induced hypotension, since indomethacin administered to dogs inhibited hypotension. Thus, participation of prostanoids in C5a-induced hypotension appears to be established and their role in the physiologic response to the anaphylatoxins is relatively well defined.

The mechanism responsible for C5a-induced hypotension has recently been shown to depend on a thromboxane-induced pulmonary vasocostruction in rabbits.[18] In this study the hypotensive response is paralleled by an increase in central venous pressure (CVP) and a decrease in cardiac output (CO), whereas the total peripheral resistance (TPR) remained unchanged. Pulmonary vasoconstruction results in decreased CO, and hypotension occurs as a consequence. When rabbits are pretreated with a thromboxane synthetase inhibitor, dazoxiben, C5a-induced hypotension was still apparent.[18] However, under this condition the hypotension is paralleled by a decrease in CVP, a less severe decrease in CO, and reduction in TRP. Therefore, in the presence of thromboxane synthetase inhibitors, the mechanism responsible for C5a-induced hypotension appears to be a prostaglandin-mediated peripheral vasodilation (Figure 1).

In conclusion, ICA-induced hypotension appears to have the same underlying mechanism as that of C5a-induced hypotension. The release of prostanoids that cause decreased arterial blood pressure is observed in both ICA and C5a reactions. Thromboxane A_2-dependent pulmonary vasoconstruction, with a corresponding decrease in cardiac output as a consequence, appear to cause the hypotension that develops from C5a challenge (see Table 1). Production and release of prostanoids appear in part mediated via histamine and its H_2-receptor, since histamine H_2-receptor antagonists significantly reduce both C5a-induced hypotension and prostanoid release (see Table 2).

B. Lung Circulation

Transient pulmonary hypertension after C5a challenge appears to occur in various species simultaneous to hypotension. Pulmonary vasoconstriction following C5a, $C5a_{desArg}$ or ICA has been reported in dogs,[17] cats,[16] sheep,[21-26] pigs,[19] and rabbits.[18,27] Corresponding to the pulmonary hypertension is a transient leukopenia with associated pulmonary leukostasis and transient hypoxemia. However, leukostasis does not appear to cause the hypertension because prolonged i.v. infusion of ZAP results in prolonged sequestration of leukocytes but with transient pulmonary hypertension.[23] Therefore, pulmonary hypertension observed after C5a challenge is caused by vasoconstriction. Pulmonary vasoconstriction also appears to mediate hypoxemia, since these two variables parallel exactly in sheep challenged with ZAP.[21]

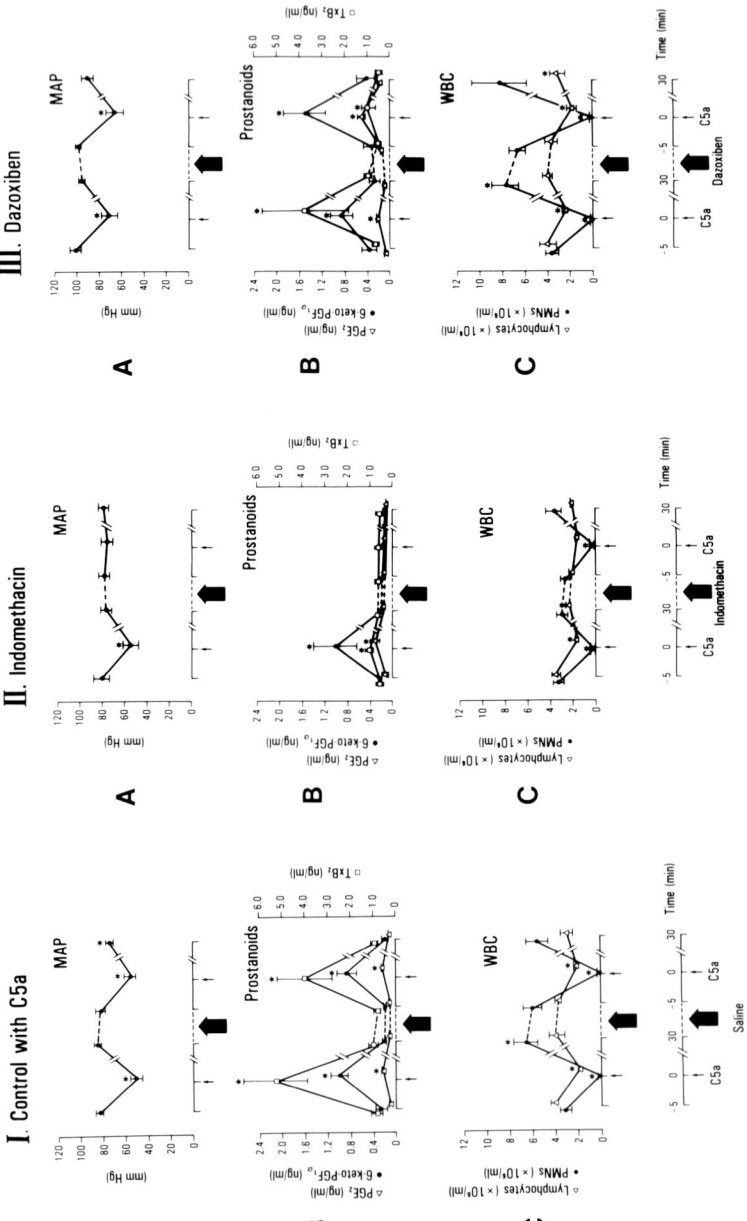

FIGURE 1.   (A) Mean arterial pressure (MAP); (B) 6-keto-PGF$_{1\alpha}$, TxB$_2$, and PGE$_2$ plasma levels; and (C) peripheral white blood cell (WBC) counts were measured prior to (−5 min), immediately after (15 to 30 sec) and 30 min after each of two consecutive i.v. injections of 10 μg porcine C5a into adult New Zealand white rabbits. The second C5a injection was performed approximately 45 min after the first one. Panel I shows the control group that received saline (i.v.) between injections of C5a. Panel II shows the result in rabbits given indomethacin (5 mg/kg, i.v.) prior to the second C5a injection. Indomethacin treatment not only inhibited C5a-induced prostanoid release (B), but also abolished the hypotensive responses (A), whereas neutropenia was not affected (C). Panel III shows the result in rabbits given dazoxiben (25 mg/kg, i.v.) prior to the second C5a injection. This treatment diminished the thromboxane release and increased the release of prostacyclin and PGE$_2$. Hypotension and neutropenia were, however, not affected, although the former was presumably caused by another mechanism as discussed in the text. Results are expressed as a mean ± SEM. Asterisks indicate significant differences compared to the respective initial values (−5 min values). (Adapted from Lundberg, C. et al., *Am. J. Pathol.*, 128, 471, 1987.)

Table 1
HEMODYNAMIC RESPONSES TO C5a, C5a$_{desArg}$,
ZYMOSAN ACTIVATED PLASMA (ZAP), OR
INTRAVASCULAR COMPLEMENT ACTIVATION (ICA)
IN VARIOUS ANIMAL SPECIES

Species	Arterial hypotension	Pulmonary hypertension	Cardiac output	Ref.
Guinea pig[a]	Yes	Not studied	Not studied	15
Cat[a]	Yes	Yes	Not studied	16
Dog[a]	Yes	Yes	Decrease	17
Rabbit[a,b,c]	Yes	Yes	Decrease	6—7,18,27
Pig[d]	Yes	Yes	Decrease	19
Sheep[d]	Yes	Yes	Decrease	21—26

[a] C5a$_{desArg}$.
[b] C5a.
[c] ICA.
[d] ZAP.

Table 2
MEDIATORS RELEASED AND HEMODYNAMIC RESPONSE OBSERVED
AFTER CHALLENGE WITH C5a, C5a$_{desArg}$, ZYMOSAN ACTIVATED
PLASMA (ZAP), OR INTRAVASCULAR COMPLEMENT ACTIVATION (ICA)
IN VARIOUS ANIMAL SPECIES

Pretreatment	Animal	PGI$_2$	PGE$_2$	TxA$_2$	Arterial hypotension/ pulmonary hypertension	Ref.
Control	Rabbit [a,b]	↑	↑	↑	Yes	7,18
	Sheep[c]	↑	NM[d]	↑	Yes	21
Cyclooxygenase	Rabbit[a]	—	—	—	No	18
inhibition	Sheep[c]	—	NM[d]	—	No	21
	Dog[e]	NM[d]	NM[d]	NM[d]	No	17
Thromboxane	Rabbit[a]	↑	↑	—	Hypotension only	18
synthetase	Sheep[c]	↑	NM[d]	—	Yes	23
inhibition						
H1-receptor	Rabbit[a,b]	↑	↑	↑	Yes	6,18
antagonist	Gunea pig[e]	NM[d]	NM[d]	NM[d]	Yes	15
H2-receptor	Rabbit[a,b]	(↑)[f]	(↑)[f]	(↑)[f]	(Yes)[f]	6,18
antagonist						

[a] C5a.
[b] ICA.
[c] ZAP.
[d] Not measured.
[e] C5a$_{desArg}$.
[f] Parentheses indicate a weaker response and arrow indicates an increase.

As in C5a-induced systemic hypotension, prostanoids appear to be of utmost importance in C5a-induced pulmonary hypertension. This is not surprising since evidence exists that C5a-induced pulmonary vasoconstriction is the mechanism underlying systemic hypotension caused by C5a. Plasma levels of thromboxane B$_2$ reportedly rise significantly and parallel exactly the rise and fall in pulmonary arterial pressure after ZAP infusion in sheep.[21] Plasma levels of TxB$_2$ as well as pulmonary hypertension were normalized after indomethacin treatment. An increase in CVP induced by C5a turned to a decrease in CVP for dazoxiben-treated rabbits, indicating that dazoxiben abolished C5a-induced pulmonary vasoconstriction.[18]

Activation of the complement system during endotoxic shock is believed to contribute to the hemodynamic changes that are observed.[28,29] Furthermore, it is well documented that pretreatment with cyclooxygenase inhibitors or selective inhibitors of thromboxane synthetase counteracts increased synthesis of prostanoids and resultant pulmonary vasoconstriction induced by endotoxin.[30-31] In addition, infusion of arachidonic acid into sheep results in pulmonary hypertension[32] preventable by indomethacin. Therefore, evidence is accumulating that supports the hypothesis that C5a-induced pulmonary vasoconstriction is mediated via thromboxane. However, thromboxane-independent pulmonary vasoconstriction has also been suggested by others. Gee et al.[23] observed that dazoxiben pretreatment did not modify pulmonary hypertension in sheep that were subjected to repeated injections of ZAP.

The source of TxA_2 in C5a- or ICA-stimulated animals is unknown. Perkowsky et al.[24] showed that ZAP failed to aggregate or release TxA_2 from sheep platelets, suggesting that TxA_2 had its origin in a cell other than the platelet. The PMN is an improbable source of TxA_2 since both neutropenic sheep and rabbits show the same increase in plasma thromboxane levels after either ZAP or C5a injection.[18,33] Both mast cells and endothelial cells release TxA_2 when stimulated and are therefore potential sources of the TxA_2 in C5a-induced pulmonary hypertension. Complement-induced thromboxane release may be an indirect phenomenon, i.e., it is not necessarily the C5a target cell that releases thromboxane.

In accordance with in vivo experiments, anaphylatoxins cause increased vascular resistance in the isolated guinea-pig lung. This pulmonary hypertension occurs simultaneously with increased levels of histamine, prostaglandins, and TxA_2 in the effluent of the challenged lung.[17,34-36] Furthermore, Friedberg et al.[37] reported that C5a was a potent effector of vasoconstriction using guinea-pig and rat pulmonary arteries. Marceau and Hugli[38] showed that the guinea-pig pulmonary artery was extremely responsive to both C3a and C5a. The C3a contractile response was inhibited most effectively by indomethacin while the C5a contractile response was maximally inhibited by a combination of indomethacin and antihistamine. These results are consistent with C3a releasing cyclooxygenase products and C5a being capable of significant histamine release as well as cyclooxygenase product formation in pulmonary vascular tissue. These results are in general agreement with C5a-induced histamine and thromboxane-dependent pulmonary vasoconstriction in the rabbit as discussed above.

The effect of ICA or purified C5a injected i.v. on permeability and injury in the lung remains controversial. It is well known that i.v. injection of endotoxin causes permeability changes that are not complement dependent and not affected by agents that interfere with prostanoid production or release.[30-31,39] Consequently, even trace quantities of endotoxin in ZAP or C5a preparations could explain discrepancies between the different studies. Intrabronchial administration of C5a is known to cause severe injury to the endothelium and prominent plasma leakage. However, there is an important difference between intra- and extravascular administration of C5a as discussed later in this chapter. It appears that ICA or intravascular C5a, as an isolated event, fail to cause permeability increases in rabbits[18,27,40] or in pigs.[19] However, in several studies where ZAP has been infused into sheep, pulmonary edema and endothelial cell damage appears to develop.[22,24-26] These permeability changes are reportedly related to the release of toxic oxygen metabolites.[22] Arachidonic acid infusion in unanesthetized sheep produced dose-related increases in pulmonary artery pressure without any alterations in permeability.[32] ZAP induces lung permeability increases in sheep, presumably via mediators other than arachidonate products.

In conclusion, pulmonary vasoconstriction in many species develops after either ICA or C5a challenge. Both in vivo and in vitro studies indicate that pulmonary hypertension is thromboxane- and histamine-dependent (see Tables 1 and 2). ICA or intravascular C5a, as an isolated event, fails to cause significant lung injury. However, sheep may be more sensitive in this respect than other species.

C. Extravascular Effect of Anaphylatoxins

Extravascular exposure to anaphylatoxins appear to cause qualitatively different responses than does intravascular exposure. We have already mentioned that neither intravascular C5a nor ICA cause permeability changes in the lung (sheep excepted), whereas intrabronchial administration of C5a definitely enhances permeability. Local accumulation of PMNs has been observed in experimental animals after intradermal injection of ZAP, C5a, or C5a$_{desArg}$ using radiolabeled PMNs and/or myeloperoxidase to monitor the cellular response.[41-43] In these inflammatory skin lesions, PMN accumulation coincides with increased permeability. Plasma leakage was also observed by Wedmore and Williams[44] after intradermal injections of C5a$_{desArg}$. In animals rendered neutropenic, no plasma leakage was observed after C5a application.[41,43,44] Furthermore, local application of C5a on the hamster cheek pouch resulted in venular margination of PMNs and migration into the extravascular tissue as observed by intravital microscopy.[45] Accumulation of PMNs and plasma leakage are both observed, and Bjork et al.[45] suggested from their experiments that the permeability observed was partly PMN dependent. When leukotriene (LT) B$_4$, a potent chemotactic factor, is injected into an arteriole in the hamster-cheek-pouch preparation, abundant venular PMN adherence occurs without extravasation of cells. Plasma leakage was not in evidence as a result of PMN sequestration. When, in the same experiment, LTB$_4$ was applied topically on the cheek pouch (i.e., extravascularly), PMNs adhered to the venules, passed through the vessel wall, and migrated into the tissue. This latter event occurred with a simultaneous plasma leakage.[72] Therefore, it appears that when the PMNs extravasate there is coincident plasma leakage. C5a administered to the extravascular compartment stimulate PMN extravasation, whereas intravascular C5a does not cause the PMN to migrate over the vascular barrier (Figure 2). As expected, a chemotactic gradient across the vessel wall is needed to cause transmigration of PMNs. In this context Williams and Morley[46] observed that vasodilatory prostaglandins markedly enhanced protein exudation caused by intradermal injections of other mediators (e.g., histamine, C5a, fMLF), an effect attributed to vasodilatory properties of these prostanoids.

Extravascular application of C3a or C5a has been shown to constrict arterioles and to cause postcapillary leakage of macromolecules using intravital microscopy.[45,47] Histamine appears to be involved both in the vasoconstriction and in enhancement of permeability.

In conclusion, effects of C5a administered to the extravascular compartment are quite different from the effect of intravascular C5a. C5a applied outside the vessels cause PMN migration into the extravascular tissue and PMN-dependent plasma leakage, whereas intravascular C5a cause PMN adherence to the endothelium without extravasation or permeability changes (see Figure 2).

III. TISSUE AND CELLULAR RESPONSES TO THE ANAPHYLATOXINS

A. Isolated Blood Vessels

Isolated blood vessels have been used to assess the vascular effects of the anaphylatoxins. Friedberg et al.[37] reported that pulmonary arteries isolated from guinea pigs and rats exhibit a contractile response to crude C5a$_{desArg}$. Guinea-pig aorta was found to contract following challenge by guinea-pig C5a and contraction was prevented by a combination of cyclooxygenase inhibitor and histamine H$_1$-antagonist.[48] Marceau and Hugli[38] and Hugli and Marceau[49] examined the effects of human C3a and C5a on isolated guinea-pig and rabbit vessels, respectively. The contactile effect of C3a on guinea-pig portal vein and pulmonary artery was weaker than C5a, and C3a is tachyphylactic while C5a is not. Pharmacologic analysis suggests that the C3a response was mainly mediated via prostanoids and leukotrienes, whereas the C5a contractile response was mediated by prostanoids and histamine. Isolated blood vessels from the rabbit exhibit less impressive responses to the anaphylatoxins com-

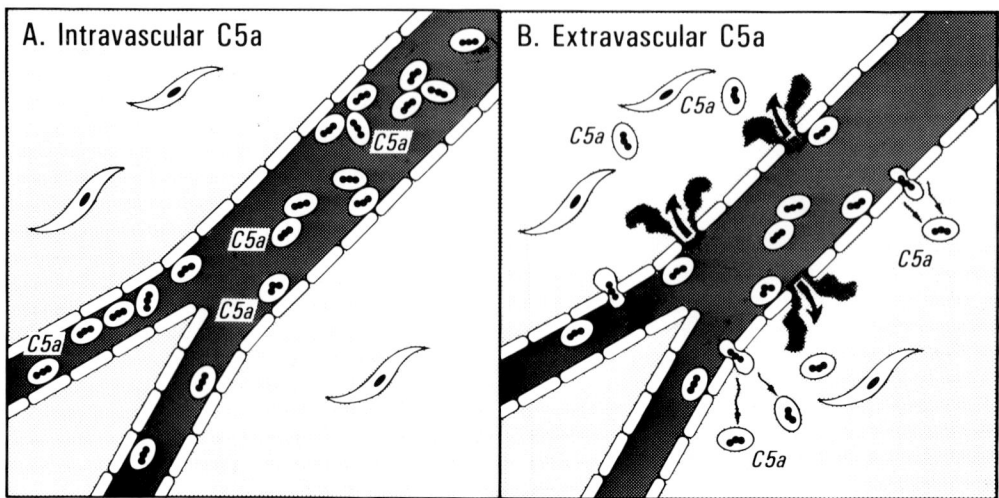

FIGURE 2.   Panel A illustrates polymorphonuclear leukocytes (PMNs) aggregating and adhering to the endothelium of postcapillary venules after intravascular administration of C5a. Note that PMN migration and plasma extravasation is minimal. Panel B indicates the effect of extravascular C5a administration. In this case, PMNs attach to the endothelium and migrate into the tissue space. Under these conditions significant plasma leakage occurs as a result of enhanced vascular permeability.

pared to the guinea-pig vessels, according to these authors.[49] Human C3a had no effect, while C5a was weakly contractile. When the portal vein from rabbits was precontracted with noradrenaline, C5a actually caused the tissue to relax. This relaxation coincided with the release of prostacyclin into the tissue bath. Both relaxation and prostacylin release was inhibited by indomethacin treatment. Purified porcine C5a, but not C3a, has been shown to release prostacylin from the intimal surface of isolated rabbit aorta,[50] This prostacylin release was greatly enhanced when rabbit $C5a_{desArg}$ and leukocytes were incubated together on the intimal surface of the aortas (discussed below).

In conclusion, responses of isolated blood vessels to C3a and C5a are in good agreement with reported effects of anaphylatoxin in vivo. The importance of prostanoids in anaphylatoxin-induced vascular effects were seemingly confirmed by these studies in a blood-free environment.

B. Polymorphonuclear Leukocytes (PMNs)

The PMN response is the best characterized of the many interactions between various cells and anaphylatoxins. Cell surface receptors that are specific for the ligand C5a were identified on the PMN.[10] Cross-linking experiments demonstrated a membrane component of 40,000 to 45,000 mol wt to which C5a could be covalently attached.[11-13] The ligand-receptor interaction on neutrophils promotes adherence, chemotaxis, oxygen burst, and granular enzyme release of the PMN.[14]

PMN adherence and chemotaxis are important early events in host defense and is prerequisite for phagocytosis and the removal of foreign substances. These two PMN functions induced by the anaphylatoxins are now well recognized. C5a induces PMN adherence and aggregation both in vitro[52-54] and in vivo.[6,7,18,55,56] The in vivo response is observed as a transient neutropenia caused either by C5a infusion or by ICA resulting from such treatments as extracorporeal circulation or cobra venom factor administration. ICA causes thrombocytopenia in addition to neutropenia,[6-7] something that is not observed after injection of purified C5a.[18] This discrepancy between ICA and purified C5a is probably attributed to the effect of C3 or C3 derivatives on platelet activation.[6]

In rabbits rendered neutropenic with nitrogen mustard, C5a is still capable of inducting hemodynamic changes of the same magnitude as observed in untreated animals.[18] Furthermore, indomethacin completely blocked the hemodynamic changes mediated by C5a, but did not affect the C5a-induced neutropenia. Therefore, C5a-induced neutropenia and hemodynamic changes appear to be independent or unrelated phenomena.

C5a is an important chemotactic stimulus not only for neutrophils but also for eosinophils, basophils, lymphocytes, and monocytes.[14] Chemotactic activity has been reported for C3a,[57,58] but is at least 1000-fold less active than C5a in this respect[59,60] and may therefore have little physiologic consequence.

PMN accumulation at an inflammatory site depends on both an adherent and chemotactic response to the stimulus. As discussed above, local application of C5a is followed by PMN accumulation at that site as well as a PMN-dependent plasma leakage. In addition to adherence, aggregation, and chemotaxis, C5a also induces a respiratory burst with release of oxygen radicals.[61] This respiratory burst has been shown to depend on PMN surface adhesion.[62] Finally, C5a is able to stimulate neutrophil granular enzyme release in cytochalsin B-treated cells or during "frustrated phagocytosis".[63]

In conclusion, the PMN response to C5a is the best characterized of interactions between cells and anaphylatoxins. Despite the profound effect C5a has on PMN activation, it appears that this response is totally unrelated and independent from the hemodynamic effects induced by C5a.

C. Mast Cells

The mast cell has long been a focus for the actions of anaphylatoxins in vivo because vasoamines are released following C3a and C5a challenge of mast cells.[64] The factors C3a and C5a stimulate mast cell release in an independent fashion, the latter being the most potent. Unlike the PMNs, receptors for the anaphylatoxins on mast cells have not been characterized. However, high levels of C3a bind to rat peritoneal mast cells. C3a binding is followed by a rapid chymase-dependent degradation and dissociation from the cell as recently reported by Gervasoni et al.[65] This degradation of the ligand on the surface of the mast cell may explain the difficulty that invstigators have had in demonstrating anaphylatoxin receptors on these cells. The mast cell is known to contain potent biologic mediators and to have a capacity to generate, de novo, active biologic material.[66] Therefore, the mast cell must be considered one of the most important cell targets for direct or indirect activation by C3a and C5a.

D. Endothelial Cells

Recently we reported that both C3a and C5a fail to stimulate cultured human endothelial cells, as measured by prostacyclin release. We also found that ^{125}I-labeled C5a did not bind to these cells.[67] However, when C5a, but not C3a, is added to a PMN-endothelial cell mixture, the endothelial cells produce and release prostacyclin (Figure 3).[67] These results were interpreted as evidence that the PMN serves as a target cell for C5a and is stimulated to release a factor(s) that in turn activates the endothelial cell. Miller et al.[68] reported that endothelial cells release prostacyclin when incubated with fMLF-stimulated PMNs. The factor from PMN that stimulates prostacyclin release from endothelial cells was either released concomitantly with the azurophilic granule or is closely related to this event. Harlan and Callahan[69] showed that hydrogen peroxide derived from activated PMNs is able to stimulate prostacyclin release from endothelial cells. In agreement with these results Rampart et al.[51] reported that leukocytes incubated with an intact endothelial cell layer on isolated rabbit aortas greatly enhanced $C5a_{desArg}$ stimulation of prostacylin release from endothelial cells. These authors also suggested that hydrogen peroxide could possibly explain the enhanced prostacyclin release stimulated by $C5a_{desArg}$ in the presence of leukocytes. Further-

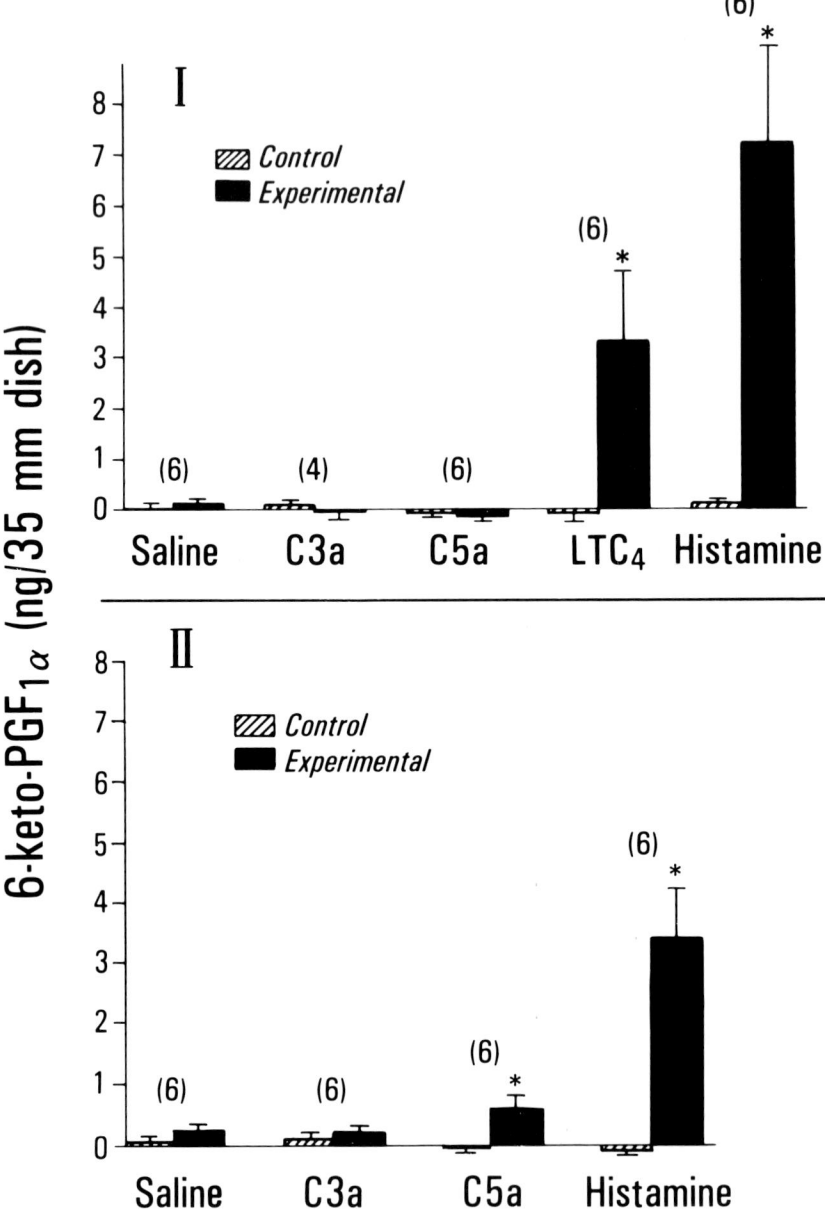

FIGURE 3.   6-keto-PGF$_{1\alpha}$ output from cultured human endothelial cells is measured for two consecutive 15-min periods (i.e., first a control and then an experimental period). Effector substances were applied at the beginning of the second (experimental) 15-min. period. Results shown in the lower panel (II) indicate release when isolated human PMNs were added to endothelial cells together with effector substances. C5a alone fail to release 6-keto-PGF$_{1\alpha}$ from the endothelial cells (Panel I), whereas C5a and PMNs together induced small but significant quantities of this prostanoid to be released from the endothelial cells (Panel II). Results are expressed as a mean ± SEM and the repetition of experiments performed are given in brackets. Asterisks indicate a significant difference as compared to the control period (P < 0.05, Students paired *t*-test). (Adapted from Lundberg, C. et al., *Immunopharmacology,* 12, 135, 1986.)

more, released leukotriene (LT) A_4 from activated PMNs is reportedly taken up by mast cells and utilized for LTC_4 production.[70] The lipoxygenase product LTC_4 has been shown to stimulate prostacyclin release from endothelial cells.[67,71]

Although the PMN may serve as a C5a target cell and products released from this cell can stimulate the endothelial cell to produce prostanoids, this is probably not the prime mechanism underlying hemodynamic effects of C5a. As discussed above, isolated blood vessels in a blood-free environment continue to release prostacyclin after C5a challenge,[49,50] and neutropenic animals also respond to C5a with hypotension and increase plasma prostanoid levels.[18] These studies, therefore, present evidence that strongly argues against a major role for the PMN in mediating vascular effects of C5a.

In conclusion, it appears that the endothelial cell is not a target cell for the anaphylatoxins. However, it seems that the endothelium functions as an intermediate, stimulated by released mediators from other anaphylatoxin target cells to release prostanoids which in turn affect the hemodynamic response.

IV. CONCLUSION

A more complete picture of the vascular effects of the anaphylatoxins, particularly C5a, is now coming into focus. Many of the biological consequences of endotoxin shock and intravascular complement activation can be attributed to effects of the anaphylatoxin C5a. The hemodynamic response following C5a appears to be mediated primarily by prostanoids and histamine. The net effect of vasodilatory prostaglandins and vasoconstrictive thromboxane release appears to vary in different organs. However, in the lung, thromboxane-dependent vasoconstriction prevails, followed by decreased cardiac output causing arterial hypotension as a consequence. Direct effect of C5a on smooth muscle cells is not likely, but rather an indirect effect depending on one or more intermediate cells. Platelets, PMNs, and endothelial cells are not directly involved in C5a-induced effects on circulation; however, the endothelium is suggested to function as an intermediate cell source. Hemodynamic effects by C5a are presumed to be targeted directly or indirectly via the mast cell, although evidence for the exact mechanism involved is still lacking. The mast cell releases potent biologic mediators, among them are histamine and TxA_2. These mediators can stimulate smooth muscle cells or activate other cells like the endothelium to produce and release the prostanoids that play a central role in eliciting the ultimate physiologic consequence (Figure 4).

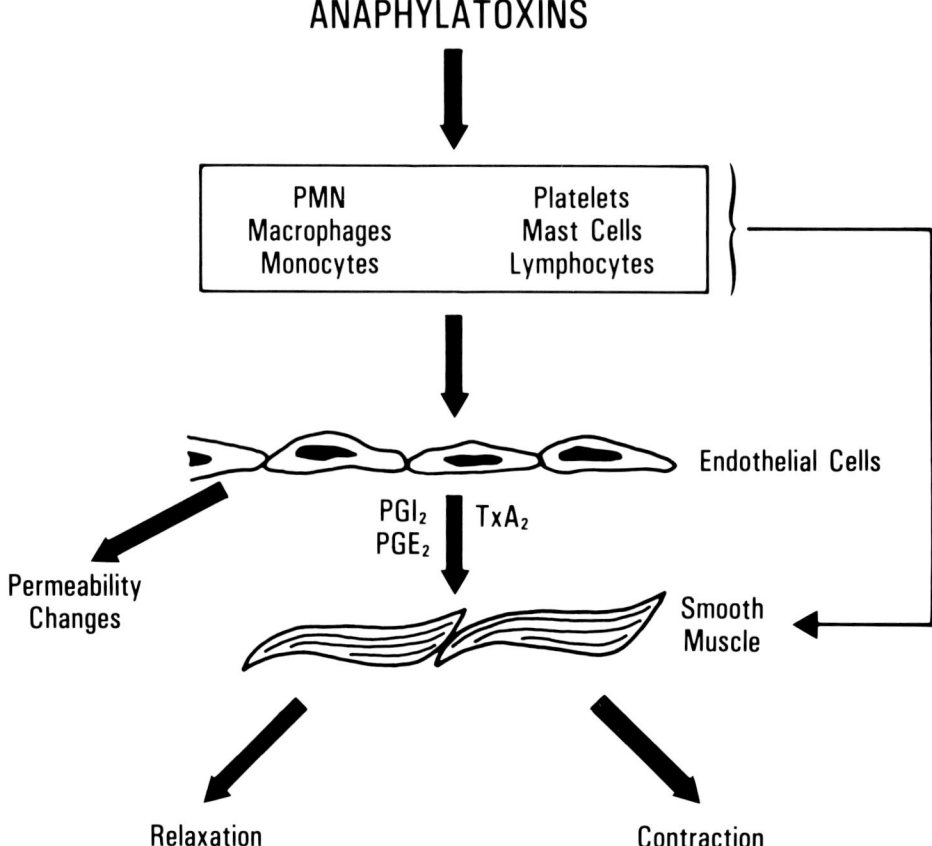

FIGURE 4.   Proposed interactions that may result in anaphylatoxin-mediated vascular responses in vivo. Anaphylatoxins activate both circulating blood cells and resident tissue cells. These activated cells in turn stimulate the endothelium via an array of cell-derived mediators. Thus C3a and C5a induce indirect effects at the endothelium resulting in the formation of intercellular gaps leading to enhanced permeability and plasma leakage. Furthermore, stimulation of endothelial cells may cause release of smooth-muscle-relaxing factors such as prostacyclin (PGI_2), prostaglandin E_2 (PGE_2), and perhaps even endothelium-derived relaxing factors (EDRF). Factors may also be released that contract smooth muscle, such as thromboxane A_2 (TxA_2). Anaphylatoxin-induced smooth muscle relaxation and contraction does not necessarily require involvement of the endothelial cell. Various mediators released from anaphylatoxin-activated circulating and tissue cells are also capable of affecting a smooth muscle tissue response.

REFERENCES

1. **Bokisch, V. A., Top, F. H., Russell, P. K., Dixon, F. J., and Muller-Eberhard, H. J.,** The potential pathogenic role of complement in dengue hemorrhagic shock syndrome,   *N. Engl. J. Med.,* 289, 996, 1973.
2. **McGabe, W. R.,** Serum complement levels in bacteremia due to gram-negative organisms, *N. Engl. J. Med.,* 288, 21, 1973.
3. **Fearon, D. T., Ruddy, S., Schur, P. H., and McGabe, W. R.,** Activation of the properdin pathway of complement in patients with gram-negative bacteremia, *N. Engl. J. Med.,* 292, 937, 1975.
4. **Gardinali, M., Cicardi, M., Agostoni, A., and Hugli, T. E.,** Complement activation in extracorporeal circulation: physiological and pathological implications, *Pathol. Immunopathol. Res.,* 5, 352, 1987.

5. **Ulevitch, R. J., Cochrane, C. G., Henson, P. M., Morrison, D. C., and Doe, W. F.,** Mediation systems in bacterial lipopolysaccharide-induced hypotension and disseminated intravascular coagulation. I. The role of complement, *J. Exp. Med.,* 142, 1570, 1975.

6. **Ulevitch, R. J. and Cochrane, C. G.,** Complement-dependent hemodynamic and hematologic changes in the rabbit, *Inflammation,* 2, 199, 1977.

7. **Bult, H., Herman, A. G., Laekeman, G. M., and Rampart, M.,** Formation of prostanoids during intravascular complement activation in the rabbit, *Br. J. Pharmacol.,* 84, 329, 1985.

8. **Hugli, T. E.,** The structural basis for anaphylatoxin and chemotactic functions of C3a, C4a and C5a, *CRC Crit. Rev. Immunol.,* 1, 321, 1981.

9. **Gorski, J. P., Hugli, T. E., and Muller-Eberhard, H. J.,** C4a: the third anaphylatoxin of the human complement system, *Proc. Natl. Acad. Sci. U.S.A.,* 76, 5299, 1979.

10. **Chenoweth, D. E. and Hugli, T. E.,** Demonstration of specific C5a receptor on intact human polymorphonuclear leukocytes, *Proc. Natl. Acad. Sci. U.S.A.,* 75, 3943, 1978.

11. **Huey, R. and Hugli, T. E.,** Characterization of a C5a receptor on human polymorphonuclear leukocytes (PMN), *J. Immunol.,* 135, 2063, 1985.

12. **Johnson, J. R. and Chenoweth, D. E.,** Labeling the granulocyte C5a receptor with a unique photoreactive probe, *J. Biol. Chem.,* 260, 7161, 1985.

13. **Rollins, T. E. and Springer, M. S.,** Identification of the polymorphonuclear leukocyte C5a receptor, *J. Biol. Chem.,* 260, 7157, 1985.

14. **Hugli, T. E.,** Structure and function of the anaphylatoxins, *Springer Semin. Immunopathol.,* 7, 193, 1984.

15. **Bodammer, G. and Vogt, W.,** Actions of anaphylatoxin on circulation and respiration of the guinea pig, *Int. Arch. Allergy,* 32, 417, 1967.

16. **Bodammer, G.,** Untersuchungen uber den Mechanismus der Blutdruckswirkung des Anaphylatoxins bei Katzen und Meerschweinchen, *Naunyn-Schmiedebergs Arch. Pharmakol. Exp. Pathol.,* 262, 197, 1969.

17. **Pavek, K., Piper, P. J., and Smedegard, G.,** Anaphylatoxin-induced shock and two patterns of anaphylactic shock: hemodynamics and mediators, *Acta. Physiol. Scand.,* 105, 393, 1979.

18. **Lundberg, C., Marceau, F., and Hugli, T. E.,** C5a-induced hemodynamic and hematologic changes in the rabbit. Role of cyclooxygenase products, and polymorphonuclear leukocytes, *Am. J. Pathol.,* 128, 471, 1987.

19. **Borg, T., Gerdin, B., Hallgren, R., Warolin, O., and Modig, J.,** Complement activation and its relationship to adult respiratory distress syndrome. An experimental study in pigs, *Acta Anaesthesiol. Scand.,* 28, 158, 1984.

20. **Hicks, R. and Sackeyfio, A. C.,** The nature of adrenergic mechanisms involved in anaphylatoxin activity in the guinea-pig, *Br. J. Pharmacol.,* 46, 260, 1972.

21. **Cooper, J. D., McDonald, J. W. D., Ali, M., Menkes, E., Masterson, J., and Klement, P.,** Prostaglandin production associated with the pulmonary vascular response to complement activation, *Surgery,* 88, 215, 1980.

22. **Gee, M. H., Perkowski, S. Z., Tahamont, M. V., and Flynn, J. T.,** Arachidonate cyclooxygenase metabolites as mediators of complement initiated lung injury, *Fed. Proc.,* 44, 46, 1985.

23. **Gee, M. H., Perkowski, S. Z., Tahamont, M. V., Flynn, J. T., and Wasserman, M. A.,** Thromboxane as a mediator pulmonary dysfunction during intravascular complement activation in sheep, *Am. Rev. Respir. Dis.,* 133, 269, 1986.

24. **Perkowski, S. Z., Havill, A. M., Flynn, J. T., and Gee, M. H.,** Role of intrapulmonary release of eicosanoids and superoxide anion as mediators of pulmonary dysfunction and endothelial injury in sheep with intermittent complement activation, *Circ. Res.,* 53, 574, 1983.

25. **Meyrick, B. O. and Brigham, K. L.,** The effect of a single infusion of zymosan-activated plasma on the pulmonary microcirculation of sheep, structure-function relationships, *Am. J. Pathol.,* 114, 32, 1984.

26. **Johnson, A., Blumenstock, F. A., Hussain, M., and Malik, A. B.,** Differential effects of complement activation induced by cobra venom factor on pulmonary transvascular fluid and protein exchange, *Am. J. Pathol.,* 114, 410, 1984.

27. **Larsen, G. L., Webster, R. O., Worthen, G. S., Gumbay, R. S., and Henson, P. M.,** Additive effect of intravascular complement activation and brief episode of hypoxia in producing increased permeability in the rabbit lung, *J. Clin. Invest.,* 75, 902, 1985.

28. **Rampart, M., Bult, H., and Herman, A. G.,** Contribution of complement activation to the rise in blood levels of 6-oxo-prostaglandin F-1a during endotoxin-induced hypotension in rabbits, *Eur. J. Pharmacol.,* 79, 91, 1982.

29. **Rampart, M., Zonnekeyn, L., Bult, H., and Herman, A. G.,** Prostacyclin biosynthesis and hypotension in relation to complement activation in rabbit endotoxic shock, *Biomed. Biochem. Acta,* 43, S191, 1984.

30. **Winn, R., Harlan, J., Nadir, B., Harker, L., and Hildebrandt, J.,** Thromboxane A2 mediates lung vasoconstriction but not permeability after endotoxin, *J. Clin. Invest.,* 72, 911, 1983.

31. **Borg, T., Gerdin, B., and Modig, J.,** Prophylactic and delayed treatment with indomethacin in a porcine model of early adult respiratory distress syndrome induced by endotoxaemia, *Acta Anaesthesiol. Scand.,* 30, 47, 1986.

32. **Ogletree, M. L. and Brigham, K. L.,** Arachidonate reaises vascular resistance but not permeability in lungs of awake sheep, *J. Appl. Physiol.,* 48, 581, 1980.

33. **Egan, T. M., Saunders, N. R., Luk, S. C., and Cooper, J. D.,** Granulocytes mediate hypoxemia in sheep infused with activated complement, *Physiologist,* 26, 36, 1983.

34. **Rocha e Silva, M., Bier, O., and Aronson, M.,** Histamine release by anaphylatoxin, *Nature (London),* 168, 465, 1951.

35. **Giertz, H., Hahn, F., Opperkuch, JW., and Schmutzler, W.,** Vergeichende Untersuchungen uber den anaphylaktischen Schock und den Anaphylatoxinschock an der isolierten Meerschweincheenlunge, *Naunyn-Schmiedebergs Arch. Pharmakol. Exp. Pathol.,* 242, 42, 1961.

36. **Sackeyfio, A. C.,** Anaphylatoxin-induced release of a substance with prostaglandin-like activity in isolated guinea-pig lungs, *Br. J. Pharmacol.,* 46, 544P, 1972.

37. **Friedberg, K., Engelhardt, G., and Meineke, F.,** Untersuchungen über die Anaphylatoxin-Tachphylaxie und über ihre Bedeutung fur den Ablaufechter anaphylaktischer Reactionen, *Int. Arch. Allergy Appl. Immunol.,* 154, 1964.

38. **Marceau, F. and Hugli, T. E.,** Effect of C3a and C5a anaphylatoxins on guinea-pig isolated blood vessels, *J. Pharmacol. Exp. Ther.,* 230, 749, 1984.

39. **Bult, H., Herman, A. G., and Rampart, M.,** Modification of endotoxin-induced haemodynamic and haematological changes in the rabbit by methylprednisolone, F(ab′)2 fragments and rosmarinic acid, *Br. J. Pharmacol.,* 84, 317, 1985.

40. **Webster, R. O., Larsen, G. L., Mitchell, B. C., Goins, A. J., and Henson, P. M.,** Absence of inflammatory lung injury in rabbits challenged intravascularly with complement-derived chemotactic factors, *Am. Rev. Respir. Dis.,* 125, 335, 1982.

41. **Issekutz, A. C.,** Quantitation of acute inflammation in the skin: recent methodological advances and their application to the study of inflammatory reactions, *Surv. Synth. Pathol. Res.,* 1, 89, 1983.

42. **Lundberg, C. and Arfors, K. E.,** Polymorphonuclear leukocyte accumulation in inflammatory dermal sites as measured by 51-Cr-labeled cells and myeloperoxidase, *Inflammation,* 7, 257, 1983.

43. **Lundberg, C., Lindbom, L., Lundberg, K., Harlan, J. M., and Arfors, K. E.,** The role of neutrophil membrane glycoprotein CDw18 (LFA) in neutrophil accumulation in vivo, *Fed. Proc.,* 45, 1156, 1986.

44. **Wedmore, C. V. and Williams, T. J.,** Control of vascular permeability by polymorphonuclear leukocytes in inflammation, *Nature (London),* 289, 646, 1981.

45. **Bjork, J., Hugli, T. E. and Smedegard, G.,** Microvascular effects of anaphylatoxins C3a and C5a, *J. Immunol.,* 134, 1115, 1985.

46. **Williams, T. J. and Morley, J.,** Prostaglandins as potentiators of increased permeability in inflammation, *Nature (London),* 246, 215, 1973.

47. **Mahler, F., Intaglietta, M., Hugli, T. E., and Johnson, A. R.,** Influences of C3a anaphylatoxin compared to other vasoactive agents on the microcirculation of the rabbit omentum, *Microvasc. Res.,* 9, 345, 1975.

48. **Regal, J. F.,** C5a-induced aortic contraction: effect on an antihistamine and inhibitors of arachidonate metabolism, *J. Pharmacol. Exp. Ther.,* 220, 102, 1982.

49. **Hugli, T. E. and Marceau, F.,** Effects of the C5a anaphylatoxin and its relationship to cyclo-oxygenase metabolities in rabbit vascular strips, *Br. J. Pharmacol.,* 84, 725, 1985.

50. **Rampart, M., Bult, H., and Herman, A. G.,** Activated complement and anaphylatoxins increase the in vitro production of prostacyclin by rabbit aorta endothelium, *Naunyn-Schmiedebergs Arch. Pharmacol. Exp. Pathol.,* 322, 158, 1983.

51. **Rampart, M., Jose, P. J., and Williams, T. J.,** Leucocytes activated by C5a des Arg promote endothelial prostacyclin (PG12) production via release of oxygen species in vitro, *Agents Actions,* 16, 21, 1985.

52. **O'Flaherty, J. Y., Kreutzer, D. L., and Ward, P. A.,** Chemotactic factor influences on the aggregation, swelling, and foreign surface adhesiveness of human leukocytes, *Am. J. Pathol.,* 90, 537, 1978.

53. **Chenoweth, D. E., Rowe, J. G., and Hugli, T. E.,** A modified method for chemotaxis under agarose, *J. Immunol. Meth.,* 25, 337, 1979.

54. **Tonnesen, M. G., Smedly, L. A., and Henson, P. M.,** Neutrophil-endothelial cell interactions. Modulation of neutrophil adhesiveness induced by complement fragments C5a and C5a des Arg and formyl-methionyl-leucyl-phenylalanine in vitro, *J. Clin. Invest.,* 74, 1581, 1984.

55. **Craddock, P. R., Hammerschmidt, D., White, J. G., Dalmasso, A. P., and Jacob, H. S.,** Complement (C5a)-induced granulocyte aggregation in vitro. A possible mechanism of complement-mediated leukostasis and leukopenia, *J. Clin. Invest.,* 60, 260, 1977.

56. **O'Flaherty, J. T., Craddock, P. R., and Jacob, H. S.,** Effect of intravascular complement activation on granulocyte adhesiveness and distribution, *Blood,* 51, 731, 1978.

57. **Ward, P.H.,** A plasmin-split fragment of C3 as a new chemotactic factor, *J. Exp. Med.,* 126, 189, 1967.

58. **Hill, J. H. and Ward, P. A.,** C3 leukotactic factors produced by a tissue portease, *J. Exp. Med.,* 130, 505, 1969.
59. **Damerau, B., Grunefeld, E., and Vogt, W.,** Chemotactic effects of the complement-derived peptides C3a, C3ai and C5a (classical anaphylatoxin) on rabbit and guinea-pig polymorphonuclear leukocytes, *Naunyn-Schmiedebergs Arch. Pharmacol.,* 305, 181, 1978.
60. **Fernandez, H. N., Henson, P. M., Otani, A., and Hugli, T. E.,** Chemotaxis response to human C3a and C5a anaphylatoxins. I. Evaluations of C3a and C5a leukotaxis and under stimulated in vivo conditions in vitro, *J. Immunol.,* 120, 109, 1978.
61. **Goldstein, I. M. and Wsissmann, G.,** Complement and immunoglobulins stimulate superoxide production by human leukocytes independently of phagocytosis, *J. Clin. Invest.,* 57, 1155, 1975.
62. **Dahinden, C. A., Fehr, J., and Hugli, T. E.,** Role of cell surface contact in the kinetics of superoxide production by granulocytes, *J. Clin. Invest.,* 72, 113, 1983.
63. **Henson, P. M., Zanolari, B., Schwartzman, N. A., and Hong, S. R.,** Intracellular control of human neutrophil secretion. I. C5a-induced stimulus-specific desenitization and the effects of cytochalasin, *Br. J. Pharmacol.,* 121, 851, 1978.
64. **Johnson, A. R., Hugli, T. E., and Muller-Eberhard, H. J.,** Release of histamine from rat mast cells by the complement peptides C3a and C5a, *Immunology,* 28, 1067, 1975.
65. **Gervasoni, J. E., Conrad, D. H., Hugli, T. E., Schwartz, L. B., and Ruddy, S.,** Degradation of human anaphylatoxin C3a by rat peritoneal mast cells: a role for the secretory granule enzyme chymase and heparin proteoglycan, *J. Immunol.,* 136, 285, 1986.
66. **Wasserman, S. I.,** The mast cell: its diversity of chemical mediators, *Int. J. Dermatol.,* 19, 7, 1980.
67. **Lundberg, C., Marceau, F., Huey, R., and Hugli, T. E.,** Anaphylatoxin C5a fails to promote prostacyclin release in cultured endothelial cells from human umbilical veins, *Immunopharmacology,* 12, 135, 1986.
68. **Miller, D. K., Sadowski, S., Soderman, D. D., and Kuehl, F. A.,** Endothelial cell prostacyclin production induced by activated neutrophils, *J. Biol. Chem.,* 260, 1006, 1985.
69. **Harlan, J. M. and Callahan, K. S.,** Role of hydrogen peroxide in the neutrophil-mediated release of prostacyclin from cultured endothelial cells, *J. Clin. Invest.,* 74, 442, 1984.
70. **Dahinden, C. A., Clancy, R. M., Gross, M., Chiller, J. M., and Hugli, T. E.,** Leukotriene C4 production by murine mast cells: evidence of a role for extracellular leukotriene A4, *Proc. Natl. Acad. Sci. U.S.A.,* 82, 6632, 1985.
71. **Cramer, E. B., Prologe, L., Pawlowski, N. A., Cohn, Z. A., and Scott, W. A.,** Leukotriene C promotes prostacyclin synthesis by human endothelial cells, *Proc. Natl. Acad. Sci. U.S.A.,* 80, 4109, 1983.
72. **Lundberg, K.,** unpublished observations.

Chapter 27

LYMPHOKINE MODULATION OF ENDOTHELIAL CELL MORPHOLOGY AND SURFACE ANTIGENS

Jordan S. Pober

TABLE OF CONTENTS

I. INTRODUCTION

Endothelium in the absence of inflammation appears as a flattened monolayer of biosynthetically inactive cells. This bland appearance led to descriptions of endothelium as a passive tissue, thought to provide a nonthrombogenic surface and to serve as an inert barrier between the cells and molecules of the blood and the cells and molecules of the underlying vessel wall and tissue space. In certain settings, such as a site of a cell-mediated immune response, endothelial cell (EC) appearance (and function) was noted to change dramatically. Specifically, the EC became plump or "hypertrophic" and filled with biosynthetic organelles; such EC were described as "activated".[1] However, neither the basal state nor the activated state could be precisely defined exclusively by histological criteria. Recently, it has been observed that injection of immune-generated mediators (i.e., a lymphokine preparation) can cause morphological changes indistinguishable from cell-mediated immunity.[2] Furthermore, it has been shown that cultured EC are directly responsive to lymphokines, an in vitro model of the changes observed in situ.[3,4] The isolation, characecization, and genetic cloning and expression of individual lymphokines in the past few years has revolutionized the study of these previously elusive molecules. In this review, the findings from my laboratory and those of my colleagues on the effects of several individual immune mediators upon EC morphology and surface antigen expression will be presented. The implications these studies have for EC activation in situ will be discussed.

II. ACTIONS OF SPECIFIC MEDIATORS ON HUMAN VASCULAR ENDOTHELIAL CELLS IN VITRO

A. Immune Interferon

Immune interferon (IFN-γ) is a protein mediator produced by activated T lymphocytes as part of the response to antigen.[5] It is encoded by a single gene and is secreted as an 18-kdalton protein core with one or two glycan moieties; the native state may be a dimer. Although it shares with leukocyte interferon (IFN-α) and fibroblast interferon (IFN-β) antiviral activties, including induction of the enzyme 2'-5' oligoadenylate synthetase, it appears uniquely active in enhancing immune responsiveness. These unique immunomodulatory effects of IFN-γ appear related to the fact that IFN-γ binds to a distinct cell surface receptor, different from the interferon receptor shared by IFN-α and IFN-β. At the cellular level, IFN-γ has been found to have multiple effects upon the cells of the immune system, including B cell differentiation and macrophage activation. IFN-γ also appears more potent than IFN-α or -β in increasing expression of cell surface major histocompatibility complex (MHC) antigen expression, especially the expression of class II antigens. Finally, IFN-γ also has some unique actions upon cells not traditionally thought of as part of the immune systems. For example, IFN-γ has been reported to be more effectively cytostatic or cytoretardive for fibroblasts than IFN-α or IFN-β,[6,7] although these experiments were performed with natural IFN-γ preparations that may well have contained other cytostatic factors (e.g., the T cell product lymphotoxin). IFN-γ also appears to induce proteins in a human lung fibroblast line not inducible with IFN-α or IFN-β. The relation of these effects on nonimmune cells to the immunomodulatory properties of IFN-γ is unclear.

The actions of IFN-γ upon MHC antigen expression is of particular interest to EC biology. Transplant immunologists had shown in the 1970s that cultured EC could induce allogeneic T lymphocytes to proliferate,[9] analoguous to the ability of peripheral blood mononuclear cells to stimulate allogeneic T cell proliferation. The major cell surface molecules on peripheral blood mononuclear cells recognized as "foreign" in this assay (i.e., the principal molecular signal seen by the responsive T cells) are class II MHC antigens (called I-region associated [Ia] antigens in mice or HLA-D related antigens [HLA-DR, DP and DQ] in humans).[10] It was subsequently shown that EC cultures could stimulate proliferation of HLA-D region-primed lymphocytes (i.e., effectively restimulate class II antigen responsive T cells) and present soluble or viral antigens to sensitized HLA-D region restricted T lymphocytes (i.e., effectively restimulate T cells that only respond to nominal antigens in combination with a self class II MHC antigen, such as HLA-DR.[12-14] These functional data strongly suggested that EC expressed class II MHC antigens. Indeed, human allosera used for HLA-DR typing were cytotoxic for cultured human EC,[15] although the patterns of reactivity were not fully concordant with HLA-DR from lymphocytes of the same donors, suggesting that other non-MHC alloantigens might be involved. Remarkably, however, none of the multiple murine monoclonal antibodies reactive with class II antigens bound to cultured EC.[16] Subsequent experiments showed that activated T cells, their products, and specifically IFN-γ (produced through recombinant gene technology and thus free of contaminating factors) all could induce class II antigen biosynthesis and expression by EC.[17] Class II MHC antigens are thus the first described EC activation antigens, induced in an unstimulated population by addition of a lymphokine. Undoubtedly, the T cells used in the proliferation assays had induced the EC to express class II molecules,[17] initiating in turn the specific T cell proliferation.

Much additional information has since emerged about the modulation of MHC antigens on EC: (1) class II antigen expression is induced *de novo,* and class I (HLA-A,B) antigen expression is also increased; (2) optimal concentrations of IFN-α and -β comparably increase class I antigen expression, but do not induce class II antigen expression; (3) surface antigen changes are first detectable at about 4 hr, reach 50% levels by 24 hr, and plateau by 4 to 6 days; (4) mRNA levels for class II antigens are undetectable in untreated cells and plateau by 48 hr, preceding the maximal change in cell surface protein; (5) both mRNA and protein levels persist as long as IFN-γ remains in the culture; mRNA drops to undetectable levels within 48 hr of removing IFN-γ, and protein levels more gradually decline; (6) all known human class II antigens (HLA-DR, -DP, and -DQ) are concomitantly induced, as well as the non-MHC encoded invariant chain, although HLA-DQ is expressed to a lower level than HLA-DR; (7) every EC in the culture coordinately responds to IFN-γ; and (8) the serologically detectable class II antigens on EC can function to activate T cells.[18,19]

The ability of EC to activate resting T cells, as measured by allogeneic T cell proliferation, appears unique among non-bone marrow-derived cells. In particular, dermal fibroblasts or vascular smooth muscle cells are much less efficient at T cell activation and often appear inhibitory.[18,20,21] We initially thought that the ability of EC to express class II antigens might account for this difference. This does not appear to be the case. In fact, cultured dermal fibroblasts and vascular smooth muscle cells express quantitatively similar amounts and densities of class II antigens in response to IFN-γ. In the case of fibroblasts, we have also shown that the induced class II molecules can be efficiently recognized by cloned (i.e., previously activated) cytolytic or helper T cells.[18,22] The unique properties of EC in inducing an unprimed allogeneic response must reside in some other property critical for activating *resting* T cells, perhaps shared with dendritic cells.[23]

IFN-γ affects EC surface molecules other than MHC antigens. Surface iodination experiments showed that both EC and fibroblasts treated with IFN-γ increased expression of a 17-kdalton polypeptide.[18] IFN-γ also increases expression of an antigen on EC[24] and fibroblasts[25] recognized by monoclonal antibody RR1/1. In the case of RR1/1 binding, the time course enhanced expression and the persistence in the presence of mediator resembles the time course of class I MHC modulation. In contrast, however, RR1/1 binding is unaffected by IFN-α or IFN-β.

IFN-γ also can induce expression of a surface antigen(s) specific for EC. This conclusion is based upon our studies on the pathogenesis of Kawasaki's Syndrome (KS), a childhood panvasculitis of unknown etiology.[26] Children in the acute phase of KS, but not the convalescent phase or age matched normals or febrile controls, have circulating IgM antibody which lyses IFN-γ-treated EC but not control EC, control or IFN-γ-treated fibroblasts, or control or IFN-γ-treated smooth muscle cells. The antibody cannot be adsorbed out with various leukocytes, further ruling out MHC antigens as the target structure. The time course of induction of the target antigen is similar to MHC antigen modulation. Neither IFN-α nor IFN-β can induce these antigens.

IFN-γ actions on EC are not confined to antigenic modulation. IFN-γ is cytostatic for EC cultures.[27] Furthermore, IFN-γ causes a profound reorganization in EC morphology.[27] In control cultures, EC are polygonal and strictly contact inhibited in migration (i.e., non-overlapping). The actin cytoskeleton is organized into dense peripheral bands, and EC elaborate an underlying fibronectin "basket-weave" matrix. Upon addition of IFN-γ, EC rapidly lose (by 24 hr) both their actin-containing dense peripheral bands and their stainable fibronectin substratum. By 48 to 72 hr, the EC become markedly elongated and reorganize their actin filaments into parallel longitudinal bundles of stress fibers. The cells in such treated cultures extensively overlap and expose focal gaps in the monolayer, revealing substratum. EC do not divide during this rearrangement, implying that although contact inhibition of migration is lost, contact inhibition of growth is not. The IFN-γ concentration dependence for EC morphological change is the same as that for MHC antigen modulation (threshhold 1 to 10 U/mℓ, maximal 100 to 200 U/mℓ). The morphological changes persist as long as IFN-γ remains in the culture medium but reverse, over several days, when IFN-γ is removed. IFN-α and IFN-β have no effect upon EC morphology. We also have not seen IFN-γ induce similar morphologic changes in fibroblast or smooth muscle cultures. Morphological changes due to IFN-γ have been described by others in keratinocyte cultures,[28] but these changes were only detectable by the ultrastructural examination of individual cells. Thus the morphological rearrangement, like the induction of antigens recognized by KS sera, appears to be an EC-specific effect of IFN-γ.

In summary, IFN-γ is a potent modulator of the EC phenotype, producing both antigenic and morphological changes. Some of these are pleiotropic whereas others appear quite EC specific. To date, all of the IFN-γ effects on EC we have observed have several common features: they are slow, reaching plateau levels in 4 to 6 days; persist in the presence of mediator; reverse slowly upon its withdrawal; and show similar concentration dependence (threshhold 1 to 10 U/mℓ, maximal 100 to 200 U/mℓ). We define these changes to constitute the "long program" of EC activation.

B. Interleukin-1

Interleukin-1 (IL-1) is a pleiotropic protein mediator of inflamation and immunity.[29] The originally defined source of IL-1 was the stimulated mononuclear phagocyte, but many cells, including EC, synthesize and secrete IL-1. The predominant active protein species isolated from human monocyte culture supernatants has a weight range of 14 to 17 kDaltons and a neutral isolectric point. Minor species of similar size but more acidic isolectric points have also been detected. This predominant species is translated from an mRNA now called IL-1 β;[30] a second mRNA (called IL-1α), encoding a distinct IL-1 protein, has also been cloned and sequenced.[31] These IL-1 messages are transcribed from different genes, and other IL-1 species derived from additional genes may also exist. All IL-1 molecules by definition have in common the ability to act as comitogens for murine thymocytes stimulated with conconvalin A or phytohemagglutinin; the murine T cell line D10.G1.4 appears to provide a more sensitive assay for the same activity.[32] Many other activities ascribed to IL-1 (e.g., fever production, stimulation of hepatic cell synthesis of acute phase reactants, etc.) may or may not be shared by all IL-1 species. Although the primordial signal for IL-1 production may have been encounter with a microorganism, IL-1 secretion is also induced by activation of the immune system by antigen.[33] In this sense, it is a lymphokine (i.e., an immune-generated mediator), and is generated locally as part of a cell-mediated immune repsonse (i.e., is present at the site of EC activation).

IL-1 has direct effects upon cultured EC. This was first demonstrated by Bevilacqua and colleagues who found that purified human monocyte-derived IL-1 induced EC to biosynthesize and express on their surface a tissue factor-like procoagulant activity.[34] These results have been subsequently confirmed with recombinant IL-1α and IL-1β.[35] IL-1 also causes EC to become adhesive for inflammatory leukocytes[36] and lymphocytes.[37] These functional actions of IL-1 can be blocked by RNA and protein synthesis inhibitors, but not cyclooxygenase inhibitors, and reach maximal expression at 4 to 6 hr. Remarkably, the activities then spontaneously decline whether or not IL-1 remains in the medium, nearing basal values by 24 hr. If IL-1 was not removed, the EC cultures become refractory to restimulation by IL-1; a "rest" in the absence of IL-1 is required before the EC can be restimulated.

To test the hypothesis that these IL-1-induced functions depend upon new EC surface proteins, we raised monoclonal antibodies against IL-1-stimulated umbilical vein EC. One such antibody, H4/18,[38] showed reactivity that paralleled the functional changes: it fails to bind to unstimulated EC; its binding is induced by IL-1; induction depends upon RNA and protein synthesis but is unaffected by cyclooxygenase inhibition; binding is maximal at 4 to 6 hr and declines spontaneously to basal levels by 24 hr; and EC maintained in the presence of IL-1 for 24 hr cannot be restimulated to bind H4/18 by IL-1. The IL-1 concentration dependence for induction of H4/18 binding also parallels the functional changes (detectable at 0.5 to 1 U/mℓ, maximal at 10 to 20 U/mℓ). Antibody H4/18 consistently blocks a portion of the IL-1-induced adhesion to EC of HL-60 cells suggesting that the antigen may serve as an inducible endothelial-leukocyte adhesion molecule (E-LAM). More recent studies with a second antibody to the same inducible protein have extended these results to identify this molecule (now termed E-LAM 1) as a ligand for neutrophil adhesion.[39]

Bacterial endotoxin is a weak inducer of both the functional changes (EC procoagulant activity and leukocyte adherence) as well as the induction of H4/18 binding.[38] Since endotoxin contamination of reagents is a common laboratory problem, the effects of IL-1 had to be separated from those of endotoxin. This was shown by selective heat inactivation of IL-1 preparations, by the selective neutralization of endotoxin by polymyxin B sulfate, by the selective neutralization of natural and recombinant IL-1 by antibody to IL-1, and by the selective loss of responsiveness of EC to one mediator or the other after 24 hr of treatment (i.e., EC treated with IL-1 for 24 hr remain sensitive to endotoxin and vice versa). Since endotoxin can induce EC to produce IL-1,[40-42] this final point is paradoxical; cultures making measurable IL-1 in response to endotoxin behave as if they have never seen IL-1 upon challenge with exogenous IL-1. Conceivably, EC-derived IL-1 differs from monocyte-derived IL-1 in its actions on endothelium. Although the structure of EC-derived IL-1 has not been fully worked out, this seems unlikely because EC-derived IL-1 can be transferred to activate other EC cultures and can be neutralized by sera reactive with monocyte-derived IL-1.[41] Furthermore, endotoxin-treated EC contain an mRNA species which hybridizes with IL-1 cDNA probes.[43] The resolution of this paradox is not yet known.

The antigen recognized by H4/18 is not the only surface change detectable in IL-1-treated EC. Antibody RR1/1 binding is also increased by IL-1.[24] The modulation of RR1/1 binding protein (also called intercellular adhesion molecule or ICAM-1) differs from the protein recognized by H4/18 by several criteria: RR1/1 binds to unstimulated EC whereas H4/18 does not; RR1/1 binding reaches maximal levels by 24 hr whereas H4/18 binding is maximal at 4 to 6 hr; and RR1/1 binding is sustained so long as IL-1 remains in the culture medium whereas H4/18 binding spontaneously declines. Furthermore, as already noted, RR1/1 binding can be increased by IFN-γ (although the rate of increase is slower than with IL-1), whereas H4/18 binding is unaffected by IFN-γ. Finally, RR1/1 binds to and is modulated on a number of cell types (e.g., dermal fibroblasts),[25] whereas H4/18 binding appears to be EC specific.[38,49] Sera from patients with KS can mediate complement-dependent lysis of IL-1-treated EC.[45] The expression of the IL-1 induced target antigen peaks at 4 hr and then declines to basal levels. Like the target antigen induced by IFN-γ, the IL-1 induced antigen appears to be EC-specific (i.e., not inducible on fibroblasts or smooth muscle cells). However, the IL-1-induced antigen clearly differs from the IFN-γ-induced antigen as shown by cross-adsorption experiments (i.e., IFN-γ-treated EC remove antibodies cytotoxic for IFN-γ-treated but not IL-1 treated EC and vice versa).

In summary, IL-1 is a potent modulator of EC phenotype. Many of its responses are rapid (compared to IFN-γ) and transient. We define these changes to constitute the "short program" of EC activation. However, as shown by the experiments with RR1/1 binding, not all IL-1-induced changes are transient or equally rapid.

C. Tumor Necrosis Factor

Tumor necrosis factor (TNF) is a pleiotropic protein mediator of inflammation and immunity.[46,47] To date, the only proven source is the activated mononuclear phagocyte, altough NK cells and other lymphoid cells may also secrete TNF. A single message (and gene) encoding TNF has been cloned, although the possibility of other natural TNFs has not been excluded. The mature protein is about 17 kdaltons with an isoelectric point of 5; it thus biochemically resembles an IL-1 species and may contaminate some natural IL-1 preparations. The native form of TNF may be oligomeric. TNF is functionally defined by its ability to produce cell death in certain tumor cell lines such as L929. The mechanism of action is not known, but cytotoxicity is slow and is enhanced by inhibitors of RNA and protein synthesis, perhaps by aborting "repair" mechanisms. Human TNF differs from IL-1 in that it does not act as a comitogen for thymocytes or the D10.G1.4 T cell line. However, TNF demonstrates many other IL-1-like properties (e.g., fever production). TNF may be the principal mediator of the mammalian response to endotoxin. Like IL-1, it is produced as part of the activation of the immune system by antigen,[48] and thus qualifies as a lymphokine (i.e., it will be present at sites of EC activation). It shows sequence homology and may share a cell surface receptor with lymphotoxin, an activated T cell product (see below).

TNF has direct actions upon EC. It shares with IL-1 the ability to induce rapid and transient expression of tissue-factor-like procoagulant activity,[35,49] leukocyte adhesion,[50] and expression of the protein recognized by H4/18.[38] These various effects have similar TNF concentration dependence (onset 1 to 10 U/mℓ, plateau 50 to 100 U/mℓ). TNF also induces the same target antigens recognized by sera from patients in the acute phase of KS as does IL-1 (i.e., TNF-treated EC can adsorb all of the cytotoxic activity against IL-treated EC and vice versa).[45] Finally, like IL-1, TNF induces rapid and sustained expression of the antigen recognized by RR1/1.[24] The effects of TNF upon these properties of EC can be distinguished from that of IL-1 by two functional criteria: (1) for some activities (e.g., procoagulant activity) the actions of TNF and IL-1 are additive at all concentrations;[35] (2) EC pretreated with ether mediator for 24 hr become selective unresponsive to that mediator, but remain fully responsive to other agent.[35-38] TNF can also be distinguished from endotoxin by the same criteria as applied for IL-1.

TNF can induce EC to secrete IL-1.[43,51] Again, this observation raises an unresolved paradox of why TNF-treated cells, which secrete IL-1, behave as if they have not been exposed to IL-1. Much of the IL-1 activity induced by TNF remains associated with the cell membrane where it may be optimally positioned to activate circulating blood cells.[65] Recent data suggest that IL-1 itself, like TNF, can induce EC biosynthesis of IL-1.[65,66] TNF, but not IL-1, leads to increased surface expression of class I MHC antigens.[52] This change is preceeded by an increase in the steady-state levels of mRNA (presumably by increased transcription). TNF shares this property with IFN-α and IFN-β and the actions of TNF on MHC antigen expression show IFN-like kinetics (i.e., maximal at day 4 to 6). The mechanism of TNF action appears different from that of IFN, as shown by the actions of the protein synthesis inhibitor cycloheximide (CHX). Specifically, we found that: (1) CHX has no direct effect upon MHC antigen mRNA levels in EC; (2) addition of CHX coincident with mediator has no effect upon the actions of IFN-γ; (3) in contrast, CHX addition markedly enhances the actions of IFN-α and IFN-β; but (4) CHX blocks the increase in class I mRNA induced by TNF. These observations suggest that TNF acts on MHC antigen expression through induction of a protein intermediate. IFN-β is an obvious candidate, but we found no effect of TNF upon the mRNA for IFN-β. Recently, we have found an increase in mRNA for another IFN-induced mRNA (2′-5′ oligoadenylate synthetase) and others have reported that TNF induces a protein in FS4 cells which copurifies and serologically cross reacts with IFN-β (called IFN-β_2).[53] A role of IFN-β_2 as the intermediate in the increased MHC antigen expression has been proposed[53] and is under current investigation.

As noted above, TNF induces some antigens in common with IL-1, some in common with IFN-γ, and some in common with IFN-α and IFN-β. It does not induce either class II MHC antigens[52] or the EC-specific antigens induced by IFN-γ recognized by KS sera.[45] However, TNF does have some additional IFN-γ-like properties. TNF is cytostatic for EC cultures.[27] In combination with IFN-γ, it can cause EC shedding and death. (Interestingly, this effect is only seen in passaged EC cultures and the increased sensitivity may represent an adaptation of EC to growth/angiogenic factor used in the culture system.) In both primary and passaged EC cultures, TNF induces morphological changes indistinguishable from those produced by IFN-γ.[27] At suboptimal doses of TNF and IFN-γ, these agents act synergystically; at optimal doses (e.g., 100 U/mℓ TNF, 200 U/mℓ IFN-γ) the combination produces unique morphological changes not seen with either mediator alone (best appreciated in primary cultures wehre EC do not shed). Specifically, doubly treated EC assume a dendritic-like morphology with extension of actin filaments into the spike-like projections. The substratum becomes exposed in numerous places. Synergy between IFN-γ and TNF has been noted in other cell types.[51] More recent studies have indicated that IL-1 species produce similar morphological changes to those induced by TNF, and that IL-1 also synergizes with IFN-γ.[67]

In summary, TNF has numerous direct actions upon EC; some are shared with IL-1, some with IFN-γ, and some are unique. In addition, TNF shows additivity with IL-1 and synergy with IFN-γ for some effects, and also can combine with IFN-γ to produce unique effects. An important implication of the sustained actions of TNF on EC is that the decline seen in the transient effects cannot be explained by loss of TNF-responsiveness of the EC or by inactivation of TNF in the medium.

D. Lymphotoxin

Lymphotoxin (LT) is a 24-kdalton protein product of activated T lymphocytes whose spectrum of activity overlaps with that of TNF.[55] These two proteins share a homologous domain[56,57] and compete for the same cell surface receptor.[58] To date, we have found that recombinant human LT shares with TNF at least five functional effects on EC: LT rapidly and transiently induces expression of E-LAM 1 rapidly but persistently elevates the antigen recognized by RR1/1, and slowly but persistently increases expression of class I MHC antigens, membrane-associated IL-1, and induces TNF-like morphological changes.[67] Like TNF, it has no effect on class II MHC antigens.

III. RELEVANCE OF LYMPHOKINE ACTIONS IN VITRO TO ENDOTHELIAL CELL ACTIVATION IN VIVO

As noted in the introduction to this review, we hypothesize that the in vitro effects of lymphokines upon cultured EC are a model for the in vivo effects of cell mediated immune responses upon the vasculature. What evidence supports this model? Class II MHC antigens, inducible by IFN-γ in vitro, are induced upon rodent vasculature in vivo at the site of an

Table 1
ENDOTHELIAL CELL MODULATION BY LYMPHOKINES

Antibody	Quantitative Binding				
	Control	IL-1	TNF/LT	IFN-α/β	IFN-γ
W6/32 (HLA-A,B)	+	+	+ + +	+ + +	+ + +
LB3.1 (HLA-DR)	−	−	−	−	+ + +
Leu 10 (HLA-DQ)	−	−	−	−	+
H4/18ᵃ (E-LAM 1)	−	+ + +	+ + +	−	−
RR1/1 (ICAM-1)	+	+ + +	+ + +	+	+ + +
KS seraᵇ	−	+ + +	+ + +	−	+ + +

ᵃ   Induction by IL-1, TNF, and LT is transient.
ᵇ   These sera were assayed by cytotoxicity; the antigens induced by IL-1 and TNF
   are different from those induced by IFN-γ.

incipient cell-mediated immune response as one of the earliest morphologically detectable events.[59] In humans, dogs, and some other species, the microvasculature appears to express class II molecules without stimulation. In dogs, cyclosporin A has been shown to eliminate this basal expression.[60] From in vitro work, cyclosporin A is known to block lymphokine synthesis by T cells but not the response of cultured EC to lymphokine.[61] The interpretation of these two cyclosporin A experiments is therefore that the basal expression of class II antigens by microvascular EC *in situ* is a response to basal levels of lymphokine and does not represent constitutive expression. Functionally, inducible EC class II molecules may (1) serve as the initial activator of circulating T lymphocytes through presentation of antigen;[62] (2) serve to select (and thus recruit) antigen specific T cells through presentation of antigen at the site of an ongoing response;[21,63] or (3) enrich a lymphocytic exudate for nonspecific T helper (i.e., CD4⁺) cells by interacting through low-affinity, antigen-independent recognition of nonpolymorphic portions of the class II molecules.[64] In any event, these *in situ* observations support in vivo relevance of the long program of EC activation. The morphological changes we have noted as part of the slow TNF and IFN-γ responses[27] (the ''long program'') may underlie the morphologic endothelial changes of activation noted *in situ*.[1] We speculate that the openings in the cell monolayer and the loss of fibronectin may represent an effort by the EC to facilitate transendothelial migration of lymphocytes into the tissues. Recently, we have begun to use antibody H4/18 as a marker for *in situ* evidence of the short program of EC activation. In an experimental delayed-hypersensitivity reaction in human skin, transient expression of H4/18 binding to endothelium was seen.[44] Our initial studies of pathological tissues have also detected H4/18 staining in settings where lymphokine production is likely (e.g., acute immune inflammation and T cell lymphoma), supporting the *in situ* relevance of the short program of EC activation as well.

In summary, therefore, several lines of investigation support the hypothesis that in vitro activation of EC by lymphokines is a good model for the in vivo events occurring in cell-mediated immunity and perhaps other inflammatory processes. Furthermore, studies with antibodies (summarized in Table 1) have shown that different lymphokines induce distinct EC responses. We conclude that EC activation *in situ* is more complex than initially expected and probably represents a cellular integration of multiple, distinct lymphokine-mediated signals. The insights gained and reagents generated from the in vitro work (especially monoclonal antibody and cDNA probes) have provided new tools for analysis of these phenomena.

ACKNOWLEDGMENT

The experiments described in this chapter were supported by grants from the N.I.H. (HL-36003 and HL-36028). The author was a Fellow of the Searle Scholars Program and is now an Established Investigator of the American Heart Association. I also wish to thank my colleagues who participated in these various studies for their many contributions.

REFERENCES

1. **Willms-Kretschmer, K., Flax, M. H., and Cotran, R. S.,** The fine structure of the vascular reponse in hapten-specific delayed hypersensitivity and contact dermatitis, *Lab Invest.,* 17, 334, 1967.
2. **Dumonde, D. C., Pulley, M. S., Paradinas, F. J., Soutcott, B. M., O'Connell, D., Robinson, M. R. G., DenHollander, F., and Schuurs, A. H.,** Histological features of skin reactions to human lymphoid cell line lymphokine in patients with advanced cancer, *J. Pathol.,* 138, 289, 1982.
3. **Montesano, R., Orci, L., and Vassalli, P.,** Human endothelial cell cultures: phenotypic modulation by leukocyte interleukins, *J. Cell. Physiol.,* 122, 424, 1985.
4. **Groenewegen, G., Buurman, W. A., and van der Linden, C. J.,** Lymphokines induce change in morphology and enhance motility of endothelial cells, *Clin. Immunol. Immunopathol.,* 36, 378, 1985.
5. **Trinchieri, G. and Perussia, B.,** Immune interferon: a pleiotropic lymphokine with multiple effects, *Immunol. Today,* 6, 131, 1985.
6. **Blalock, J. E., Georgiaoles, J. A., Langford, M. P., and Johnson, H. M.,** Purified human immune interferon has more potent anticellular activity than fibroblast or leukocyte interferon, *Cell Immunol.,* 49, 390, 1980.
7. **Rubin, B. Y. and Gupta, S. L.,** Differential efficacies of human type I and type II interferons as antiviral and antiproliferative agents, *Proc. Natl. Acad. Sci. U.S.A.,*
8. **Weil, J., Epstein, C. J., Epstein, L. B., Sedmak, J., Sabran, J. L., and Grossberg, S. E.,** A unique set of polypeptides is induced by γ interferon in addition to those induced in common with α and β interferons, *Nature (London),* 301, 437, 1983.
9. **Hirshberg, H., Evensen, S. A., Henriksen, T., and Thorsby, E.,** The human mixed lymphocyte-endothelium culture interaction, *Transplantation,* 19, 495, 1975.
10. **Dupont, B., Hansen, J., and Yunis, E. J.,** Human mixed lymphocyte culture reaction: genetic specificity and biological implications, *Adv. Immunol.,* 23, 188, 1976.
11. **Moen, T., Moen, M., and Thorsby, E.,** HLA-D region products are expressed in endothelial cells. Detection by primed lymphocyte typing, *Tissue Antigens,* 15, 112, 1980.
12. **Hirschberg, H., Braathen, L. R., and Thorsby, E.,** Antigen presentation by vascular endothelial cells and epidermal Langerhans cells: the role of HLA-DR, *Immunol. Rev.,* 66, 57, 1982.
13. **Wagner, C. R., Vetto, R. M., and Burger, D. R.,** The mechanism of antigen presentation by endothelial cells, *Immunobiology,* 168, 453, 1984.
14. **McCarron, R. M., Kempski, O., Sptaz, M., and McFarlin, D. E.,** Presentation of myelin basic protein by murine cerebral vascular endothelial cells, *J. Immunol.,* 134, 3100, 1985.
15. **Hirschberg, H., Moen, T., and Thorsby, E.,** Complement and cell-mediated specific destruction of human endothelial cells treated with anti-DRw antisera, *Proc. Natl. Acad. Sci. U.S.A.,* 11, 776, 1979.
16. **Pober, J. S. and Gimbrone, M. A., Jr.,** Expression of la-like antigens by human vascular endothelial cells is inducible *in vitro:* demonstration by monoclonal antibody binding and immunoprecipitation, *Proc. Natl. Acad. Sci. U.S.A.,* 79, 6641, 1982.
17. **Pober, J. S., Gimbrone, M. A., Jr., Cotran, R. S., Reiss, C. S., Burakoff, S. J., Fiers, W., and Ault, K. A.,** Ia expression by vascular endothelium is inducible by activated T cells and by human γ interferon, *J. Exp. Med.,* 157, 1339, 1983.
18. **Pober, J. S., Collins, T., Gimbrone, M. A., Jr., Cotran, R. S., Gitlin, J. D., Fiers, W., Clayberger, C., Krensky, A. M., Burakoff, S. J., and Reiss, C. S.,** Lymphocytes recognize human vascular endothelial and dermal fibroblast la antigens induced by recombinant immune interferon, *Nature (London),* 305, 726, 1984.
19. **Collins, T., Korman, A. J., Wake, C. T., Boss, J. M., Kappes, D. J., Fiers, W., Ault, K. A., Gimbrone, M. A., Jr., Strominger, J. L., and Pober, J., S.,** Immune interferon cultivates multiple class II major histocompatibility complex genes and the associated invariant chain gene in human endothelial cells and dermal fibroblasts, *Proc. Natl. Acad. Sci. U.S..A.,* 81, 4917, 1984.

20. **Geppert, T. D. and Lipsky, P. E.,** Antigen presentation by interferon-γ-treated endothelial cells and fibroblasts: differential ability to function as antigen-presenting cells despite comparable la expression, *J. Immunol.,* 135, 3750, 1985.

21. **Pober, J. S., Collins, T., Gimbrone, M. A., Jr., Libby, P., and Reiss, C. S.,** Overview: inducible expression of class II major histocompatibility complex antigens and the immunogenicity of vascular endothelium, *Transplantation,* 41, 141, 1986.

22. **Umetsu, D. T., Pober, J. S., Jabera, H. J., Fiers, W., Yunis, E., Burakoff, S. J., Reiss, C. S., and Geha, R. S.,** Human dermal fibroblasts present tetanus toxoid antigen to antigen specific T cell clones, *J. Clin. Invest.,* 76, 154, 1985.

23. **Inaba, K. and Steinman, R. M.,** Resting and sensitized T lymphocytes exhibit distinct stimulatory (antigen-presenting cell) requirements for growth and lymphokine release, *J. Exp. Med.,* 160, 1717, 1984.

24. **Pober, J. S., Gimbrone, M. A., Jr., Lapierre, L. A., Mendrick, D. L., Fiers, W., Rothlein, R., and Springer, T. A.,** Overlapping patterns of activation of human endothelial cells by interleukin 1, tumor necrosis factor and immune interferon, *J. Immunol.,* in press.

25. **Dustin, M. L., Rothlein, R., Bhan, A. K., Dinarello, C. A., and Springer, T. A.,** Induction of IL-1 and interferon-γ, tissue distribution, biochemistry and function of a natural adherence molecule (ICAM-1), *J. Immunol.,* 137, 245, 1986.

26. **Leung, D. Y. M., Collins, T., Lapierre, L. A., Geha, R. S., and Pober, J. S.,** IgM antibodies in the acute phase of Kawasaki syndrome lyse cultured vascular endothelial cells stimulated by gamma interferon, *J. Clin. Invest.,* 77, 1428, 1986.

27. **Stolpen, A. H., Guinan, E. C., Fiers, W., and Pober, J. S.,** Recombinant tumor necrosis factor and immune interferon act singly and in combination to reorganize human vascular endothelial cell monolayers, *Am.J. Pathol.,* 123, 16, 1986.

28. **Nickoloff, B. J., Mahrle, G., and Morhern, V.,** Ultrastructural effects of recombinant gamma-interferon on cultured human kertinocytes, *Ultrastruct. Pathol.,* 10, 17, 1986.

29. **Powanda, M. C.,** The role of interleukin-1 in homeostasis, in *The Physiologic, Metabolic, and Immunologic Actions of Interleukin-1,* Kluger, M. J., Oppenheim, J. J., and Powanda, M. C., Eds., Alan R. Liss, New York, 1985, 535.

30. **Auron, P. E., Webb, A. C., Rosenwasser, L. J., Mucci, S. F., Rich, A., Wolff, S. M., and Dinarello, C. A.,** Nucleotide sequence of human monocyte interleukin 1 precursor cDNA, *Proc. Natl. Acad. Sci. U.S.A.,* 81, 7907, 1984.

31. **March, C.J., Mosley, B., Larsen, A., Cernetti, D. P., Braedt, G., Price, U., Gillis, S., Henney, C. S., Kronheim, S. R., Grabstein, K., Conlon, P. J., Hopp, T. P., and Cosman, D.,** Cloning, sequence, and expression of two distinct human interleukin-1 complementary DNAs, *Nature (London),* 315, 641, 1985.

32. **Kaye, J., Parcelli, S., Tite, J., Jones, B., and Janeway, C. A.,** Both a monoclonal antibody and antisera specific for determinants unique to individual cloned helper T cell lines can substitute for antigen and antigen-presenting cells in the activation of T cells, *J. Exp. Med.,* 158, 836, 1983.

33. **Durum, S. K., Schmidt, J. A., and Oppenheim, J. J.,** Interleukin 1: immunological perspective, *Annu. Rev. Immunol.,* 3, 263, 1985.

34. **Bevilacqua, M. P., Pober, J. S., Majeau, G. R., Cotran, R. S., and Gimbrone, M. A., Jr.,** Interleukin 1 (IL-1) induces biosynthesis and cell surface expression of procoagulant activity in human vascular endothelial cells, *J. Exp. Med.,* 160, 618, 1983.

35. **Bevilacqua, M. P., Pober, J. S., Majeau, G. R., Fiers, W., Cotran, R. S., and Gimbrone, M. A., Jr.,** Recombinant tumor necrosis factor induces procoagulant activity in cultured human vascular endothelium: characterization and comparison with the actions of interleukin 1, *Proc. Natl. Acad. Sci. U.S.A.,* 83, 4533, 1986.

36. **Bevilacqua, M. P., Pober, J. S., Wheeler, M. E., Cotran, R. S., and Gimbrone, M. A., Jr.,** Interleukin-1 acts on cultured human vascular endothelial cells to increase the adhesion of polymorphonuclear leukocytes, monocytes and related leukocyte cell lines, *J. Clin. Invest.,* 76, 2003, 1985.

37. **Cavender, D. E., Haskard, D. O., Joseph, B., and Ziff, M.,** Interleukin 1 increases the binding of human B and T lymphocytes to endothelial cell monolayers, *J. Immunol.,* 136, 203, 1986.

38. **Pober, J. S., Bevilacqua, M. P., Mendrick, D. L., Lapierre, L. A., Fiers, W., and Gimbrone, M. A., Jr.,** Two distinct monokines, interleukin 1 and tumor necrosis factor, each independently induce biosynthesis and transient expression of the same antigen on the surface of cultured human vascular endothelial cells, *J. Immunol.,* 136, 1680, 1986.

39. **Bevilacqua, M. P., Pober, J. S., Mendrick, D. L., Cotran, R. S., and Gimbrone, M. A., Jr.,** Identification of an inducible endothelial leukocyte adhesion molecule, E-LAM 1, *Proc. Natl. Acad. Sci. U.S.A.,* in press.

40. **Wagner, C. R., Vetto, R. M., and Burger, D. R.,** Expression of I-region associated antigen (Ia) and interleukin 1 by subcultured human endothelial cells, *Cell Immunol.,* 93, 91, 1985.

41. **Stern, D. M., Bank, I., Nawroth, P. P., Cassimeris, J., Kisiel, W., Fenton, J. W., Dinarelio, C., Chess, L., and Jaffe, E. A.,** Self-regulation of procoagulant events on the endothelial cell surface, *J. Exp. Med.,* 162, 1223, 1985.

42. **Miossec, P., Cavender, D., and Ziff, M.,** Production of interleukin 1 by human endothelial cells, *J. Immunol.,* 136, 2486, 1986.

43. **Libby, P., Ordovas, J. M., Auger, K. R., Robbins, A. H., Birinyl, L. K., and Dinarello, C. A.,** Endotoxin and tumor necrosis factor induce interleukin-1 gene expression in adult human vascular endothelial cells, *Am. J. Pathol.,* 1986, in press.

44. **Cotran, R. S., Gimbrone, M. A., Jr., Bevilacqua, M. P., Mendrick, D., and Pober, J. S.,** Induction and detection of a human endothelial activation antigen in vivo, *J. Exp. Med.,* 164, 661,1986.

45. **Leung, D. Y. M., Geha, R. S., Newburger, J. W., Burns, J. C., Fiers, W., Lapierre, L. A., and Pober, J. S.,** Two monokines, interleukin 1 and tumor necrosis factor, render cultured vascular endothelial cells susceptible to lysis by antibodies circulating during Kawasaki syndrome, *J. Exp. Med.,* in press.

46. **Old, L. J.,** Tumor necrosis factor (TNF), *Science,* 230, 630, 185.

47. **Beutler, B. and Cerami, A.,** Cachectin and tumour necrosis factor as two sides of the same biological coin, *Nature (London),* 320, 584, 1986.

48. **Nedwin, G. E., Svedersky, L. P., Bringman, T. S., Palladino, M. A., Jr., and Goeddel, D. V.,** Effect of interleukin 2, interferon-γ, and mitogens on the production of tumor necrosis factors α and β, *J. Immunol.,* 135, 2492, 1985.

49. **Nawroth, P. P. and Stern, D. M.,** Modulation of endothelial cell hemostatic properties by tumor necrosis factor, *J. Exp. Med.,* 163, 740, 1986.

50. **Gamble, J. R., Harlan, J. M., Klebanooff, S. J., and Vadas, M. A.,** Stimulation of the adherence of neutrophils to umbilical vein endothelium by human recombinant tumor necrosis factor, *Proc. Natl. Acad. Sci. U.S.A.,* 82, 8667, 1985.

51. **Nawroth, P. P., Bank, I., Handley, D., Cassimeris, J., Chess, L., and Stern, D.,** Tumor necrosis factor/cachectin interacts with endothelial cell receptors to induce release of interleukin 1, *J. Exp. Med.,* 163, 1363, 1986.

52. **Collins, T., Lapierre, L. A., Fiers, W., Strominger, J. L., and Pober, J. S.,** Recombinant tumor necrosis factor increses HLA-A,B antigens in vascular endothelial cells and dermal fibroblasts *in vitro,* *Proc. Natl. Acad. Sci. U.S.A.,* 83, 446, 1986.

53. **Kohase, M., Henriksen-DeStefano, D., May, L. T., Vilcek, J., and Sehgal, P. B.,** Induction of β$_2$-interferon by tumor necrosis factor: a homeostatic mechanism in the control of cell proliferation, *Cell,* 45, 659, 1986.

54. **Williamson, B. D., Craswell, E. A., Rubin, B. Y., Predergast, J. S., and Old, L. J.,** Human tumor necrosis factor produced by human B cell lines: synergistic cytotoxic interaction with human interferon, *Proc. Natl. Acad. Sci. U.S.A.,* 80, 5397, 1983.

55. **Ruddle, N.,** Lymphotoxin redux, *Immunol. Today,* 6, 156, 1985.

56. **Gray, P. W., Aggarwal, B. B., Benton, C. V., Bringman, T. S., Henzel, W. J., Jarrett, J. A., Leung, D. W., Moffat, B., Ng, P., Svedersky, L. P., Palladino, M. A., and Nedsin, G. E.,** Cloning and expression of cDNA for human lymphotoxin, a lymphokine with tumour necrosis activity, *Nature (London),* 312, 721, 1984.

57. **Pennica, D., Nedwin, G. E., Hayflick, J. S., Seeburg, P. H., Derynck, R., Palladino, M. A., Kohr, W. J., Aggarwal, B. B., and Goeddel, D. V.,** Human tumour necrosis factor: precursor structure, expression and homology to lymphotoxin, *Nature (London),* 312, 724, 1984.

58. **Aggarwal, B. B., Eessalu, T. E., and Haas, P. E.,** Characterization of receptors for human tumour necrosis factor and their regulation by γ-interferon, *Nature (London),* 318, 665, 1985.

59. **Sobel, R. A., Blanchette, B. W., Bhan, A. K., and Colvin, R. B.,** The immunopathology of experimental allergic encephalomyelitis. II. Endothelial cell Ia increases prior to inflammatory cell infiltration, *J. Immunol.,* 132, 2402, 1984.

60. **Groenewegen, G., Buurman, W. A., and van der Linden, C.J.,** Lymphokine dependence of *in vivo* expression of MHC class II antigens by endothelium, *Nature (London),* 316, 361, 1985.

61. **Groenewegen, G., Buurman, W. A., Jeunhomme, G. M. A. A., and van der Linden, C. J.,** Cyclosporin A affects MHC-class II antigen expression by arterial and venous endothelium *in vitro, Transplantation,* 40, 21, 1985.

62. **Burger, D. R. and Vetto, R. M.,** Vascular endothelium as a major participant in T-lymphocyte immunity, *Cell Immunol.,* 70, 357, 1982.

63. **Pober, J. S., Gimbrone, M. A., Jr., Collins, T., Cotran, R. S., Ault, K. A., Fiers, W., Krensky, A. M., Clayberger, C., Reiss, C. S., and Burakoff, S. J.,** Interactions of T lymphocytes with human vascular endothelial cells: role of endothelial cell surface antigens, *Immunobiology,* 168, 483, 1984.

64. **Masuyama, J.-I., Minato, N., and Kano, S.,** Mechanisms of lymphocyte adhesion to human vascular endothelial cells in culture, *J. Clin. Invest.,* 77, 1596, 1986.

65. **Kurt-Jones, E. A., Fiers, W., and Pober, J. S.,** Membrane IL-1 induction on human endothelial cells and dermal fibroblasts, *J. Immunol.,* 139, 2317, 1987.

66. **Warner, S. J. C., Auger, K. R., and Libby, P.,** Interleukin 1 induces interleukin 1. II. Recombinant human interleukin 1 induces interleukin 1 production by adult human vascular endothelial cells, *J. Immunol.,* 139, 1911, 1987.

67. **Pober, J. S., Lapierre, L. A., Stolper, A. H., Brock, T. A., Springer, T. A., Fiers, W., Bevilacqua, M. P., Mendrick, D. L., and Gimbrone, M. A., Jr.,** Activation of cultured human endothelial cells by recombinant lymphotoxin: comparison with tumor necrosis factor and interleukin-1 species, *J. Immunol.,* 138, 3319, 1987.

Index

INDEX

A